Risk Analysis in the Private Sector

ADVANCES IN RISK ANALYSIS

This series is edited by the Society for Risk Analysis.

Volume 1 **THE ANALYSIS OF ACTUAL VERSUS PERCEIVED RISKS**
Edited by Vincent T. Covello, W. Gary Flamm, Joseph V. Rodricks, and Robert G. Tardiff

Volume 2 **LOW-PROBABILITY/HIGH-CONSEQUENCE RISK ANALYSIS**
Issues, Methods, and Case Studies
Edited by Ray A. Waller and Vincent T. Covello

Volume 3 **RISK ANALYSIS IN THE PRIVATE SECTOR**
Edited by Chris Whipple and Vincent T. Covello

A Continuation Order Plan is available for this series. A continuation order will bring delivery of each new volume immediately upon publication. Volumes are billed only upon actual shipment. For further information please contact the publisher.

PREFACE

The theme of this volume--risk analysis in the private sector--reflects a changing emphasis in risk analysis. Until recently, attention has been focused on risk analyses conducted in support of federal regulatory decision making. Such analyses have been used to help set safety standards, to illuminate issues of regulatory concern, and to evaluate regulatory alternatives. As this volume indicates, however, risk analysis encompasses a broader set of activities. Analyses performed by private sector institutions aimed at preventing or reducing potential adverse health or environmental effects also play an important part in societal risk management. In virtually all societies, there have been strong incentives for the private sector to conduct such analyses. These incentives range from moral or altruistic norms and values to simple self-interest based on fear of monetary loss, possible civil or criminal litigation, or punitive or restrictive government action.

The papers in this volume address the overall theme from a variety of perspectives. Specifically, the papers represent contributions from such diverse fields as toxicology, epidemiology, chemistry, biology, engineering, statistics, decision analysis, economics, psychology, philosophy, and the law.

All papers were presented at the third annual meeting of the Society for Risk Analysis held in New York City. Support for the meeting was provided by grants from the American Petroleum Institute, the Chemical Manufacturers Association, Shell Oil Corporation, Mobil Corporation, Exxon Corporation, Union Carbide Corporation, the U.S. Department of Energy, and the U.S. Environmental Protection Agency. We would also like to thank the following individuals for their contributions to the meeting above and beyond the call of duty: Richard Schwing, Robert Tardiff, Richard Burk, Robert Johnson, Diane Taub, and Pauline Hostettler.

C. Whipple and
V. Covello

CONTENTS

SECTION 1: CHEMICAL RISK MANAGEMENT

CHEMICAL RISK MANAGEMENT: INTRODUCTION

Rae Zimmerman

Graduate School of Public Administration
New York University
New York, NY

INTRODUCTION

The level of concern over the environmental and occupational risks from chemicals is often measured in terms of the vast number of chemicals in circulation, many of which are suspected of having adverse human health effects. The specific numbers often cited in connection with the volume of chemicals potentially in the environment are: the over 5.3 million chemicals listed in Chemical Abstracts, the 70,000 synthetic chemicals currently in circulation, and the average rate of over 1,000 chemicals per year introduced into the market place.[1] The chemical industry, which did an estimated volume of sales of $162 billion in 1980, accounts for five percent of the gross national product.[2] As a result of these concerns, the number of environmental laws that deal with chemical risks has grown exponentially over the past decade. Several hundred chemicals are closely monitored under these laws.

In light of the centrality of the risk problem to the chemical industry, its suppliers, clients, and regulators, an understanding of the nature of the industry and the trends in the risk problems it faces is critical. An understanding of the potential approaches used by, or of use to, the industry in identifying risks and the reactions of the chemical industry to these risks and the regulations that control them, is also important if risk reduction measures are to be integrated effectively into chemical industry operations. The papers presented in this session begin to explore these various aspects of chemical risk management as it pertains to the private sector.

Price identifies and compares the underlying level of risk used in developing standards for drinking water and workplace conditions.

Standards for 55 chemicals are compared. He finds that risk, expressed as permitted dosage, was many times greater for the workplace than for drinking water. He indicates that this may be the result of underlying differences in the safety factors used, i.e., the level of acceptable risk.

Rosenblum, et al. present a method of rapidly ranking chemicals with respect to their environmental, health, and safety risks. The Integrated Risk Index allows the variables that go into the index to be weighted by such factors as annual production of the chemical and the potential population exposed. The index is much more multidimensional than those generally in use for hazardous waste site ranking. In contrast, McKone, et al. develop a chemical indexing system based upon biogeochemical cycles for toxic chemicals generated by energy systems. The index, which incorporates a spatial mapping component for chemicals, estimates the incremental additional risk posed by increases in the concentrations of selected toxic chemicals. The model is used to compare the ranking of chemicals with respect to their toxicity in soil and water.

Sueishi and Ueta explore chemical usage patterns for one toxic metallic substance, lead, in Japan. The analysis attempts to link usage and waste generation patterns as a means of assessing overall changes in the potential environmental impact posed by lead in Japan's commerce.

Finally, Zimmerman looks at the response of major corporations which have faced risks from the chemicals they manufacture, use, or discard. These responses are portrayed in terms of market and product diversification, organizational diversification, and several unclassifiable responses that aim at risk reduction or risk avoidance.

Thus, chemical risk management in the private sector has to be looked at from many perspectives. The risks posed by their manufacture, usage, and disposal need to be understood in terms of market mechanisms, organizational behavior, and global biogeochemical impacts. Indexing systems provide one way of simplifying the risk component in order to look more closely at the relationship of these other factors to chemical risk.

FOOTNOTES

1. Committee on Chemical Environmental Mutagens, National Research Council. Identifying and Estimating the Genetic Impact of Chemical Mutagens. Washington, DC: National Academy Press, 1983. Page 17.
2. Council on Environmental Quality. Environmental Quality--1981. Washington, DC: CEQ, 1982. Page 115.

A GENERIC ENVIRONMENTAL HEALTH RISK MODEL FOR RANKING THE TOXIC BY-PRODUCTS OF ENERGY SYSTEMS

Thomas E. McKone*
David Okrent**
William E. Kastenberg**

*Lawrence Livermore National Laboratory, L-140
Livermore, CA 94550

**School of Engineering and Applied Science
University of California
Los Angeles, CA 90024

ABSTRACT

This paper describes and illustrates an approach for indexing the long-term health risks of toxic chemicals and radionuclides using crustal chemical cycles. The external environment is expressed in a pathways model which uses a "landscape prism" as a tool for mapping toxic material cycles on and near the crustal surface. Average concentrations in air, water, and food are derived and translated into a collective population dose field expressed as daily intake. The population dose is converted to population risk, expressed as lifetime cancer risk per individual or as a range of blood levels for non-stochastic agents. The models are used to index the health risk added to a hypothetical landscape from additions of arsenic, lead, uranium-238, and radium-226 to groundwater and soil. It is found that each species, by its behavior in the total environment, imposes a hazard that does not correlate to current measures of toxicity such as drinking water standards.

KEY WORDS: Chemical Cycles, Health Risks, Toxic Elements, Radionuclides

INTRODUCTION

In designing and operating energy systems, decisionmakers in the private sector face the task of balancing engineering, economic, environmental, and social considerations. Certainly, one decision involves the appropriate distribution of risk management resources among several toxic substances that will be released to the environment. Such a decision requires assessment of both the inherent toxicity of these substances and the availability of the toxic risk through environmental transport. Every energy system involves alterations of the geochemistry of a local, regional, or global environment. As an example, each year coal mining moves 20,000 tons of uranium from the earth's crust to its surface and atmosphere.[1] Coal fly ash is often highly enriched in uranium, mercury, lead, and other toxic elements. Such geochemical mobilizations must be analyzed in a way that integrates knowledge from several disciplines. However, within the fields of geochemistry and ecology, there have been rapid advances in our ability to observe and describe natural systems interactions at all scales. This information makes it possible to map the history of chemical elements as they participate in terrestrial processes.

This paper describes and illustrates an approach for indexing the long-term health risks of toxic chemicals and radionuclides using crustal chemical cycles. This approach considers chemical distributions within, and interactions between, the human biological system and the external biogeochemical environment. The result is a two-component model consisting of an environmental pathways model and an exposure/dose-response model. The pathways model uses a "landscape prism" as a tool for mapping toxic material cycles on and near the crustal surface. The landscape prism is divided into a set of compartments consistent with patterns of element circulation observed in the global environment. General geochemical data or site-specific monitoring data are used to calculate transfer coefficients for the landscape systems model. A system of coupled first-order linear differential equations is used to describe the dynamics of an element or a radionuclide chain within the components of the near-surface earth system. Average environmental concentrations in air, water, and food are derived and translated into a collective population dose field expressed as daily intake. The collective population dose is converted to population risk, which is expressed as lifetime cancer risk per individual for carcinogens and as range of blood levels for other toxins. The models are used to investigate and index the impact of additions of arsenic, lead, uranium ore, uranium-238, and radium-226 to the groundwater and soil of a generic landscape. It is found that when one considers environmental chemical cycles, the hazard indexes of these species fail to correlate with traditional measures of toxic hazard such as drinking water standards.

ELEMENT CYCLES AND HEALTH RISK

The components of the earth's surface are linked by chemical cycles to form a system in which there is chemical balance. Each element has an "environmental chemical cycle"[2] which can be mapped in terms of local, regional, or global fluxes. The chemical cycle of most elements begins in the crystalline rock that is at the base of the upper crust. Groundwater dissolution, erosion, uplift, and volcanoes transfer elements from this zone to the surface environment. At the surface, elements are distributed among sediments, soils, flora, fauna, rivers, lakes, and oceans. These cycles provide chemical stock for the biosphere, including humans. The ancient Greeks first noted that overall health is influenced by the chemistry of the environment. In Air, Water, and Places, Hippocrates demonstrated that the well-being of individuals is influenced by quality of air, water, and food; the topography of the land; and general living habits.

For this paper the methods of "environmental geochemistry" provide a robust tool for mapping element cycles in landscape. According to Fortescue,[3] "environmental geochemistry is concerned with the chemistry of the earth's component parts at or near the daylight surface." It is a science suited to evaluating the abundance and movement of toxic elements in the earth system. The tool used in environmental geochemistry to visualize the flow of elements near the earth's surface is the "landscape prism." Figure 1 illustrates a general landscape prism and shows the relationship among environmental compartments. Landscape prisms can be drawn as an accurate cross-section in order to represent a specific local zone or they can be used to represent compositions and interactions common to a region of, or to the whole of, the earth's surface.

CALCULATION OF ELEMENT AND NUCLIDE DISTRIBUTIONS

The calculation of element and nuclide distributions within environmental compartments is composed of four steps. First, one constructs the landscape prism. Second, the landscape is divided into compartments and the mass of each compartment and the mass exchange among the compartments is obtained. Third, for each element or nuclide, transfer coefficients between each set of compartments are determined. Fourth, these transfer coefficients, together with nuclear decay constants, are used to set up a system of first-order ordinary differential equations which define the time-dependent distribution of a radionuclide series or stable element in the landscape. This set of equations is solved numerically. The health risk experienced by the population of the landscape, which must be tailored for each nuclide or element, is taken up in the next section.

In this paper the landscape prism is viewed as a lumped parameter system composed of eight compartments. These compartments and element

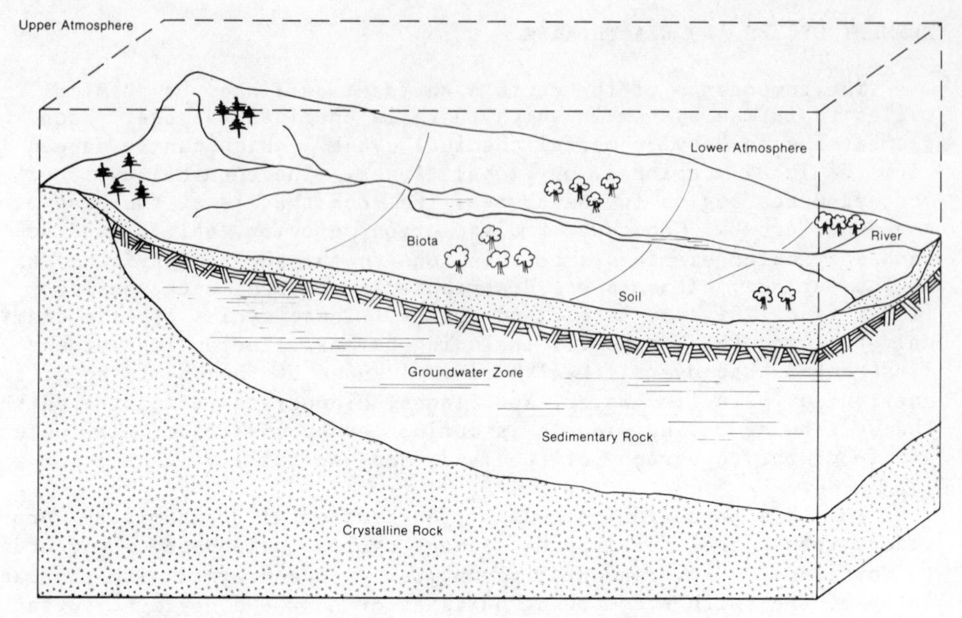

Figure 1. The General Landscape Prism

pathways among them are shown in Figure 2. The flux rates within the landscape prism are derived by considering mass and element flows of the overall earth system.

Table I provides a list of both inventories and mass fluxes for the global sedimentary cycle.[4] In each compartment except groundwater, the entries for gains, losses, and fluxes are summed over all individual elements but exclude H, O, C, N occurring as H_2O, CO_2, NO_2, and O_2. Gains, losses, and fluxes for groundwater include all elements. Values for inventories include all elements within a compartment. The initial inventory of element i in a compartment k is obtained from the product of the total mass X_k of the compartment and the abundance of the element. The mass flux $S^i_{k,m}$ of element i from one compartment to another is obtained from the expression,

$$s^i_{k,m} = S_{k,m} K^i_{k,m} a^i_K \qquad (1)$$

where $S_{k,m}$ = total mass flux k to m,

$K^i_{k,m}$ = distribution coefficient, which expresses the abundance of i in the mass stream from k to m relative to abundance of i in k, and

a^i_k = abundance of i in k.

TABLE I

LANDSCAPE CELL INVENTORIES AND MASS FLUXES DERIVED FROM THE GLOBAL SEDIMENTARY CYCLE[a]

Compartment	Inventory kg/m^2	External Gains $kg/m^2/yr$	External Losses $kg/m^2/yr$	Flux[b] in $kg/m^2/yr$ to:					
				Sedimentary Rock	Ground-water	Soil	Surface Water	Biota	Lower Atmosphere
Sedimentary Rock	8.5×10^6	0	0	0	-	0.14	0	0	0
Ground[b] water	3.5×10^4	0	0	-	0	0	90.	0	0
Soil	1.2×10^3	0.03	0	0	-	0	0.17	0.023	0.38
Surface Water	0.15	0	0.17	0	0	-	0	0	0
Biota	0.4	0	0	0	0	0.017	0	0	6.3×10^{-3}
Lower Atmosphere	1.0×10^{-2}	1.0×10^{-5}	0	0	0	0.38	0	0	0

[a] The use of a hyphen in the table indicates that the gross mass flux for a pathway is not known. For these pathways the mass flux of a specific element is derived using other mass fluxes and the partitioning coefficients discussed in the text.

[b] The groundwater flux to surface water is the amount of water mass whereas all other fluxes are mineral mass.

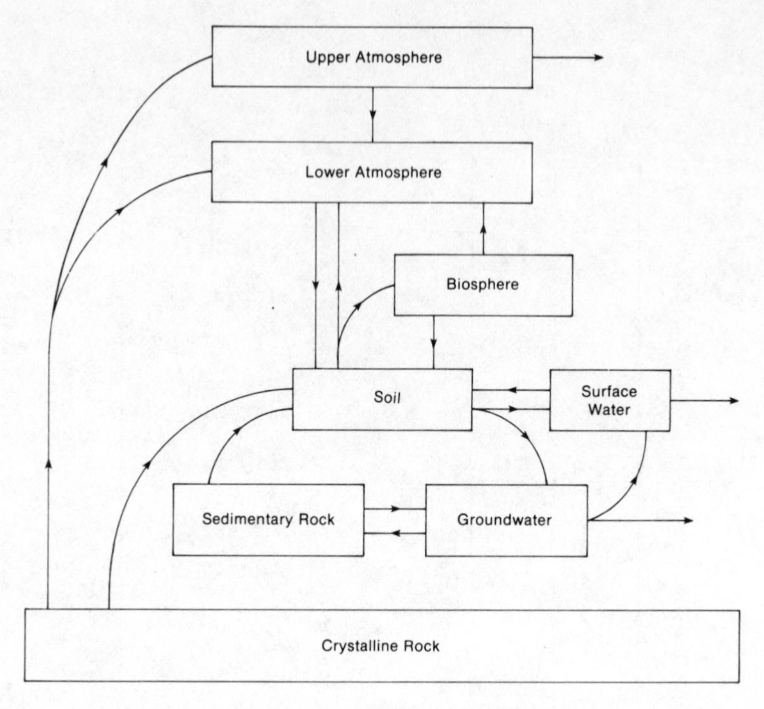

Figure 2. Element and Radionuclide Flow Paths in the Landscape Prism.

The set of differential equations describing the inventory of a set of radionuclides in the compartments of a landscape prism have the general form:

$$\frac{dx_k^i(t)}{dt} = -\lambda^i x_k^i(t) + \lambda^{i-1} x_k^{i-1}(t) - \sum_{\substack{m=1 \\ m \neq k}}^{8} T_{k,m}^i x_k^i(t)$$

$$+ \sum_{\substack{n=1 \\ n \neq k}}^{8} T_{n,k}^i x_n^i(t) + q_k^i(t) \tag{2}$$

where

$x_k^i(t)$ = inventory of nuclide i in compartment k at time t (moles),

λ^i = radioactive decay constant of i (yr^{-1}),

$T_{k,m}^i = s_{k,m}^i / x_k^i$ = transfer coefficient for i from compartment k to m (yr^{-1}),

$q_k^i(t)$ = source term for the introduction of nuclide i into compartment k.

This system of equations is solved using the GEARB[5] numerical package. Initial inventories are determined using a source term that is characteristic of steady-state landscape element sources such as element abundances in rocks. A transient source is used to simulate anomalous geochemical events such as ore bodies, volcanoes, and human activities--events which modify the steady-state landscape element or radionuclide cycles within a landscape.

CALCULATION OF DOSES AND HEALTH RISKS

This section describes models used to translate environmental concentrations into doses and health risks. Detrimental effects on human health caused by toxic species are classified as "stochastic" and "non-stochastic" effects.[6] Stochastic effects are those for which the probability of an effect occurring rather than the severity of the effect, is proportional to dose, without threshold. Non-stochastic effects are those for which severity of effect is a function of dose and for which a threshold may exist . The health effects of carcinogens are usually assumed to be stochastic, whereas the health effects of neurotoxins such as lead and mercury are non-stochastic. In calculating doses, it is assumed that an average member of the landscape population breathes 20,000 liter/day of air, drinks 2 liter/day of water, and consumes 1 kg/day of food. The assumed values are for an adult member of the population. Half of the drinking water is assumed to come from groundwater and the remainder from surface water.

Radiation doses from the uranium series can be attributed to external and internal exposures. Linear dose-conversion factors translate the inventory of a radionuclide in an environmental compartment into dose rates in specific organs or tissues. The external whole-body dose from the uranium series is 30 mrad/yr per pCi of uranium-238 in soil.[4] Table II lists dose conversion constants based on UNSCEAR models,[7] that were used for internal exposure to the uranium series.

One multiplies exposure rates in rad/yr by a quality factor to convert to dose equivalent rates in rem/yr. The ICRP[6] risk factors

TABLE II

EQUILIBRIUM DOSE CONVERSION CONSTANTS FOR THE URANIUM SERIES

Table II. Equilibrium Dose Conversion Constants for the Uranium Series.

Nuclide	Internal Dose Conversion Factors (mrad/yr per pCi/day ingested)			
	Lung	Bone Marrow	Bone Lining Cells	Other Tissues
^{238}U (α)	.05	.05	.3	.05
^{234}Th (β,γ)	.0005	.0015	.0075	.0005
^{234m}Pa (β,γ)	.0075	.05	.100	.0075
^{234}U (α)	.05	.075	.35	.05
$^{230}Th^{*}$ (α)	40.	50.	800.	4.0
^{226}Ra (α)	.03	.10	.76	.03
^{226}Ra (β,γ)	.0014	.013	.032	.0014
$^{222}Rn^{*\#}$ (α)	3.5	.031	.031	.024
^{210}Pb (β,γ)	.0002	.00017	.00027	.0002
^{210}Bi (β,γ)	.013	.067	.13	.013
^{210}Po (α)	.10	.23	1.0	.20

*Dominant intake pathway is inhalation.
#Includes its four short-lived decay products.

provide the estimated likelihood per unit exposure of inducing fatal malignant disease, non-stochastic changes, or substantial genetic defects. Multiplying the population risk by the expected lifetime (70 years) of the population cohort provides an estimate of an individual's lifetime risk of incurring disease per unit of continuous annual exposure.

Ingestion of inorganic arsenic compounds has been linked to induction of skin and lung cancer in humans.[8] According to the EPA[9] an intake of 50 ng/day corresponds to a lifetime cancer risk of 10^{-5} per individual. Leaching by ground and surface water, plant uptake, and volcanic activity maintain the natural distribution of arsenic in the environment. Arsenic is a by-product of coal combustion. Before the advent of organic pesticides, lead-arsenate was widely used as a pesticide. Arsenic is also present in phosphate rock used to manufacture fertilizer and detergent.

Even though lead has not been linked to cancer induction, it is a documented neurotoxin.[10] Human uptake of lead results in both acute and chronic health risks. Health effects from lead intake are correlated to blood levels in μg per 100 ml blood. Acute lead exposures correspond to blood levels greater than 100 μg/100 ml. Symptoms of acute lead poisoning include colic, severe abdominal pain, neurologic injury, and convulsions.[10] Lead levels in the blood of 50 to 100 μg/100 ml result in chronic exposure and impact the hemopoietic system and the central nervous system. The EPA drinking water standard for lead is 50 μg/l.[9] Lead concentration in blood is 5.3 μg/100 ml for every 100 μg/day of lead ingested and 2.6 μg/100 ml for every 100 μg/day inhaled.[11] Lead enters the atmosphere and surface waters from natural sources. However, the major source of lead in urban landscapes is anthropogenic--resulting from metal smelting, combustion of leaded gasoline, combustion of coal, and the ingestion by children of paint containing lead pigments.

APPLICATIONS OF THE MODEL

In this section the models described above are used to rank the risks induced by continuous additions of uranium, uranium ore, radium, arsenic, and lead to soil or groundwater. Figure 1 provides an illustration of the general landscape used for these calculations. In this landscape there are surface out-croppings of both crystalline and sedimentary rock. The surface is covered by soil averaging 60 cm in depth and having a density of 2.0 g/cc. In 80 percent of this landscape the soil is underlain by a 500m zone of sedimentary rock which is saturated with groundwater; the remainder is underlain by igneous rock. The mean density of the rock is 2.5 g/cc and the average porosity is 0.07. The inventory of groundwater is 3.5×10^{13} g/km^2. Elements flow out of the landscape by water erosion and to a lesser extent by groundwater discharge. The actual area of the landscape is arbitrary, because all calculations are carried out on a unit area basis.

The four toxins and uranium ore are ranked in terms of the steady state health risk imposed on residents of the landscape prism from continuous inputs to soil and groundwater. The results, given in Figure 3, provide the source flux in g/km^2/yr to soil or groundwater that results in an incremental increase in lifetime health risk of 10^{-5}. For lead, the corresponding blood-level increase is taken 5 μg/100 ml. The toxic rank in Figure 4 is obtained by determining the steady-state flux of a given species required to equal the population risk of 1 g/km^2 per year of radium-226 similarly introduced. The results provide a rough measure of the quantity of a toxin such as arsenic that provides the equivalent detriment of a unit quantity of radium. The basis of measurement is the steady-state change of population health risk within a physical region, such as a river basin, as a result of continuous additions to soil and groundwater. The toxic rank is similar for the four species in soil and groundwater. However, toxic rank based on these asymptotic distributions do not compare to toxic rank obtained from drinking water standards.[12,13] For arsenic, uranium-238,

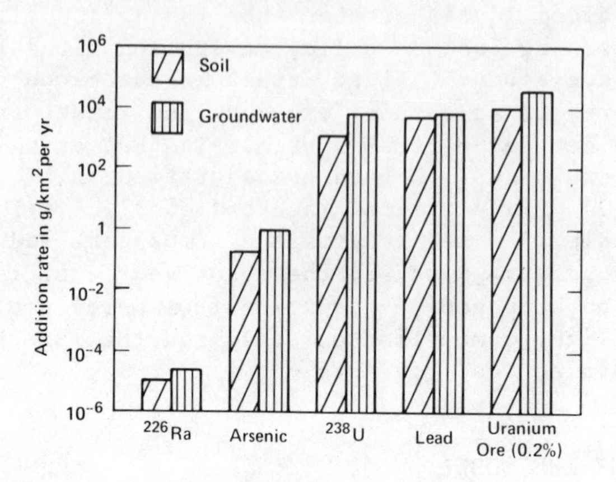

Figure 3

Additions to Soil and Groundwater That Increase Lifetime Cancer Risk (or Equivalent Detriment) by 10^{-5}

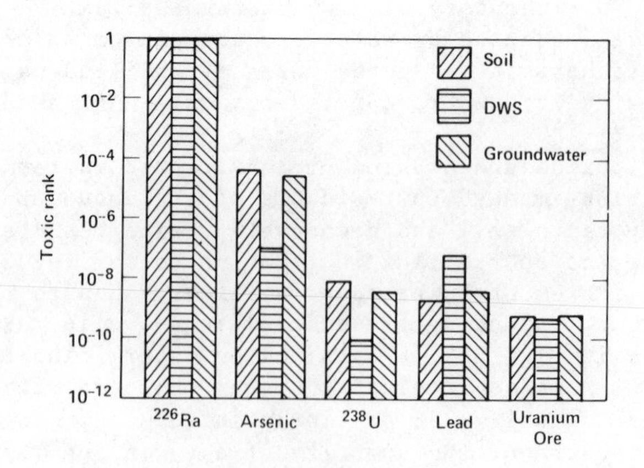

Figure 4
Toxic Ranks Derived from Continuous Release Studies Compared to Rank Derived from Drinking-Water Standards

and lead there is roughly an order of magnitude separating the rank obtained from drinking water standards relative to the rank obtained from our generic pathways model.

SUMMARY AND OBSERVATIONS

This paper uses geochemical systems models to describe the temporal distribution of chemical elements and radionuclide series and the resulting health risks. The landscape of interest is divided into eight homogeneous compartments between which nuclides are exchanged. The time-dependent distribution of elements and decay chains is described by a set of first order ordinary differential equations which are solved using integration methods. Health risks to humans are derived from compartment inventories using models for human interaction with their environment.

There are both advantages and disadvantages associated with the use of compartment models. The major advantages are that these models are flexible and can be matched to the data available; changes in regional characteristics can be represented by a suitable choice of compartments. These types of models are widely studied and numerical methods exist to solve the equations that arise. The resulting computer codes are usually inexpensive to run and well-suited to parametric studies. Among the disadvantages are the treatment of nuclide or element inventories within a compartment as uniform concentrations. Appropriate compartmentalization is often difficult and a compartment structure suited to one problem is not necessarily appropriate to others.

Biogeochemical cycles can either enhance or diminish the potential health risks of a toxic substance introduced to the environment. Failure to understand the ultimate fate of a toxic substance in the general environment can result in a variety of unpleasant surprises. Nonetheless, failure to recognize the role that environmental systems play in the attenuation of health risks can result in inflexible, inappropriate, and expensive standards. This paper provides an approach which uses chemical cycle models together with dose-response models to rank the relative availability of toxic risk in the general environment. Such a method can be used as an input to the standards process or as a screening tool for selecting hazardous species which require additional research.

FOOTNOTES

1. W. S. Fyfe, The environmental crisis: Quantifying geosphere interactions, Science 213:105-110 (1981).
2. H. J. M. Bowen, Environmental Chemistry of the Elements, Academic Press, London (1979).

3. J. A. C. Fortescue, Environmental Geochemistry - A Holistic Approach, Springer-Verlag, New York (1971).
4. T. E. McKone, Chemical Cycles and Health Risks of Some Crustal Nuclides, Ph.D. dissertation, University of California, Los Angeles (December, 1981).
5. A. C. Hindmarsh, "GEARB: Solution of Ordinary Differential Equations Having a Banded Jacobian," Lawrence Livermore Laboratory Report UCID-30059 Rev. 1 (March 1975).
6. International Commission on Radiological Protection, "Recommendations of the International Commission on Radiological Protection," ICRP Publication 26, Pergamon Press, New York (1977).
7. United Nations Scientific Committee on the Effects of Atomic Radiation (UNSCEAR), Sources and Effects of Ionizing Radiation, United Nations, New York (1977).
8. International Agency for Research on Cancer, IARC Monographs on the Evaluation of Carcinogenic Risk of Chemicals to Man, Vol. 2, "Some Inorganic and Organometallic Compounds," IARC, Lyon, France (1973).
9. U. S. Environmental Protection Agency, Notice of Water Quality Criteria Documents, Federal Register 45, 79318-79341 (November 28, 1980).
10. National Academy of Sciences, Lead, Airborne Lead in Perspective, Report of the Committee on Biological Effects of Atmospheric Pollutants, National Academy of Sciences, Washington, DC (1972).
11. R. M. Garrels, F. T. MacKenzie and C. Hunt, Chemical Cycles and the Global Environment: Assessing Human Influences, William Kaufmann, Inc., Los Altos, California (1975).
12. U. S. Environmental Protection Agency, National Interim Drinking Water Regulations, 40 CFR 141, U.S. Government Printing Office, Washington, DC.
13. U. S. Nuclear Regulatory Commission, Standards for Protection Against Radiation, 10 CFR 20, U.S. Government Printing Office, Washington, DC.

PRIVATE SECTOR RESPONSE PATTERNS TO RISKS FROM CHEMICALS

Rae Zimmerman

Graduate School of Public Administration
New York University
New York, NY

INTRODUCTION

Over the past decade, corporations that use or market chemicals have had to manage a wide range of risks posed by these chemicals. This management problem is a result of a considerable amount of uncertainty in the external organizational environments these corporations face, particularly in the regulatory environment. The uncertainties that exist relate to identifying the kinds of chemicals that pose risks, establishing the level of risk reduction necessary, and determining the acceptability of different forms of risk reduction and risk aversive behavior.[1]

To examine the responses of corporations to their regulatory environment and the risks in general that chemicals pose (in a nonregulatory framework), a number of cases were reviewed involving the manufacture, use, or disposal of hazardous substances in which various corporations played a substantial role. These cases are listed and summarized in Table 1. The firms involved in each of the cases were classified by the type of activity, manufacture, use, or disposal, that they were engaged in involving chemical risks. Manufacturers included Union Carbide (Aldicarb), Occidental (DBCP), Allied Chemical (Kepone), and B.F. Goodrich (Vinyl Chloride). Corporations categorized as users were General Electric (using PCBs in the manufacture of capacitors), Ventron Chemical (using mercury in its role as a chloralkali plant), and Marathon Battery (using cadmium in the manufacture of nickel-cadmium batteries). Companies operating disposal sites were Texaco (the Positive Chemical site on Staten Island), Hooker Chemical (Love Canal), and the Chemical Control Corporation (Elizabeth, New Jersey, site).

TABLE I

LIST OF CASES AND SUMMARY OF SELECTED CHARACTERISTICS

Case	Location	Technology	Source		Pathway/Receptor	Toxic Effects	Estimated Population at Risk	Estimated Land Area Directly Affected (Acres)	Estimated Quantity of Contaminant (Tons)
			Agent	Usage					
Cadmium	Foundry Cove, New York	Industrial/electrical equipment	Marathon Battery	Manufacture of batteries, subsequent seepage into surface water	Seepage to surface water and into aquatic life	Bioaccumulation, bioconcentration; carcinogenicity	2,000	400	20
Mercury	Berry's Creek, Hackensack Meadowlands, New Jersey	Mercury processing	Ventron Corp.	Miscellaneous industrial usages	Industrial waste discharge into surface water and soil	Bioconcentration, bioaccumulation; neurological disorders	15,000	40	200
PCBs	Hudson River Valley, New York	Electronics	General Electric	Capacitors	Industrial waste discharge into surface water	Bioaccumulation and bioconcentration; carcinogenicity	44,000	360	250
PCBs	College Point, New York	Industrial waste oil disposal	Durante Bros.	Not known	Industrial waste discharge into surface water from waste disposal	Bioaccumulation and bioconcentration; carcinogenicity	19,774	10	600
Aldicarb (Temik)	Suffolk County, Long Island, New York	Agricultural production (pest control)	Union Carbide	Control of Colorado potato beetle and golden nematode	Seepage to potable water supplies	Cholinesterase suppression in humans	3,100	25,000	1,050
Dibromo-chloropropane (DBCP)	San Joaquin Valley and Lathrop, Calif.	Agricultural production (pest control)	Amvac Chemical, Occidental, Shell, Dow	Nematode control: citrus fruits, tomatoes	Seepage to potable waters via irrigation wells and percolation	Reduced fertility; suspected carcinogenicity	56,750	335,600	6,800
Kepone	James River-Chesapeake Bay, Virginia	Agricultural production (pest control)	Allied Chemical	Pesticide against banana root borer; ant and roach poison	Industrial spills to surface water affecting aquatic life	Bioaccumulation and biomagnification; carcinogenicity	25,000	5,888	51
Trichoro-ethylene (TCE)	Southeastern Pennsylvania	Machining	Multiple industries	Chemical degreasing agent	Seepage to surface and groundwater to potable water supplies	Animal carcinogen	20,000	128,000	Not available
Chlorinated organics (Trihalo-methanes)	Dade County, Florida	Water purification	Miami-Dade Water and Sewer Authority	Chlorination of naturally occurring organic material in groundwater	Chlorination of potable water supplies with high natural organic content	Carcinogenicity	1,300,000	1,000,000	Not available
Chlorinated organics	North Miami Beach, Florida	Miscellaneous industrial technologies	North Miami Beach Water and Sewer Authority	Use of well water contaminated with industrial wastes	Seepage of industrial waste into groundwater used as potable water	Suspected carcinogenicity	38,000	3,264	Not available
Chemical Control Corp.	Elizabeth, New Jersey	Miscellaneous industrial operations	Chemical Control Corp.	Industrial waste disposal	Seepage of miscellaneous toxic substances into surface and groundwater; explosion releasing contaminants into the air	Suspected carcinogenicity	104,406	4	8,800
Love Canal	Niagara, New York	Chemical manufacture	Hooker Chemical	Industrial waste disposal	Seepage of miscellaneous toxic substances into surface and groundwater	Suspected carcinogenicity	1,000	300	21,800
Positive Chemical	Travis, Staten Island, New York	Miscellaneous industrial operations	Positive Chemical/Chelsea Terminal	Industrial waste disposal	Seepage of miscellaneous toxic substances into surface and groundwater	Suspected caroinogenicity	130	6	1,200

Source: R. Zimmerman, "The Management of Risk". Report to the National Science Foundation, TARA group. 1982.

In these cases, corporations involved in the marketing or use of risky chemicals displayed discrete response patterns to uncertainties in the regulatory and risk environments. The complexity of the regulatory environments experienced by companies in each of the cases is shown in Table 2. The major programs included permits to discharge wastewater or dredge spoil into natural waterways and permits to manufacture, use, or dispose of hazardous substances.

CORPORATE RESPONSES TO CHEMICAL RISKS

Corporate responses to risks posed by chemicals fall into a number of distinct categories. These responses may be input-related (e.g., sources of raw materials), output-related (in terms of product markets and clients), or internal (in terms of internal organizational characteristics and behavior patterns).[2]

Corporate responses to risks from chemicals are more apparently related to organizational outputs and internal environments. Responses in the form of manipulating the output environment tend to fall within several of the categories suggested in the Strategic Diversity Classification System developed by Nathanson and Cassano,[3]

TABLE II

REGULATORY PROGRAMS APPLICABLE TO SELECTED HAZARDOUS WASTE PROGRAMS

	Regulatory Program										
	Indirectly Related to Human Health							Directly Related to Human Health			
	CWA-404 Permit	Wetland, Stream Permit	NPDES Permit	RCRA Permit	Superfund	EIS	CAA Permit	OSHA	Pesticides		Public Water Supply Permit
									Registration	RPAR	
Union Carbide (Aldicarb)	–	–	–	–	–	–	–	–	X	–	X
Occidental (DBCP)	–	–	–	–	–	–	–	X	X	X	X
B. F. Goodrich (VC)	–	–	–	–	–	–	–	X	–	–	–
Allied (Kepone)	–	–	X	–	–	–	–	–	X	X	–
G.E. (PCB)	X	(X)	X	(X)	(X)	X	–	X	–	–	–
Ventron (Mercury)	(X)	(X)	X	(X)	–	X	–	X	–	–	–
Marathon Battery (Cadmium)	X	X	X	–	X	–	–	X	–	–	–
Texaco	–	–	–	X	–	–	–	–	–	–	–
Hooker	–	–	X	–	X	–	–	–	–	–	–
Chemical Control	–	–	–	(X)	–	–	–	–	–	–	–

Abbreviations: PBCP = Dibromochloropropane
VC = Vinyl Chloride
PCB = Polychlorinated Biphenyl

Note: X's in parentheses indicate programs that are under consideration for pending plans.

namely, market diversity and product diversity. In addition, some response patterns tend to focus upon internal organizational environments, such as various forms of organizational diversification. Finally, there are simple response categories such as negotiated cleanups, property transfers, or cessation of operations. Table 3 lists the cases organized by activity and major corporate entity, and the responses of these corporate entities to the societal risks their activities posed.

MARKET DIVERSIFICATION

Nathanson and Cassano define market diversity as the degree to which one or more unrelated markets are pursued or "the number and relatedness of . . . 'strategic business units' - business groups with distinct customer needs, competitors, and channels of distribution."[4] The companies categorized by Nathanson and Cassano as having highly diverse markets that were involved in the cases reviewed are B.F. Goodrich (Vinyl Chloride case), Union Carbide (Aldicarb), and General Electric (PCBs). Several of the other companies reviewed here, but not categorized by Nathanson and Cassano, can be viewed as highly diverse from a market point of view (as indicated by their descriptions in industrial directories); namely, Texaco (by virtue of being a strongly vertically integrated firm) and Occidental. Ventron and Marathon Battery had fairly small markets, and thus had a low degree of market diversity. The relative degree of market diversity of these companies is summarized in the first column of Table 4.

In two of the cases involving pesticides, the response to societal pressures to reduce risk as measured by the intensity of regulatory action occurred in the form of the reduction in the firms' market diversification for the product considered as having a high risk. In the case of Aldicarb, the Aldicarb market on Long Island was eliminated i.e., the registration of the pesticide was changed to exclude its use on Long Island, though it continued to be used in other parts of the country. In the case of Dibromochloropropane (DBCP), the market in California was eliminated, though its use continued outside of the United States and for emergency purposes within the U.S. (e.g., Florida). Thus, at least in the pesticide area, firms with a high degree of market diversity already, were able to further manipulate the diversity of their markets in order to avoid regulatory sanctions associated with the marketing of risky chemicals.

PRODUCT DIVERSIFICATION

Nathanson and Cassano defined product diversity as the number of "manufacturing nexuses" or raw material, process, and technology sets.[5] Thus, if a company uses the same inputs, processes, and technologies to produce many different outputs, it will only have a low

TABLE III

PRIVATE SECTOR ORGANIZATIONAL RESPONSE PATTERNS TO RISKS FROM HAZARDOUS SUBSTANCES FOR SELECTED CASES

Cases: By Activity and Major Corporate Entity	Responses					
	(I)	(II)	(III)	(IV) Site-Related Actions		
	Market Diversification	Product Diversification	Organizational Diversification	Negotiated Clean-Up	Cessation of Operations	Sale of Site
Manufacturing (Product is sold directly)						
Union Carbide (Aldicarb)	X	–	–	–	–	–
Occidental (DBCP)	X	–	–	–	–	–
B.F. Goodrich (VC/PVC)	–	–	–	X	–	–
Allied Chemical (Kepone)	–	–	X	–	–	–
Use (In manufacturing another product)						
General Electric (PCB)	–	X	X	X	–	–
Ventron (Mercury)	–	–	X	–	X	–
Marathon Battery (Cadmium)	–	–	–	–	X	X
Disposal						
Texaco	–	–	–	X	–	–
Hooker Chemical	–	–	–	–	–	X
Chemical Control	–	–	–	–	X	–

Abbreviations: DBCP = Dibromochloropropane
VC/PVC = Vinyl Chloride/Polyvinylchloride
PCB = Polychlorinated Biphenyl

TABLE IV

MARKET AND PRODUCT DIVERSIFICATION IN SELECTED INDUSTRIAL FIRMS

Firm	Market Diversi-fication	SIC Codes as Indicators of Product Diversity: 2-digit codes	Total Number of 4-digit codes	4-digit categories within SIC 28	Overall Product Diversity
Allied Chemical	Medium	14	27	8	High
General Electric	High	3	6	0	High
B.F. Goodrich	High	7	11	2	Medium
Hooker Chemical	Medium	3	6	4	Low
Occidental Petroleum	High	4	7	4	High
Texaco	High	5	10	0	Medium
Union Carbide	High	10	30	9	High
Ventron	Low	2	3	2	Low
Marathon Battery	Low	-	-	-	Low

Note: Source of SIC classifications by firm is Standard & Poor's Register of Corporations, Directors and Executives, Volume 1. 1983. New York, N.Y.: Standard & Poor, 1983. Product diversity is a combination of the Nathanson and Cassano classification and the SIC codes.

Market diversity is based on extrapolations from the Nathanson and Cassano classification.

degree of product diversity. The companies in the cases reviewed that were categorized as having very high levels of product diversity were Union Carbide and General Electric. Standard Industrial Classification (SIC) codes, as they are defined by the U.S. Department of Commerce, are also indicative of product diversity. Table 4 lists the companies involved in the cases reviewed and the number of two and four digit SIC codes typically assigned to them. In addition, the number of four digit SIC codes within the chemical and allied products group (SIC 28), in which many of the chemical risks are concentrated, is also listed. On the basis of the SIC code listing, Allied Chemical could also be classified as having a high degree of product diversity. Texaco and B.F. Goodrich have only a moderate degree of product diversity, according to the Nathanson and Cassano study, even though they have a high degree of market diversity.

Only one of the cases involved a diversification of its product as a result of risk regulatory actions. General Electric, a company already characterized by a high degree of product diversity, diversified its product even more by modifying the manufacture of its capacitor using a substitute for PCBs.

ORGANIZATIONAL DIVERSIFICATION

Diversification is a multidimensional concept, referring to the degree to which organizations have or add specialized units, the nature and location of these units (Eads, 1982; Freedman, 1981), and changes in ownership patterns.

Specialized Units - Trends and Location

The most common organizational response to environmental uncertainty is the formation of specialized units to respond to the sudden recognition of risk from a particular product or in anticipation of risks ensuing from the use of existing products. In the area of chemical risk management, this usually refers to units specializing in product safety, industrial hygiene, environmental health, and similar concerns. Eads,[6] adapting the data from McGuire, 1979, points out the following trend in the formation of specialized product safety units:

	Non-Divisionalized	Divisionalized		
		Corporate	Group	Major Division
Time period:				
Early 1970s	26%	25%	22%	33%
Late 1970s	79	73	58	74

Percentages indicate the percentage of responding firms in the particular category that had organizational entities located in the particular functional area.

Regardless of the form that specialized units took, the percentage across all categories has increased dramatically over the decade of the 1970s. In divisionalized companies, the most dramatic change has been at higher levels of management, i.e., major divisions and corporate levels, rather than group levels.

These specialized units generally take a number of different forms, and follow a pattern similar to that occurring in the early 1970s when specialized untis emerged to cope with general environmental pollution issues rather than environmental health issues. As a matter of fact, a number of the environmental policy or environmental control groups formed in the early 1970s were transformed into or supported the environmental health groups that emerged a decade later.

According to Eads, the form product safety units very much reflect the philosophy of the organization. These units, he argues, can have an operational orientation (i.e., to improve the design and operation of a product in order to reduce risks), a legal orientation (functioning to help the company cope with liability defenses), or a regulatory orientation (as an interface with regulatory agencies).[7] Eads generally favors the former as being more responsive to environmental health issues, though he understands the need for the latter two orientations as well. He recognizes further that the operational orientation can be in conflict with the goals of the legal and regulatory approaches. He concludes that this form of corporate response has not been successful overall in responding to product safety issues: "...as an alternative to regulation, they appear to have proved an extremely blunt instrument, at least in the current legal environment."[8]

In addition to the existence of these units and the type they assume, is the issue of where in the organizational structure these units that deal with risk should be located. A recent study by The Conference Board explored some corporate case studies in this area for very large corporations (Freedman, 1981). Most of the units reported to vice presidents rather than to senior vice presidents directly. To a large extent this may be a function of the size of the organization rather than its philosophy. As important as what level of officer is in charge of the environmental health function is what type of officer. In some cases, e.g., Johns Mansville, officers are named specifically in the environmental health or safety areas. In other cases, e.g., Atlantic Richfield and Monsanto, the officer is in the engineering professions or technology or, as in the case of Allied Chemical, in the area of operations. Depending upon where this function is placed in the corporation, these professionals may not have the authority to overrule or influence corporate decisions. The system of reporting and authority for the firms covered in the Conference Board study is summarized in Table 5.

The sizes of these units depend substantially on the types of products, company resources, traditions, and visibility of the risks. Many of the larger corporations involved in the cases reviewed had corporate or divisional environmental units as far back as the early 1970s. General Electric did create corporate level units to deal specifically with the PCB issue, not only in the Hudson River but nationwide as well. Smaller companies, such as Marathon Battery and Ventron never had such units nor were known to have developed any. Firms like these tend to rely upon engineering and environmental expertise outside of the firm.

Transformation of Corporate Ownership Patterns in Response to Regulation

Companies typically change ownership (including the transfer of property) in response to regulatory pressures, in an attempt to transfer liabilities to companies with smaller assets. This takes a number of different forms and occurred in several of the cases.

TABLE V

SPECIALIZED UNITS IN SELECTED CORPORATIONS FOR ENVIRONMENTAL HEALTH AND SAFETY

Company	Reporting Unit	Officer
Johns Mansville	1 manager, 1 chief environmental scientist (VP), 1 medical director	Sr. VP for Health, Safety, & Environment
	5 managers	VP, Chf. Env. Scientist
DuPont	Medical Division, Safety & Fire Protection Division	General Manager who reports to VP
Exxon	Medicine & Environmental Health Dept., Science & Technology Dept.	Senior Vice President
General Electric	Clinical Medicine; Occupational Medicine & Environmental Health; Safety & Environmental Quality	VP for Corporate Health & Safety; Medical Dir.
	Environmental Audit Team in Environmental Protection	VP for Corporate Operating Services
Atlantic Richfield	Occupational & Environmental Protection	VP, Technology
Allied	VP for Environmental Affairs (with 6 directors reporting)	Sen. VP, Operations

Summarized from: A. Freedman, "Industry Response to Health Risk." New York, N.Y.: The Conference Board, 1981.

Ventron/Velsicol-Mercury in Berry's Creek. For over 45 years a number of corporate transformations occurred while the chlor-alkali plant was operating in the Hackensack Meadowlands. F.W. Berk & Co., the initial owner of the property and plant, sold the property to Velsicol in 1960 as the New Jersey Department of Environmental Protection (NJDEP) began increasing its concern over violations of water quality standards. Velsicol, in turn, formed a wholly-owned subsidiary, the Wood-Ridge Chemical Corporation (WRCC). Seven years later, WRCC sold 33 acres, including the plant, to Velsicol, making WRCC a wholly-owned subsidiary of Ventron. The plant ceased operations in 1974 at which time Ventron had sold the property to Wolf, while WRCC retained 7.1 acres of land. This was after Ventron had unsuccessfully tried to negotiate a number of wastewater treatment arrangements with municipalities and to design an on-site system as well. The plant closed after attempts at meeting the requirements of the new National Pollutant Discharge Elimination System (NPDES) for wastewater discharge permits were unsuccessful. It was under this program that mercury per se was listed as a wastewater discharge parameter rather than just alkalinity. When the firms were ultimately brought to court, Velsicol tried to argue that stock ownership in a subsidiary was not enough to make it liable "for the torts of the latter," citing Mueller v. Seaboard Commercial Corporation, 5 NJ 28, 34 (1950). The court ruled that while personal liability cannot be avoided through the corporate form:

> The court does not, however, conceive the governing standard to be the same for stockholders seeking to avoid the usual contact or tort liability as for a 100% stockholder, who with knowledge, allows the operating corporation to violate environmental standards, create or continue a public nuisance, or in such a manner allow that subsidiary to act in the face of the policy of the State. Both Velsicol and Ventron had the ability through its 100% stock ownership to control its subsidiary. Liability is justified where the parent, with knowledge, fails to act. The corporate shell may not be used as a means to evade the thrust of an environmental control statute.[9]

In spite of these strong declarations about the improper use of the corporate form, the court ruled that the companies did not act negligently. It declared that: "The conduct of those defendants was reasonable in light of the state of knowledge as it then existed."[10] These transformations are shown in Figure 1.

<u>Marathon Battery-Cadmium</u>. A similar, but simpler, set of circumstances involving extensive corporate transformations occurred in the Marathon Battery case. Gould, Inc., was a parent company of Clevite, Inc., who owned 100 percent of Sonotone stock. It was during the period that the Sonotone Corporation operated the plant that the worst discharges are said to have occurred, in the opinion of government officials. It was during that period that NJDEP issued orders to the company to stop wastewater discharges. In the late 1960s, after unsuccessfully dealing with the water pollution problem, Sonotone sold the plant to Marathon Battery, after the State told Sonotone to clean up its alkaline waste discharges into Foundry Cove.[11] The only other property transfer that occurred was well after the cadmium had been dredged from Foundry Cove and entombed on the site (but was found to be leaking back into the cove). As a result of leakage from the dredge spoil site, the site became a Superfund site around 1980. About that time the site was sold to Merchandise Dynamics who used the area as a warehouse for books. Occupational hazards were investigated as a result of allowing that use to exist. Prior to the sale to Merchandise Dynamics, Marathon Battery had unsuccessfully tried to deed

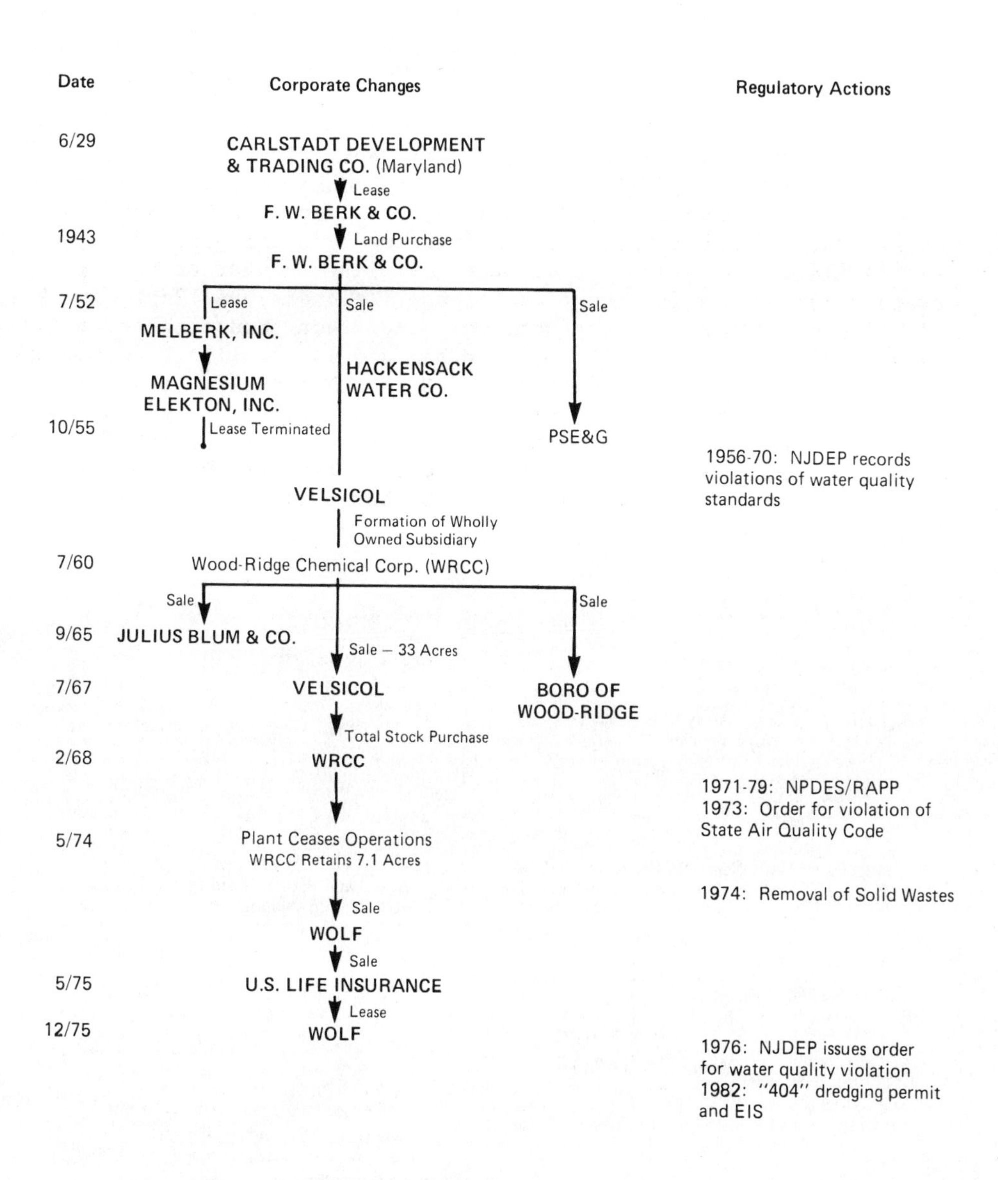

Source: Compiled from McCormick, 1977: 5-6; Tasher, 1981, 209+.

Figure 1

Corporate Developments and Property Transfers in the Ventron/Velsicol-Mercury Case Berry's Creek, New Jersey

the property to the Village of Cold Spring. In spite of these changes in ownerships, Gould, Inc., the parent company, was held responsible for the site clean-up. Figure 2 shows the course of events in this case with respect to corporate changes.

Allied Chemical-Kepone. The Allied Chemical production of Kepone represents another set of corporate arrangements that were somewhat unique. Allied Chemical used "tolling" agreements with other companies, rather than outright stock or property transfers, to reduce its liability. Goldfarb[12] tells the following account of the corporate arrangements for the production of kepone since 1949. In 1949, Allied Chemical's Hopewell plant, which had been producing fertilizers, began producing kepone. Allied entered into two "tolling" agreements to produce the chemical, one with the Nease Chemical Company from 1958 to 1960 and the other with Hooker after 1960. "Tolling is a common arrangement in the chemical industry whereby another company performs processing work for a fee or 'toll' and then returns the final product. The keynote of a tolling arrangement is that during the processing

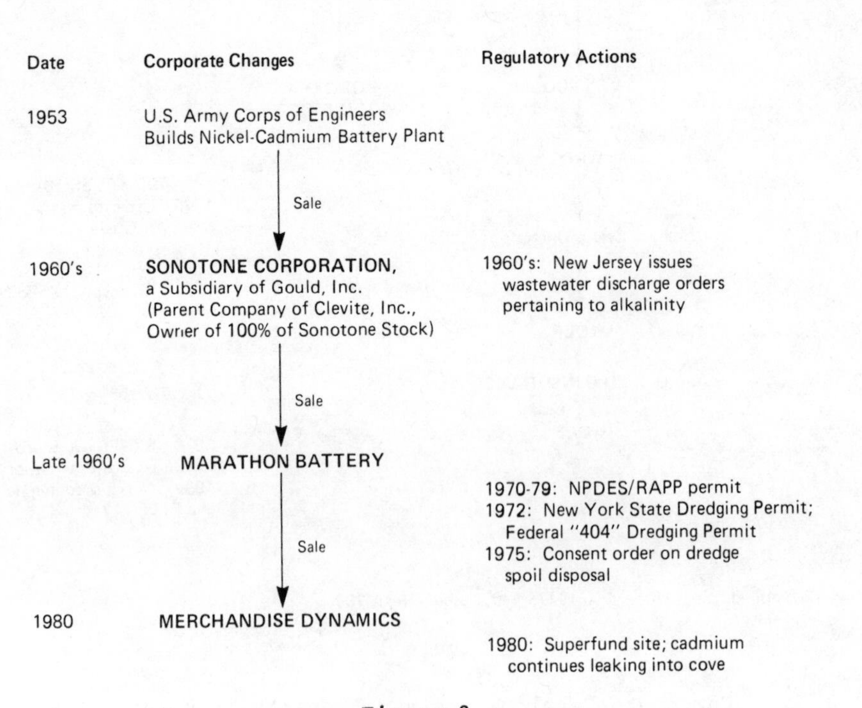

Figure 2

Corporate Developments in the Marathon Battery-Cadmium Case
Foundry Cove, New York

period legal title to the materials and product remains in the supplier, in this case Allied."[13] In the mid-1960s, Allied's Semi-Works facility in Hopewell produced the pesticide under the Agricultural Division first, but then transferred it to the Plastics Division in 1973. This transfer was to shift the priorities for manufacturing the pesticide relative to other products. In 1973, another pesticide, THEIC, was favored over kepone, and it was decided to toll kepone. Because of some personal locational needs on the part of the two Allied executives that had been in charge of kepone production, they decided to form Life Sciences Products (LSP) Co., and to toll kepone from Allied through this corporation. Allied was ultimately fined $13.2 million in the case, LSP was fined in accordance with its assets, and the owners were personally fined as well. Allied was exonerated from charges of conspiracy with LSP, since in this case the two companies were considered one and the same, and a corporation cannot conspire with itself. For an analogous reason, Allied was also exonerated from the vicarious liability charge, i.e., of aiding and abetting LSP to commit a crime. These events are outlined in Figure 3.

Positive Chemical-Hazardous Waste Disposal. Positive Chemical, a waste disposal case, was involved in a similar set of occurrences with the transition from Positive Chemical to Chelsea Terminals, Inc., to prevent the potential problems of litigation when the NYS DEC took action against the site (see Figure 4). Texaco, the site owner, was ultimately responsible for the cleanup.

Public utilities show organizational diversification patterns similar to those of private corporations, though they theoretically operate under very different circumstances from those of private corporations. The organizational responses by some of the utilities operating nuclear power plants where risks have come close to being realized particularly bear some similarities to the corporate situation described above for chemicals. Martin,[14] in his description of the Three Mile Island (TMI) accident, goes into a lengthy description of the utility's organizational arrangements as it pertained to TMI (see Figure 5). The General Public Utilities (GPU) Corporation headquartered in New York is the holding company for Jersey Central Power and Light (since 1949), Pennsylvania Electric Company, and Metropolitan Edison (the operator of TMI). In 1967 GPU formed a subsidiary, the Nuclear Power Activities Group, which was given the responsibility for planning and building plants. The three other companies were to retain the responsibility of running the plants. The construction of TMI II was a result of the failure of GPU to build a second unit at Oyster Creek during the late 1960s because of labor corruption practices. In order not to lose its investment in the design and purchase of equipment, it decided to move the plant to Pennsylvania and make Met Ed half owner and the other two companies quarter owners.[15] Because of the corporate structure, in spite of the fact that Met Ed

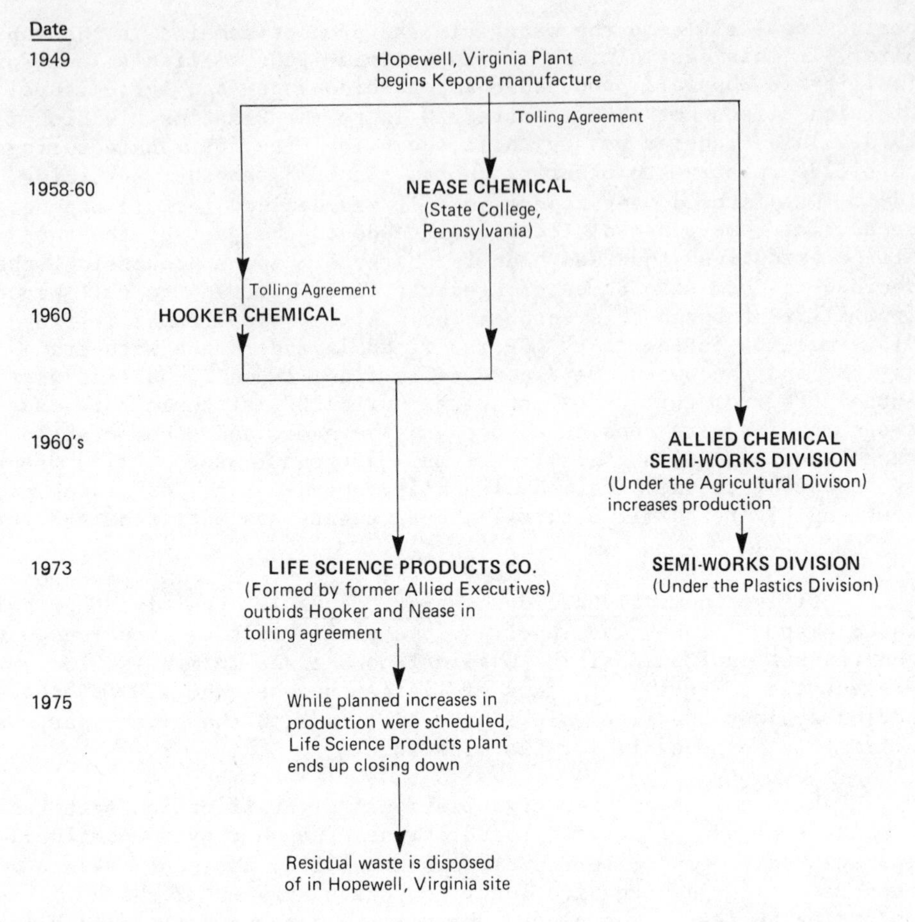

NOTE: Compiled from Goldfarb, 1980.

Figure 3

Corporate Developments in the Allied Chemical-Kepone Case

owned half of the plant, it was unable to make certain changes with regard to the construction of the plant. For instance, they wanted to make the two central control rooms for TMI I and II analogous. That and other changes were difficult to make, given the degree to which the planning for the Oyster Creek unit had advanced. Several of these factors, according to Martin, contributed significantly to the accident and the recovery phase.

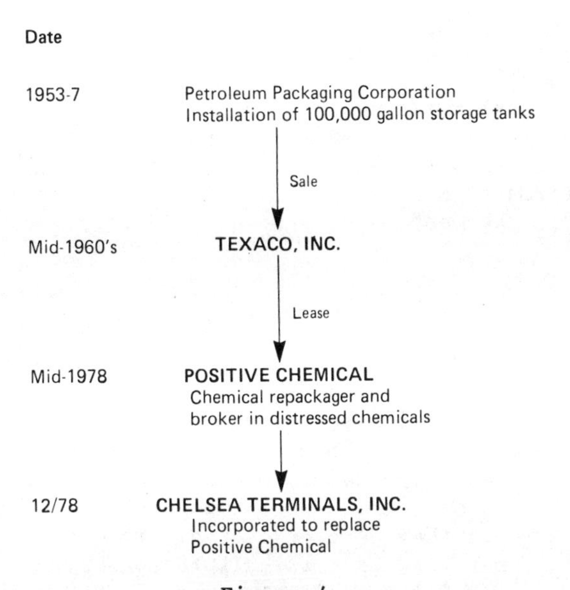

Figure 4

Corporate Developments in Positive Chemical Waste Disposal Case
Travis, Staten Island, New York

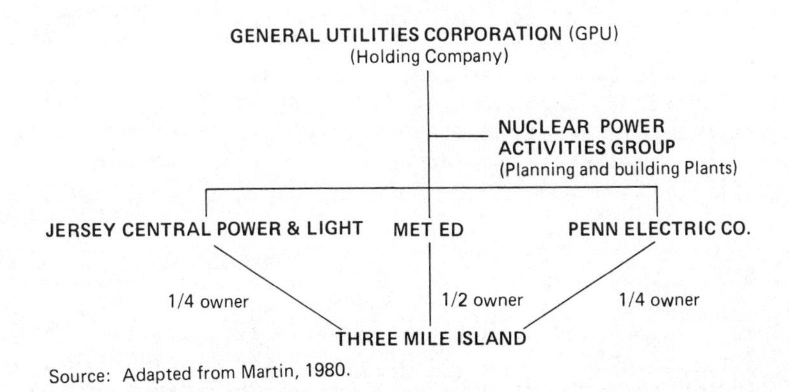

Source: Adapted from Martin, 1980.

Figure 5

Utility Arrangements in the Three Mile Island Case

OTHER RESPONSES

Other responses or strategies that do not clearly fall into any neat category (or may accompany the three major strategies outlined above) are cessation of operations, sale of the site, and negotiated cleanups. The former two might be considered a type of market diversification in reverse, i.e., market consolidation. This response is typical of chemical disposal operations, where there is little room to maneuver in terms of market diversity (in this context, finding another location for the wastes) or product diversity (transformation of wastes).

CONCLUSION

When corporations involved in the manufacture, use, or disposal of chemicals are faced with the management of risks from these chemicals, they appear to react in a number of discrete ways. The firms that generally have a high degree of market diversity rely upon that strategy to cope with a stringent regulatory environment regarding the risks from chemicals they were producing. Product diversity may act in a similar way but this is difficult to confirm, since only one firm used a product diversification approach to its risk problem.

Organizational diversification is the most complex of the strategies for coping with risk. This strategy generally takes two forms: the formation of organizational entities to deal with chemical risk, and various corporate transformations involving the sale or transfer of property and holdings to reduce financial risks that are foreseen from the manufacture, use, or disposal of chemicals. The formation of corporate or divisional entities to deal with environmental and occupational health risks seems popular among the larger firms, and is a more long-term approach to general firmwide risks, rather than being linked specifically to a particular risk issue. Except in the case of one large chemical company, Allied, the strategy of organizational diversification in the form of corporate transformations appears to be popular among smaller firms marked by a very low degree of market and product diversification.

REFERENCES

G.C. Eads, "Increased Corporate Product Safety Efforts: A Substitute for Regulation?", in Social Regulation. Strategies for Reform, Institute for Contemporary Studies, San Francisco, CA (1982).

A. Freedman, "Industry Response to Health Risk," The Conference Board, New York, NY (1981).

W. Goldfarb, "Kepone: A Case Study," in Water Quality Management, B. Lamb, ed., Ann Arbor Science, Ann Arbor, MI (1980).

B. Hazen, "Cadmium in an Aquatic Ecosystem," Ph.D. dissertation, Institute of Environmental Medicine, New York University, Sterling Forest, NY, February 1981.

D. Martin, Three Mile Island: Prologue or Epilogue?, Ballinger, Cambridge, MA (1980).

D.A. Nathanson and J.S. Cassano, "Organization, Diversity, and Performance," Wharton Magazine, Vol. 6 (Summer, 1982), pp. 19-26.

D. Nelkin, and M. Pollak, "Consensus and Conflict Resolution," in Technological Risk, M. Dierkes, S. Edwards and R. Coppock, eds., OG&H, Cambridge, MA (1980).

S.A. Tasher, "Technical Elements of the Government's Case," Ventron Case, in Hazardous Waste Litigation, R.M. Mott, ed., Practising Law Institute, New York, NY (1981).

NOTE: Part of this research was supported by a grant from the National Science Foundation, Division of Policy Research and Analysis, TARA group (Grant No. PRA-8209795).

FOOTNOTES

1. To a large extent the problems in the regulatory environment that corporations are faced with today with respect to the marketing of chemicals can be looked at in terms of the typical responses of organizations to changes in their external environments in general, about which there is a large empirical and theoretical literature.
2. Models of corporate response to risk tend to parallel and enrich the models that have developed primarily in the field of conflict resolution to explain multiple party responses to conflict in general and environmental problems in particular (see for instance, Gladwin, 1981, and Nelkin and Pollak, 1981).
3. D.A. Nathanson and J.S. Cassano, "Organization, Diversity, and Performance," Wharton Magazine, Vol. 6 (Summer 1982), pp. 19-26.
4. Ibid., p. 21, The specific categories they use range from one business market to seven or more unrelated business markets.
5. Ibid.
6. G.C. Eads, "Increased Corporate Product Safety Efforts: A Substitute for Regulation?", in Social Regulation, Strategies for Reform, Institute for Contemporary Studies, San Francisco, CA (1982).
7. Ibid., p. 298.
8. Ibid., pp. 312-313.
9. S.A. Tasher, "Technical Elements of the Government's Case," Ventron Case, in Hazardous Waste Litigation, R.M. Mott, ed., Practising Law Institute, New York, NY (1981), p. 258.
10. Ibid., p. 270.

11. B. Hazen, "Cadmium in an Aquatic Ecosystem," Ph.D. dissertation, Institute of Environmental Medicine, NYU, Sterling Forest, NY (February, 1981), p. 3.
12. W. Goldfarb, "Kepone: A Case Study," in Water Quality Management, B. Lamb, ed., Ann Arbor Science, Ann Arbor, MI (1980), pp. 130-131.
13. Ibid., p. 131.
14. D. Martin, Three Mile Island: Prologue or Epilogue?, Ballinger, Cambridge, MA (1980), pp. 1-9.
15. Ibid., p. 6.

THE CONTROL OF THE USE PATTERN AND THE LIFE OF METALS FOR RISK MANAGEMENT: THE CASE OF LEAD

Kazuhiro Ueta*
Tomitaro Sueishi**

*Environmental Economics Unit
Kyoto Institute of Economic Research, Kyoto University
Yoshioda-Homachi
Sakyo-ku
Kyoto, 606, Japan

**Department of Environmental Engineering
Osaka University
Yamadakami, Suita
Osaka, 565, Japan

INTRODUCTION

This paper proposes a new paradigm of socio-academic study for achieving a balance between the utility of metals and the hazards of their waste. One of the major dilemmas of our time is how to make decisions about dealing with unknowable and yet conceivable environmental risks. Uncertainty about risks of pollution and uncertainty about costs of reducing pollution allow enormous scope for distortion in representing what can or should be done. Risk assessment based on currently available information can lead to over-confidence about the knowledge of ways in which accidents can occur. Most of the data on the risks of metal waste are not connected with those on the benefits of metal products. It may well be that people, in assessing the risks, are also analyzing the social worth of the technology using metals, the credibility of the science behind that technology, and the promoting and governing institutions that foster and create it. An assessment, therefore, must be made of the total metal cycle and not just an isolated part of it. Risks are involved at all stages--mining, the manufacture of metal, and metal waste management processes.

It is very difficult to estimate the risk of metal wastes when they have gradually accumulated in the environment, because in the field of dose-response relationship, the effect of long-term accumulation of heavy metals in the environment has not been resolved. Taking into account the interactions among many substances in the environment, it is difficult to clarify the cause-effect relationship between heavy metals and the ecological system. Therefore, in reducing risks it is more important to make clear the decisionmaking and the using process of metals than to adhere to the development of new cause-effect relationship models. The main factors that influence the frequency and the magnitude of risk caused by heavy metals are:

- The time between the utilization of heavy metals by human society and the disposal or discarding of those materials
- The use pattern of heavy metals
- The spatial distribution of heavy metal products and their waste
- The organization of and the institutions concerning the metal recycling system.

In spite of the presupposition that the recycling of heavy metal waste reduces the metal pollution and environmental risks, the organization of an adequate recycling system cannot be achieved simply by increasing the price of primary metals and the cost of pollution control without a social assessment of metal use. A recycling plan should not only accelerate the improvement of recycling technology, but also establish an evaluation and policymaking system for using and disusing metals. An attempt has been made to establish the planning theory of a totalized social-environmental metal recycling system in the following study which will be referred to as a "socio-metallic study" conducted under the following headings (Sueishi and Ueta, 1981):

- A study of the socio-economic assessment of metal use
- A study of the social cost of discarded metals
- A study on the optimal design of a metal recycling system

In order to develop such a theory, using lead as an indicative metal, the effects of structural changes in the use pattern and the life of lead upon environmental risks were analyzed as a part of the above study and are presented in the following sections.

THE STRUCTURAL CHANGE OF THE USE PATTERN OF LEAD AND ITS EFFECT UPON THE PROBLEM OF WASTE

The average consumption growth rate of non-ferrous metals during the 10 years from 1963 to 1973 in Japan was as follows: lead, 6.8 percent; zinc, 9.8 percent; copper, 12.8 percent; and aluminum, 20.8

percent. The historical change in the value of the substantial GNP (the price of 1970 fiscal year) and the amount of the domestic demand for copper, lead, and zinc is given in Figure 1. Supposing that the values of these indices in 1955 were normalized to 100, the values of the amounts of domestic demands of copper, lead, and zinc were 1078, 558, and 748 respectively in 1973 when the first oil shock occurred. These figures are higher than those for the United States, West Germany, and other industrialized countries and also the growth rate is higher than that of the GNP in Japan during the same period. The industrial structure of Japan during this period of high economic growth became increasingly resource-consumption oriented. The increased amounts of used metals have, in turn, led to various types of social cost concerning the environment.

The demand structure for lead has been drastically changed together with the rapid increase of the amount of lead consumption through the process of high economic growth.

The main uses of lead, the lead content of products and the estimated life of these products, are shown in Table 1. Figure 2 illustrates the annual change in lead supply for the main use categories and the Recovery Effort Index (REI) (Pearce, 1976),

where,

$$REI_a = \frac{SC + RSC + X_s - M_s}{PC + SC + RSC} \quad (1)$$

$$REI_b \quad \frac{SC + RSC + SI + X_s - M_s}{PC + SC + RSC} \quad (2)$$

where,

PC = primary lead consumption,
SC = secondary lead consumption,
RSC = remelted secondary lead consumption,
X_s = exports of scrap,
M_s = imports of scrap, and
SI = scrap input to primary lead refining.

REI_a does not include the Recovery Effort Index with lead scrap input for primary lead refining, while REI_b does. By comparison of REI_a and REI_b, the annual variation shows not only the recovery ratio of scrap but also the distribution of scrap utilization between primary and secondary refining. The total amount of the demand for lead has been increasing except in 1973 when Japan experienced the first oil shock and consequently the demand for lead was less than during the previous year. The increase in the total demand for lead

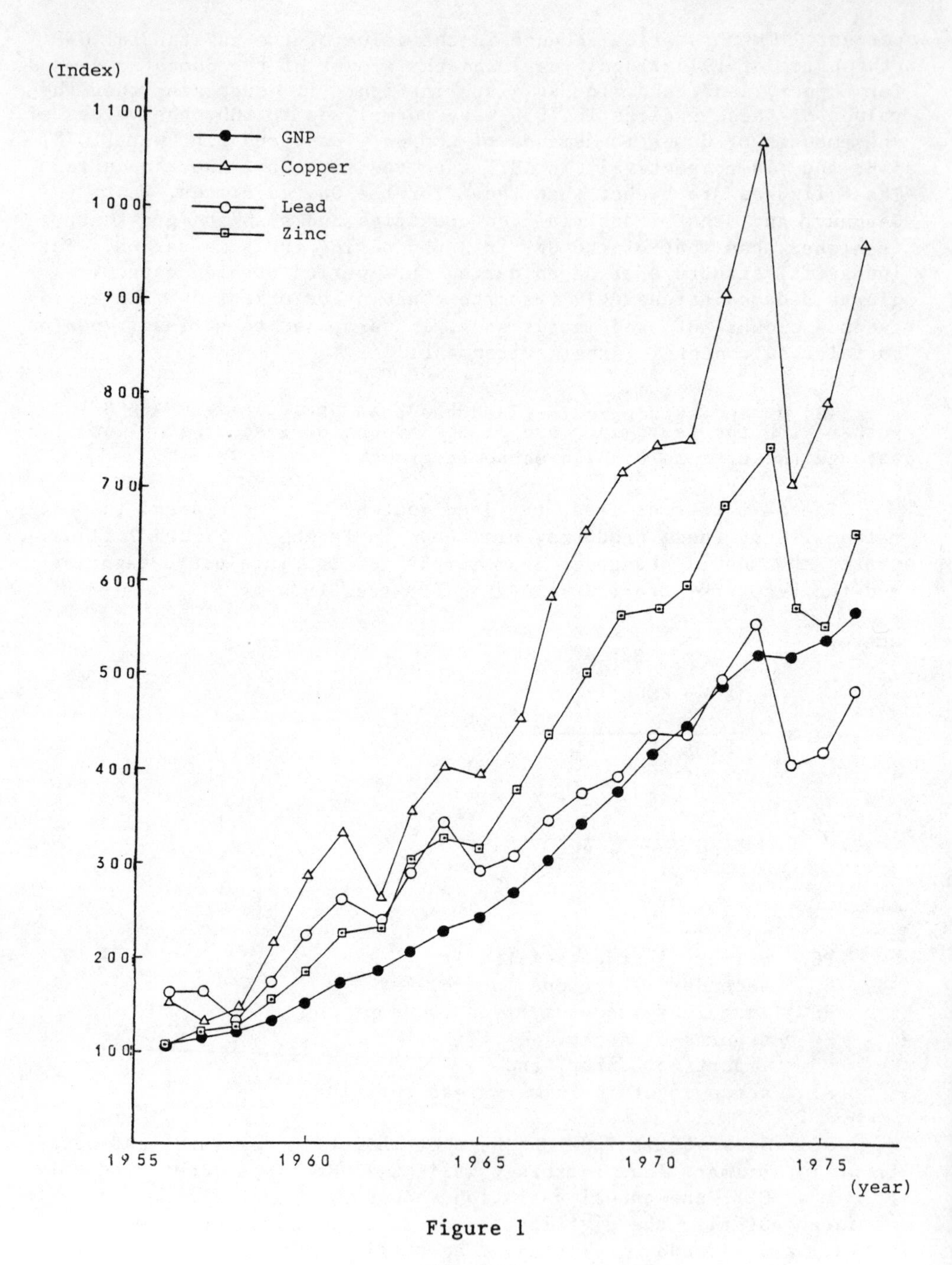

Figure 1

The Annual Change of Substantial GNP and Domestic Demand for Copper, Lead, and Zinc

TABLE I

MAIN USES AND AVERAGE LIFE OF LEAD IN JAPAN

Category		Lead Content (%)	Main Uses	Estimated Life (years)
I Batteries		70	Automobile, Traction, Fixed	2 ~ 3
II Inorganic Chemicals	Litharge	93	Stabilizer for Vinyl Chloride, Drier	1 ~ 6
	Chrome Yellow	90	Yellow Pigment	1 ~ 6
	Red Lead	90	Enamel, Dye for Rubber	1 ~ 6
III Lead Pipes and Sheets		90 ~	Water Pipe, for Chemical Industry	5 ~ 35
IV Antifrictions Metals and Solders		40 ~ 80	Cans, for Machinery Industry	3 ~ 10
V Type Metals		60 ~ 80	Sheet for Printing	2
VI Cables		90 ~	Cable Sheathing	5 ~ 20

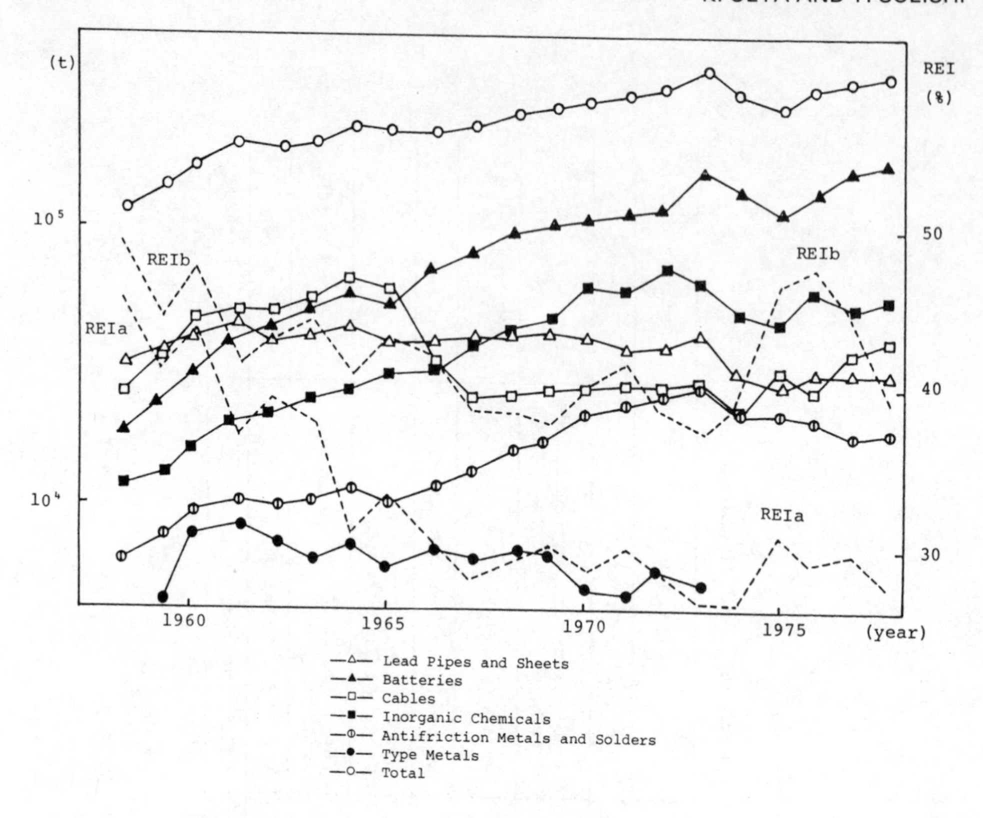

Figure 2

Annual Change of Lead Supply for the Main Use Categories and Recovery Effort Indices

was caused by the increase in the demand for lead for batteries and inorganic chemicals. As can be seen in Figure 2, during the latter half of the 1950s, the demand for lead increased mainly for use in cables and batteries, inorganic chemicals, and lead pipes and sheets. In the 1960s, there was no increase in the amount of lead consumed for lead pipes and sheets because PVC began to replace lead in water pipes. Therefore, the main uses of lead were for batteries and cables. During the latter half of the 1960s, the amount of lead consumed for cables was reduced by almost 50 percent because lead cable sheathing was rapidly substituted with aluminum. On the other hand, the amount of lead consumed for batteries increased with the growth of the automobile industry, and the amount of lead consumed in inorganic chemicals, whose use reflects the increasing sophistication of human life, also rose. Therefore, the amount of lead consumed for both batteries and inorganic chemicals has accounted for 60 to 70 percent of the total demand for lead. One characteristic of the demand structure of lead in the 1970s was the fact that the amount of lead consumed in inorganic chemicals decreased and did not return to the peak of 1972. This phenomenon was caused by (1) the slow recovery of the demand for lead which followed the sharp decrease caused by the first oil shock in 1973 and (2) the decrease in the amount of lead for paint owing to environmental pollution problems. In effect, the main uses of lead changed from goods used by industry in prewar days and immediately postwar to consumer goods afterwards.

The effects of this structural change of the use pattern of lead upon the obsolescent process of lead were as follows:

First, there has been a shortening in the useful life of lead. The average disuse function of lead is:

$$f(\tau) = \frac{\sum_{i=1}^{n} I_i f_i(\tau)}{\sum_{i=1}^{n} I_i} \qquad (3)$$

where $f(\tau)$ = average disuse function,
$f_i(\tau)$ = the disuse function in the use i of lead,
τ = the time elapsed from the input of lead,
I_i = lead consumption for use i, and
i = 1,.....,n.

The disuse function shows the relationship between input and output of material shown in Figure 3. The disuse function of each use of lead is shown in Figure 4. Supposing that the form of the disuse function $f_i(\tau)$ had not changed from the postwar days to 1980, the form of the average disuse function $f(\tau)$ would have changed as shown in Figure 5 according to the change in the use pattern of lead. The peak year in the average disuse function has not changed, but the peak value has increased remarkably. The average life of lead has changed as shown in Figure 6, according to the following definition of the average life of lead:

$$\tau_m = \int_0^\infty \tau f(\tau) d\tau \qquad (4)$$

where, τ_m = average life.

The average life of lead decreased remarkably during the period of high economic growth. This change corresponds to the change in the use pattern of lead as shown in Figure 2.

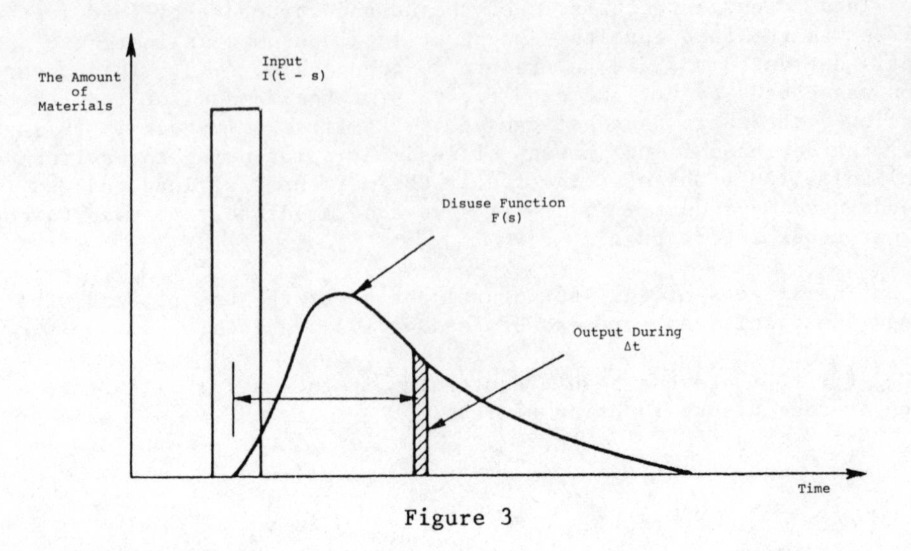

Figure 3

Relation between Input and Output of Materials

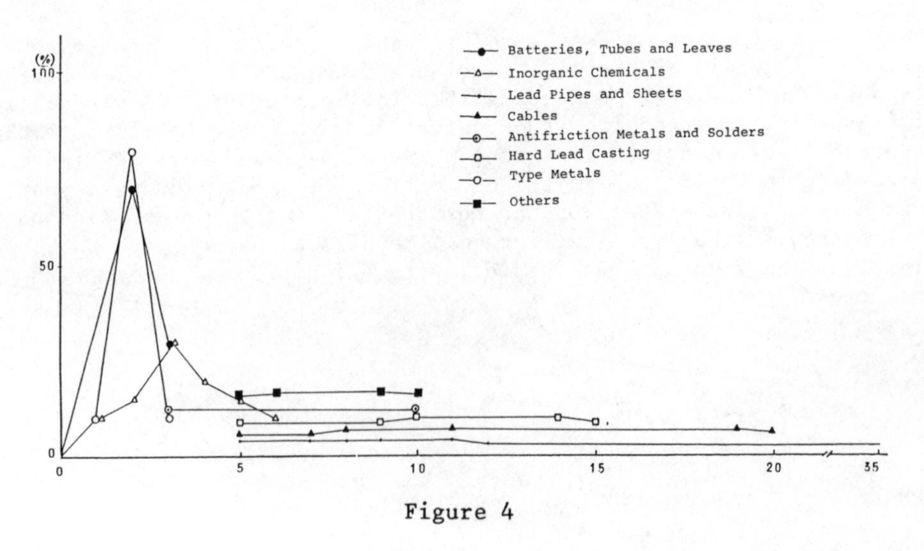

Figure 4

Disuse Function of Lead in Each Use

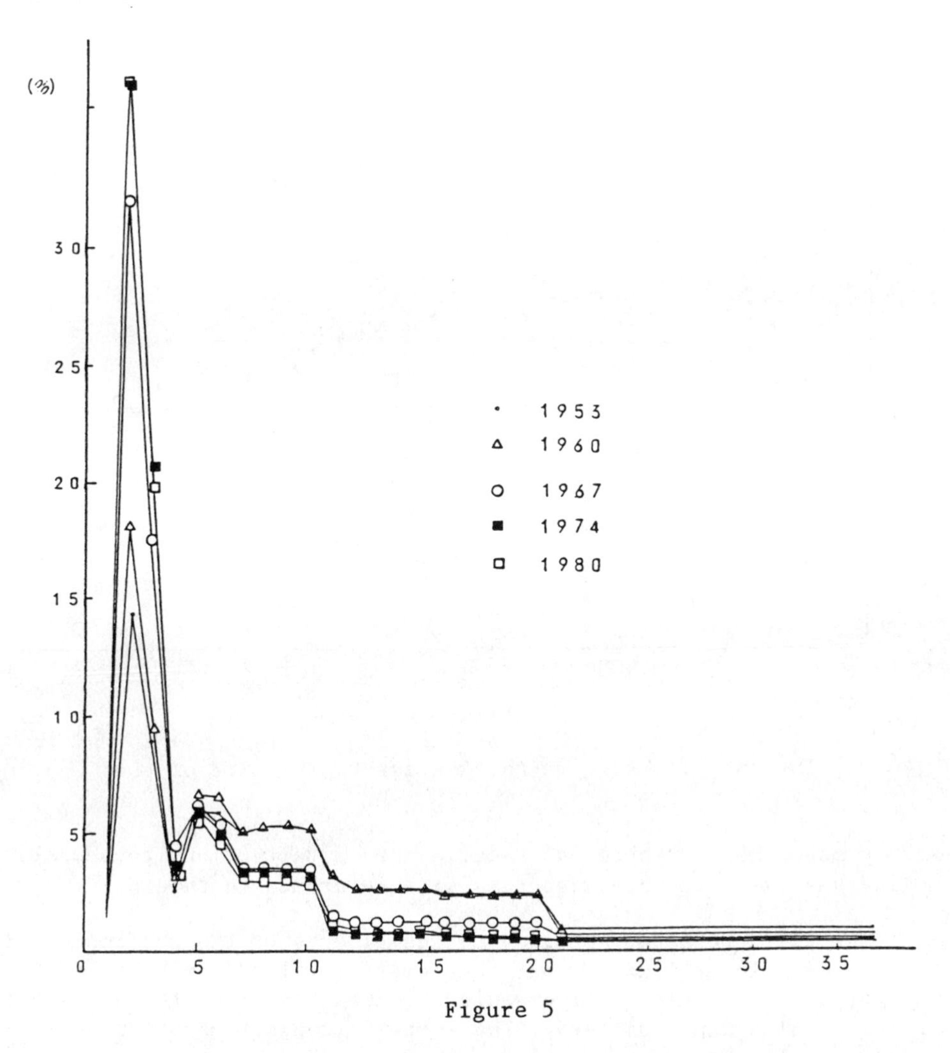

Figure 5

The Change in the Disuse Function of Lead

The second point to notice is an expansion of the spatial distribution of lead with variety in the use phase of lead. It has become difficult to recognize and control the spatial distribution of lead. The increase in the total amount of lead consumption is due not only to the increase in the amount of lead consumed in traditional ways but also to the expansion of new uses of lead. For example, the increase in the total amount of metals consumed for inorganic chemicals caused the increase in the number and the kinds of chemical products that contain metals. The number of kinds of chemical products that contain more than one of the 41 main metal elements increased from about 320 in 1968 to about 720 in 1982 as shown in Figure 7. The

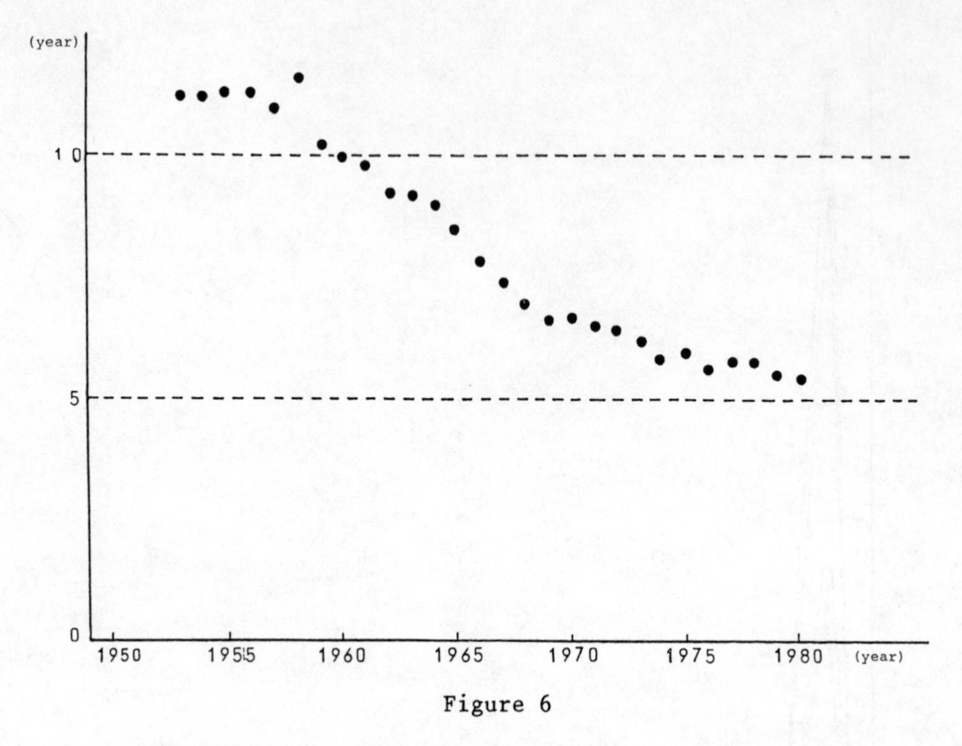

Figure 6

The Annual Change in the Average Useful Life of Lead

safety management of chemical materials as a part of environmental policy has been conducted recently with reference to this.

The third point with regard to the effects of the shortening of the useful life of lead is that the demand for lead for various unrecyclable uses such as inorganic chemicals amounted to 35 percent of the total demand for lead. The economic conditions of realizing a waste recycle system are as follows:

1) A large quantity of waste generation
2) Useful property of waste
3) Recycling technology
4) Demand for the recycled product

Moreover, it is necessary to meet these four conditions simultaneously. Strictly speaking, the possibility of recycling cannot be decided only with reference to what the metals are used for. However, the state of lead wastes can be classified by usage because lead products are used in similar places and similar forms according to each use.

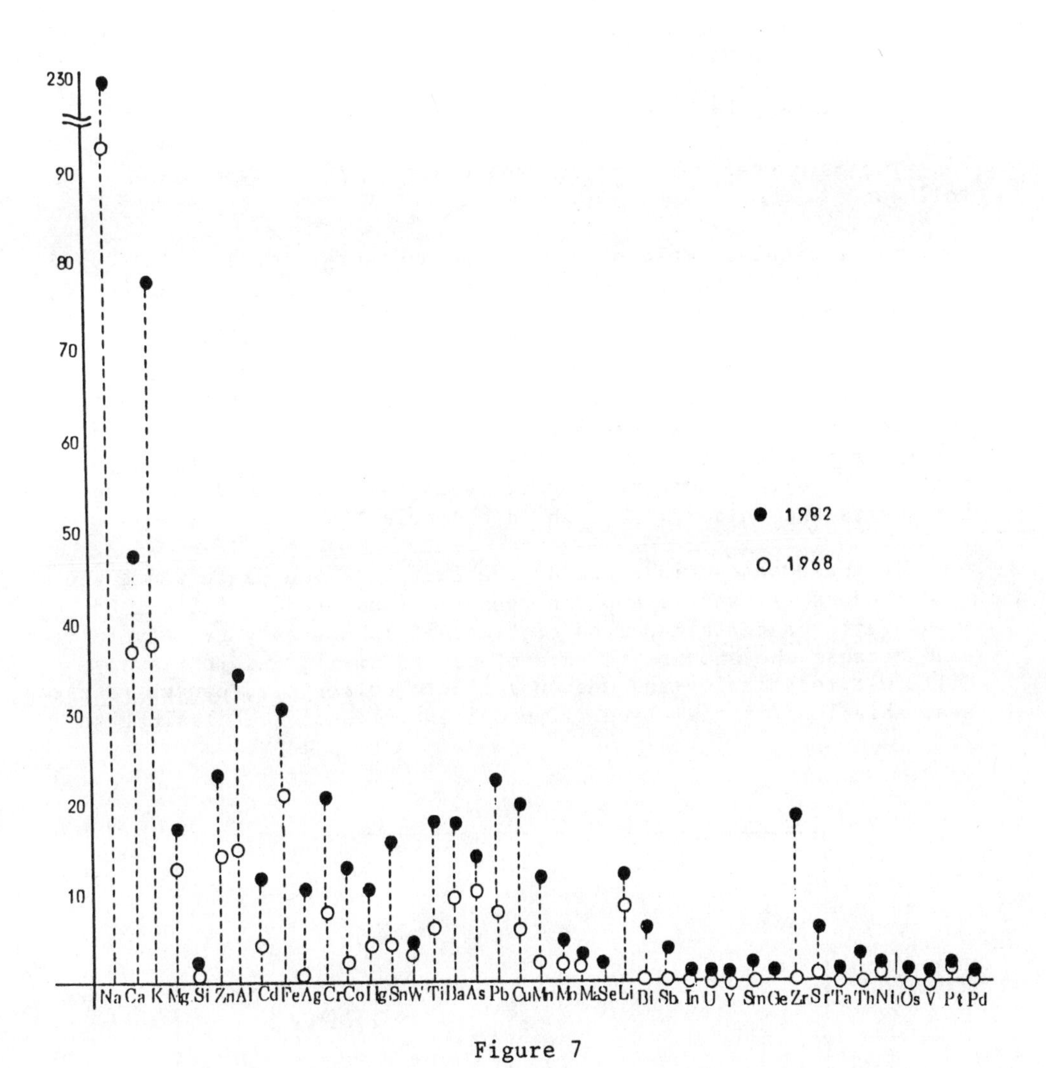

Figure 7

The Change in the Number of Chemical Products Including Metals, by Metal

We classify the pattern of lead waste disposal into the following four types:

1) Recyclable disposal type
2) Non-disposal type
3) Dispersive disposal type
4) Dissipative disposal type

The main uses relating to each waste disposal type are as follows:

The typical example of a recyclable disposal type is lead used for car batteries and for the greater part of lead pipe and sheet. A typical example of a non-disposal type is the lead consumed for radioactive screening sheets over which our society has to have control for extremely long periods of time. Another example of a non-disposable type is the lead used for art objects and cultural artifacts. In these two uses, lead wastes are stored in a particular space and are rarely recycled. Typical examples of a dispersive disposal type are the lead used for inorganic chemicals in paint and lead sound screening sheets and solder in household electrical appliances.

In these cases, in spite of the fact that the place where the lead product is used is apparent and the lead product falls into disuse after a certain period of time, it is not easy to recycle the lead because the amounts in each place are small, and it is very difficult technically and uneconomical to collect and concentrate the available lead in one place. Typical examples of the dissipative

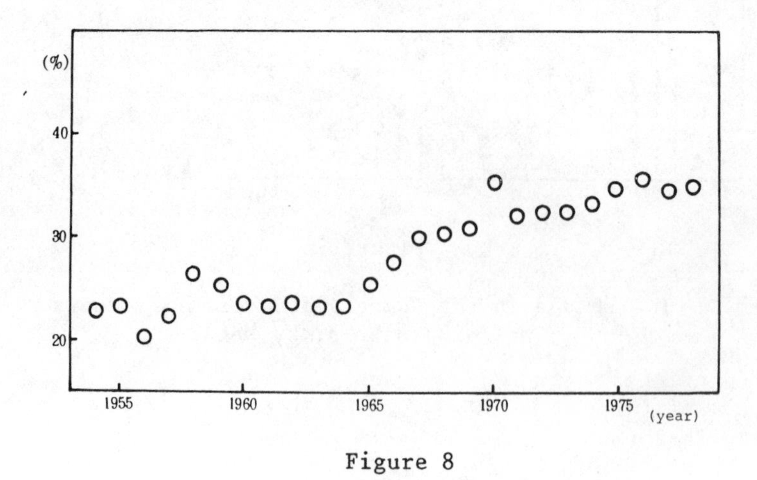

Figure 8

Annual Change of the Ratio of the Demand for Lead in Unrecyclable Use disposal type are tetraethyl lead consumed as an antiknock compound in automobile gasoline, and lead arsenate used in pesticides. In these cases, it is impossible to recycle the wastes because lead is discharged and dispersed into the environment simultaneously with the consumption of the lead product. Lead wastes are not recycled now in the cases of lead consumed in non-disposal type uses, dispersive disposal type uses, and dissipative disposal type uses. The distribution ratio of the demand for lead for these various unrecyclable uses in the total demand for lead has increased gradually in the postwar period as shown in Figure 8 (Ueta, 1981). This is one of the effects caused by the structural change in the demand for lead.

THE ENVIRONMENTAL IMPACT OF LEAD USE AND THE CONSERVATION EFFECTS OF PRODUCT LIFE EXTENSION

The total amount of lead in the use phase having a potential environmental impact is evaluated in the following way (Sueishi, Morioka, and Ueta, 1981), on condition that the annual change in the lead supply in each use category is given. The waste amount in each use category is expressed by the convolutional integral

$$O_i(t) = \int_0^{\infty} I_i(t - s) f_i(s) ds \tag{5}$$

where,

O_i = the waste amount in use i,
I_i = the total consumption of lead in use i, and
$f_i(s)$ = the disuse function in use i.

This equation shows the relationship between the input of lead product and the output of its waste as shown in Figure 3.

The output to a unit input and the indicial admittance can be used to obtain the output to a force that is an arbitrary function of time by application of the principle of superposition. The principle of superposition can be applied to any linear system. However, the application of the method mentioned above is subject to another restricting condition. Let us assume that a unit input is applied at the time s, and that we observe the output at the instant t; then we also assume that this output is only a function of the elapsed time t-s and does not depend either on t or on s separately. This will be the case if the coefficients of the differential equations of the system are constants; it will not be the case, in general, if the coefficients are functions of the time. These two conditions are known as Duhamel's integral (Karman and Biot, 1940). Strictly speaking, it is not easy to check whether these conditions are satisfied in input-output system of lead use, because the disuse function of each commodity has not been investigated enough and because of the planned obsolescence of products. There are, however, two advantages to this method. One is that the state of future waste generation can be predicted with reference to the estimated value of the annual lead supply in each use category corresponding to the estimated period, based on past records. The other advantage is that we can compare the obsolescence process in each use category.

I_i is calculated by the following equation:

$$I_i = PC_i + SC_i + RSC_i \tag{6}$$

where,

PC_i = primary lead consumption in use i,
SC_i = secondary lead consumption in use i, and
RSC_i = remelted secondary lead consumption in use i.

The amount of recovered lead in use i is calculated by:

$$RA_i = SC_i + RSC_i + X_{si} - M_{si} + SI_i \tag{7}$$

where,

RA_i = amount of recovered lead in use i,
X_{si} = exports of scrap in use i,
M_{si} = imports of scrap in use i, and
SI_i = scrap input to primary lead refining.

The value of X_{si}, M_{si}, and SI_i are not reported in governmental statistical reports in each use category, and so, we can estimate only the total amount of exports of scrap, imports of scrap, scrap input to primary lead refining, and also the amount of recovered lead.

The total amount of lead having a potential environmental impact EI can be calculated by

$$EI = O - RA \tag{8}$$

where,

$$O = \sum_{i=1}^{n} O_i \tag{9}$$

$$RA = \sum_{i=1}^{n} RA_i = SC + RSC + X_s - M_s + SI \tag{10}$$

$$SC = \sum_{i=1}^{n} SC_i \tag{11}$$

$$RSC = \sum_{i=1}^{n} SC_i \tag{12}$$

$$X_s = \sum_{i=1}^{n} X_{si} \tag{13}$$

$$M_s = \sum_{i=1}^{n} M_{si} \tag{14}$$

$$SI = \sum_{i=1}^{n} SI_i \tag{15}$$

Substituting the values of Figure 2 and Table 1 in Eqs. (1), (2), and (8), the annual change in the total amount of demand for lead, the amount of lead wastes, and the lead discharged into the environment is estimated as shown in Figure 9. The annual change in the value of Recycling Ratio a (RRa) and Recylcing Ratio b (RRb) as shown in the following equation is also shown in Figure 9.

$$RRa = \frac{RA(t) - SI(t)}{o(t)} \tag{16}$$

$$RRb = \frac{RA(t)}{o(t)} \tag{17}$$

Both the amount of cumulative lead waste and the amount of cumulative lead discharged into the environment have been increasing annually, but the rate of increase is slowing down. The amount of lead waste in 1978 was 8.32 percent of the amount of cumulative lead wastes for the period from 1960 to 1978. The amount of lead discharged into the environment was 9.39 percent of the amount of cumulative lead discharged into the environment for the period from 1960 to 1978. The amount of lead discharged into the environment in 1978 can be broken down as follows: inorganic chemicals, 38.6 percent; cable, 18.1 percent; batteries, 14.1 percent; lead pipes and sheets, 13.8 percent, antifriction alloy and solder, 11.5 percent. As lead wastes generated from inorganic chemicals cannot be recycled, the amount of inorganic

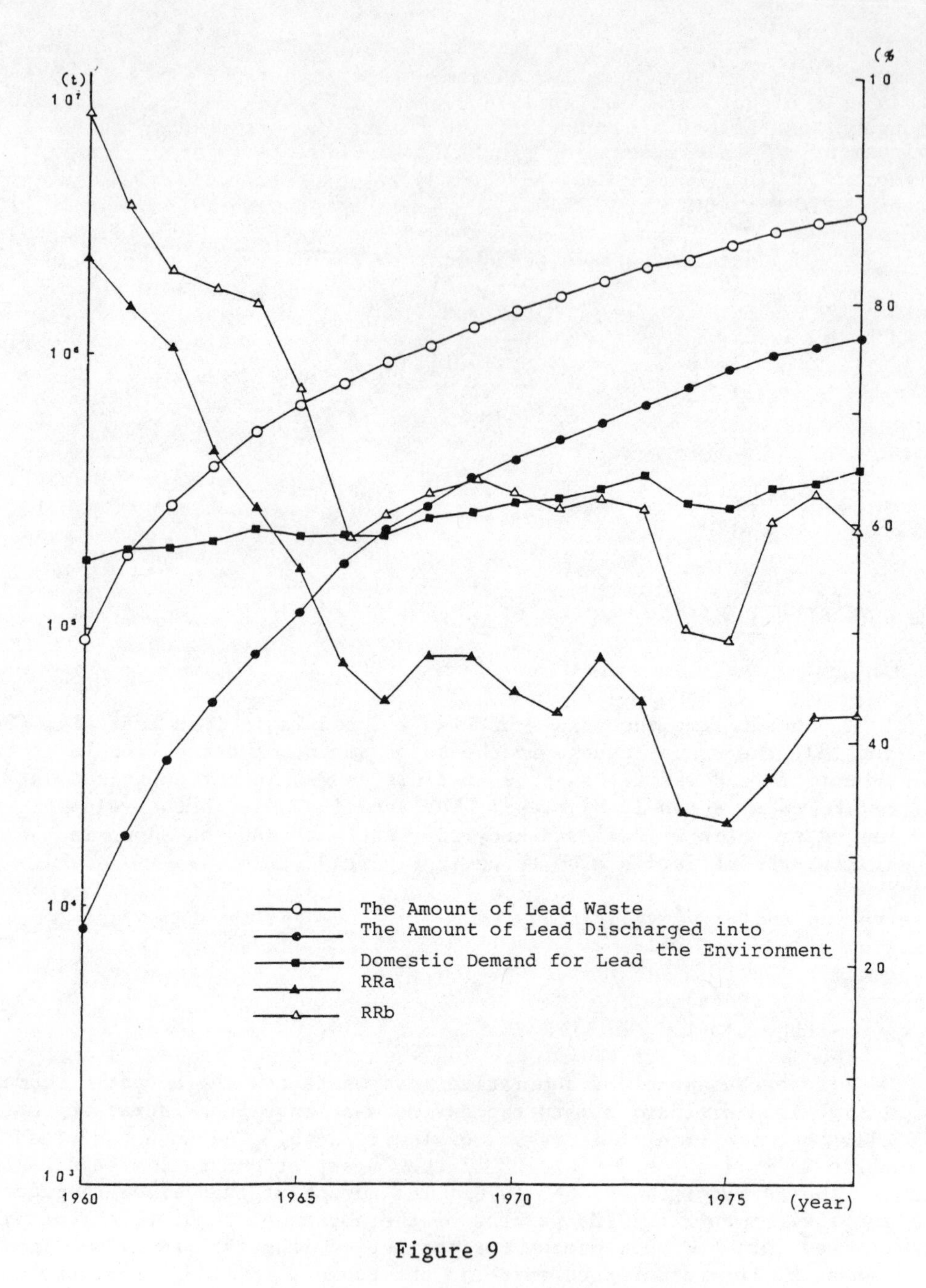

Figure 9

The Annual Change in the Demand for Lead, the Amount of Lead Waste, the Amount of Lead Discharged into the Environment, and the Recycling Ratio

chemical waste accounts for more than 30 percent of lead wastes discharged into the environment. The amount of lead cable waste is more than that of battery waste in 1978 because the life of a cable is longer than that of a battery and the cumulative effect upon the environment of waste generation of cables used in the 1960s cannot be ignored. The amount of lead discharged into the environment had been increasing annually until 1975 but dropped after that. In contrast, the amount of lead waste has been increasing annually. The increasing rate of the amount of lead discharged into the environment is less than that of lead waste because the Recycling Ratio has been rising since 1975. Both RRa and RRb have changed in a similar way to REIa and REIb respectively, but the tendency since 1973 has been considerably different.

The estimated total amount of lead discharged into the environment for the period from 1960 to 1978 is 1,194,000 tons. If the lead discharged into the environment of Japan accumulated evenly in the outer layer of the soil 10 cm in depth, the lead content in the soil would have risen 32.3 ppm, which is about three times the normal lead content of 10 ppm. The additional amount of 59,000 tons of tetraethyl lead from 1969 to 1978 having a potential environmental impact should also be considered in the above estimate. Statistics for lead production and consumption do not include tetraethyl lead because is is not manufactured in Japan.

We regard product life extension as a useful policy instrument to solve resource, energy, and environmental problems. This applies not only when seen from the point of view of the acceptability of product life extension for society in the moral sense; but also because of the value of the recognition that the useful life of a product is not physically fixed but is a socio-economic operational variable of the society. The effect of lead product life extension upon resource conservation and waste generation can be investigated through a comparison of resource consumption between the case of extending the lead product life by 20 percent and the case of retaining the existing product life. This comparison is implemented under the assumption that the stock of products in actual use phase is constant. If the utility contributed by the product to society is independent of the life and the use years, the constant stock of the products means that the utility contributed by the stock of products and the stock of lead in the products to society is also constant.

The disuse function of lead products in the case of extending the product life 20 percent is shown in Figure 10. The decreasing rate of the demand for lead for the period from 1960 to 1978 by extending the product life 20 percent is shown in Figure 11. The effects on the demand for lead have been increasing annually. The decreasing rate of the demand for lead in lead pipes and sheets is higher than that of the total demand after 1971. In contrast, the decreasing rate is higher for lead batteries than it is for the total from 1960 to 1965;

but, as in the 13 years from 1966 to 1978, there are only four years in which the decreasing rate is higher than that of the total figure. The range of the decreasing rate in the demand for lead varies widely, and it is difficult to recognize a consistent change that relates to all uses. It is thought that the decreasing rate in the demand for lead is affected by some factors other than product life extension. Considering the three main uses having a different useful life--batteries, antifriction metals and solders, and lead pipes and sheets--and the five cases of demand growth rates--20 percent, 10 percent, 5 percent, 0 percent, and -5 percent--the decreasing rate in the demand for lead is compared in each case. The different factors affecting the demand for lead in the various uses are only the growth rate of the demand for lead and the disuse function for each use. On the relation between the demand for lead and the stock of lead, it is found that the annual growth rate of the demand for lead is equal to the annual growth rate of the stock of lead.

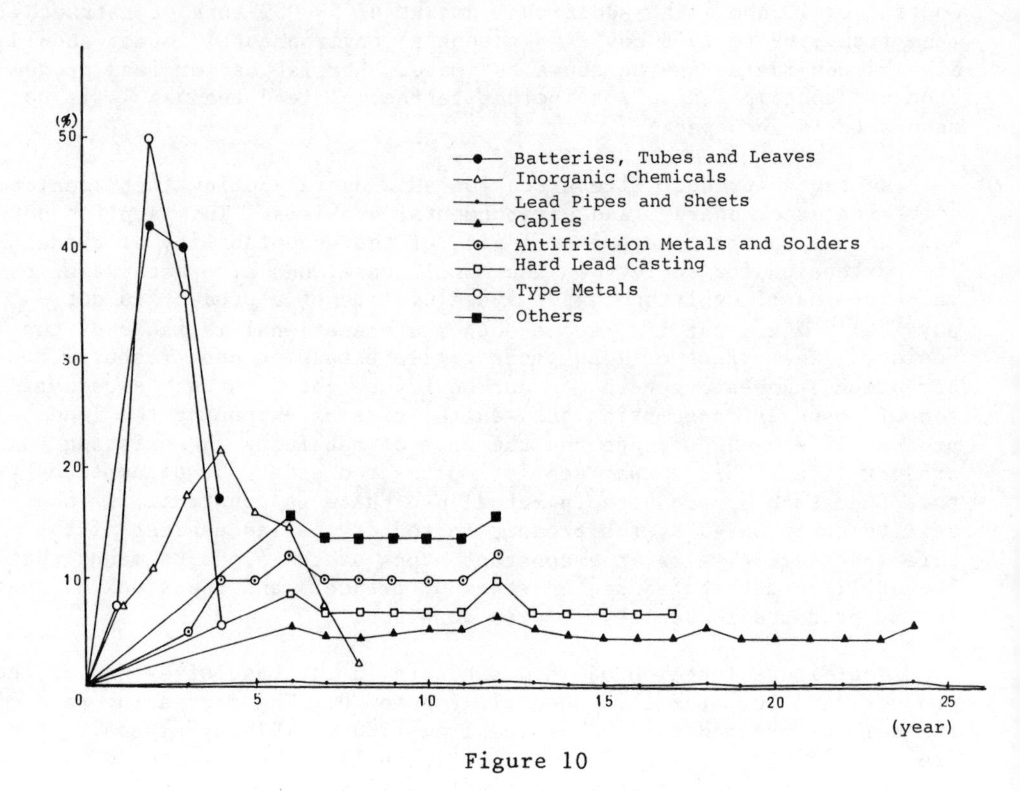

Figure 10

The Disuse Function of Lead by Use

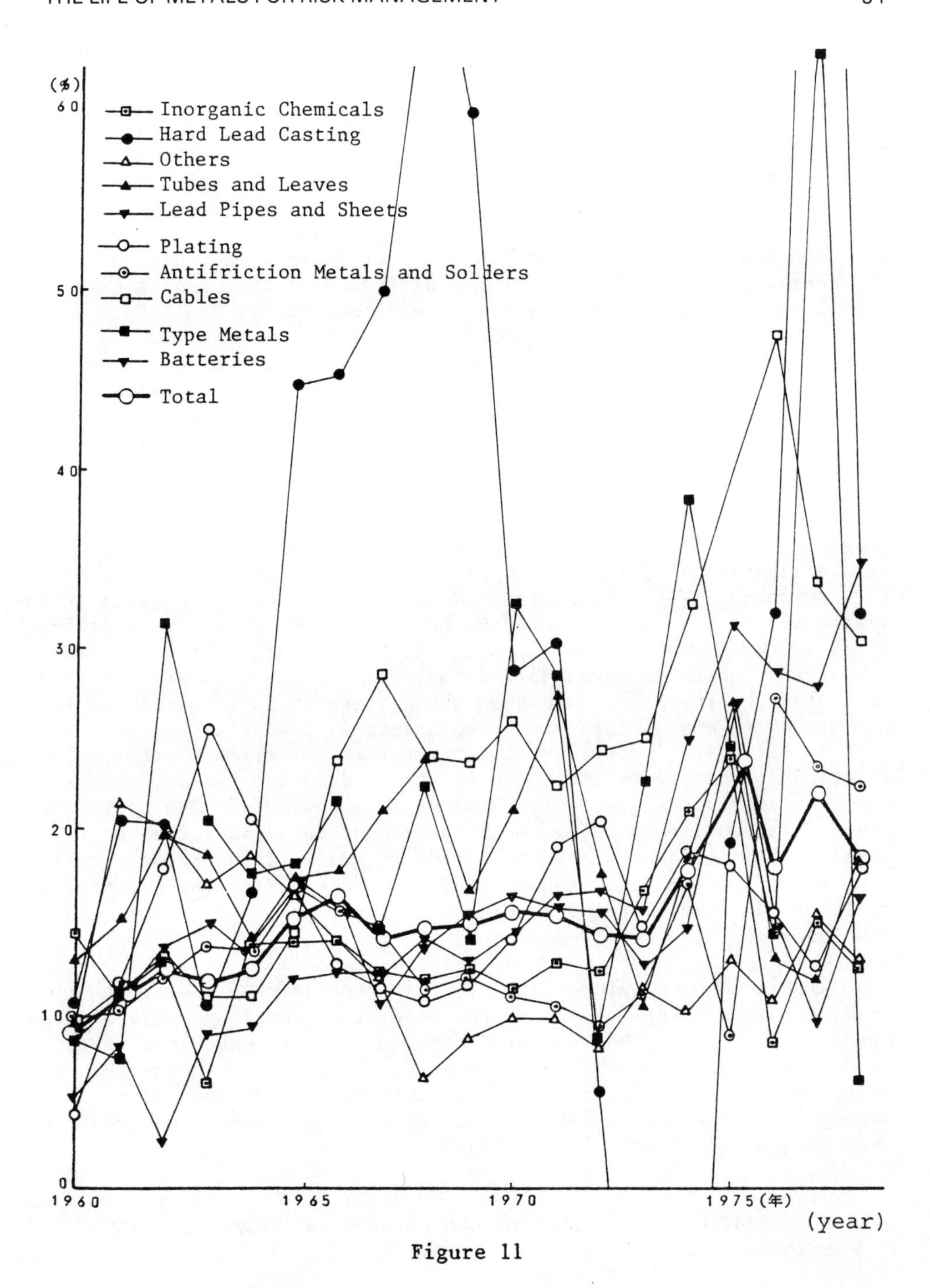

Figure 11

The Decreasing Rate of the Demand for
Lead by Product Life Extension

The decreasing rate of the demand for lead is calculated under the following conditions: the stock of lead in the three main use categories at the starting point is supposed to be 0. The demand for lead in the three uses at year 1 is assumed to be 10,000 tons. The stock of lead after that is estimated on the assumption that the stock of lead changes based on the growth rate of the demand for lead and the disuse function in each use. Subsequently, if the product life is extended by 20 percent, the demand for lead will change in order to maintain the stock of lead at the same level had it not been extended. The decreasing rate of demand for lead from when the product life is not extended is calculated as shown in Figure 12.

In the relation between the annual growth rate of the stock of lead and the decreasing rate in the demand for lead, the value of the latter becomes lower as the value of the former goes up. The amount of lead saved by product life extension is very small compared with the stock of lead because the higher the growth rate of the demand for lead, the higher the growth rate of the stock of lead. The rate of difference between the decrease in the demand for lead, when the growth rate of the stock of lead is 0 percent and that when the growth rate of the stock of lead is 20 percent becomes larger as the product life shortens. With lead pipes and sheets, the decreasing rate in the demand for lead goes down to 42.64 percent as the growth rate of the stock of lead goes up from 0 percent to 20 percent. For batteries, the decreasing rate in the demand for lead goes down to 81.85 percent. It is thought that this is due to the difference in the stock of lead. The stock of lead in lead pipes and sheets is 2.64 times higher than that in batteries. Moreover, in contrast to batteries where the effect of product life extension is realized during two or four years, the process takes 42 years for lead pipes and sheets. Therefore, in order to accomplish a decrease in the demand for lead for a short period, it is more effective to extend the life of the product having a shorter life than the product having a longer life. This is true in spite of the fact that the stock of lead in the former is less than in the latter.

With lead manufactured goods, it is necessary to take into account the growth rate of demand for lead, the recycling ratio of lead waste, and the stock of lead products, in order to make lead product life extension effective for resource, energy, and environmental conservation. It is clear that lead product life extension, without taking into consideration those factors mentioned above, may require more energy than before (Ueta, 1983).

THE PHILOSOPHY OF LEAD POLLUTION CONTROL FOR ENVIRONMENTAL RISK MANAGEMENT

With the hazards of metals that are poisonous in the environment, the time delay between the benefits gained by their use and the risks

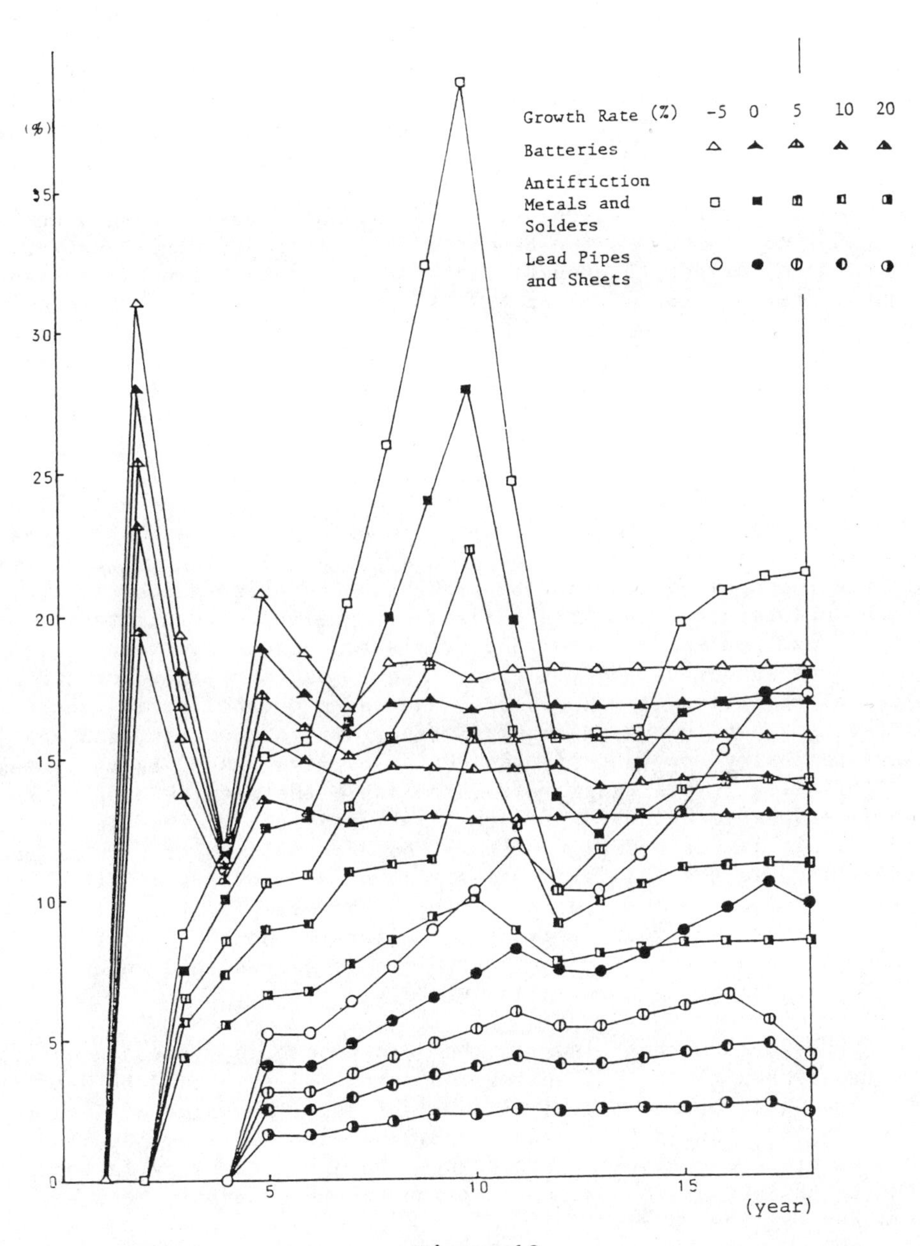

Figure 12

The Annual Change of the Reduction Rate of the Demand for Lead by Product Life Extension in the Three Main Uses

from their waste vary according to use. The benefits and the risks may fall on different groups of people, so that the areas in which risks are concentrated may be different from those of the benefits. Therefore, the control of the use pattern, and the life and spatial distribution of metals for balance between the utility of metals and the hazards of their waste, is indispensable in metal risk management.

The estimation of the benefits of hazardous waste management is in its infancy. Because the benefits of pollution control--savings in health costs, damage to property and the environment generally--cannot be quantified accurately, society has to set the level of acceptable risk in relation to the likely costs involved in reducing the hazard. Controlling the risks of toxic substances is especially difficult both because the exact nature of the hazard has yet to be fully recognized by regulatory authorities and because of certain characteristics displayed by these pollutants which make control by traditional policy instruments fairly ineffective. Risk management basically involves four phases: risk identification, risk estimation, risk evaluation, and risk control, which by necessity are closely related. Risk evaluation is really the central task of risk management for it involves not only the social judgement of risks as identified and estimated, but the balancing of such risks against perceived or estimated social gains. Lead pollution discharged from automobile exhaust has been shown to be dangerous and toxic. It should be noted, however, that potential pollution from lead scrap and waste metabolism is even greater as shown in the previous section. It is not easy, and not always necessary, to identify the degree of risk caused by the potential pollution, since exaggerating the scale and magnitude of risks beyond reasonable compass for investigation is not worthwhile for society. It is not useful to attempt to ascertain the risks too precisely, because the full risk assessment of all toxic products is extremely complicated and time consuming. For example, it will take at least ten years and cost millions of dollars to do the full risk assessment of all toxic products required to be examined under the U.S. Toxic Substances Control Act (1976).

Although it is not clear if the lead discharged into the environment causes health damage, an environmental policy should be determined with foresight, erring on the side of safety. The value of lead in consumer goods should be reconsidered. We have already found that the order of preference in lead use changes when the recovery factor is added to the production and consumption scheme (Sueishi, Morioka, Ueta, and Migita, 1979).

In the context of environmental risk problems, policymakers must decide what degree of risk is acceptable. However, there is a general lack of data on pollution generation, transfer functions and dose-response relationships, the potential for catastrophic costs, and relatively modest benefits in terms of environmental risk. Therefore, the very great uncertainty surrounding the potential costs and benefits, and the severe institutional problems created by the extended delay between the hazard occurrence and the manifestation of its damage effects and the irreversibility of the effect, cannot be clearly determined. Because of the fair distribution of costs and risk over time and the latency and irreversibility characteristics of risk, the redesign of lead pollution control strategies is necessary in order to be able to anticipate adverse effects rather than merely react to existing, known effects. The redesign of pollution-management programs should be focused on the control of the use pattern and the life of lead. Therefore, an adequate recycling plan for resource, energy, and environmental conservation cannot be realized without the redesign of the use structure of lead.

SUMMARY AND CONCLUSION

In this paper, characteristics of the change of lead use structure in Japan relating to waste problems after the Second World War were summarized as follows:

1) The useful life of lead has been shortened from more than 11 years in 1955 to about 5.5 years in 1980.
2) The use of lead has been expanded from goods used in production to consumer goods; therefore, we have not only to control enterprises but also to warn households--the subjects of the use of lead for risk management.
3) The demand for lead for various unrecyclable uses amounts to 35 percent of the total demand for lead in 1980, increasing from about 20 percent in 1955.

The estimated total amount of lead having a potential environmental impact for the period from 1960 to 1978 is 1,194,000 tons. It is considered that the environmental risk derived from lead has increased because of the sudden catastrophic change of the demand structure of lead without adequate management of the use pattern and the life of lead.

Early warning of potential pollution hazards is a major concern of the public and governments. The methods for assessing the chronic effects of pollutants are also in need of improvement so as to provide workable techniques for pollution control. However, since such assessments would impose impossible organizational and financial burdens on government authorities and the public--in the framework of existing control structures--the control of the use structure of lead for environmental risk management should be realized without waiting for the development of such assessment techniques. An adequate recycling system could be established through the redesign of the use pattern and the life of lead in the direction of a closed system without increasing social cost such as an extravagant consumption of energy.

REFERENCES

Karman, T. and M. Biot (1940), Mathematical Methods in Engineering; An Introduction to the Mathematical Treatment of Engineering Problems, McGraw-Hill.

Pearce, D.W. (1976), "Environmental Protection, Recycling and the International Materials Economy," in I. Walter (ed.), Studies in International Environmental Economics, p. 319, John Wiley & Sons.

Sueishi, T., T. Morioka, K. Ueta, and J. Migita (1979), "A Utility Evaluation of Lead Use by Means of Multi-Attribute Utility Function Method," 15th Symposium on Sanitary Engineerng, JSCE, (in Japanese), p. 121.

Sueishi, T. and K. Ueta (1981), "Proposal of Socio-Metallic Study," Bulletin of the Japan Institute of Metals, Vol. 20, No. 6, (in Japanese), pp. 177-184.

Sueishi, T., T. Morioka and K. Ueta (1981), "Use of Secondary Lead for Resource, Energy and Environmental Conservation," Conservation and Recycling, Vol. 4, No. 3, pp. 177-184.

Ueta, K. (1981), "The Economic Principle of Recycling," Research on Environmental Disruption Toward Interdisciplinary Cooperation, Vol. 11, No. 2, (in Japanese), pp. 2-10.

Ueta, K. (1983), A Structural Change of the Use of Metals and Disposal of Resultant Wastes, with Particular Reference to the Case of Lead and the Conservation Effects of Product Life Extension, KIER Discussion, Paper No. 185.

ACCEPTABLE LEVELS OF RISK IN SETTING CHEMICAL STANDARDS

Paul S. Price

Office of Toxic Substances
U.S.E.P.A.*

ABSTRACT

The setting of a chemical standard is a function not only of the toxic properties of the chemical but also the level of risk under which the standard is established. This study demonstrates the quantitative effects of different levels of acceptable risk for two groups of standards, the drinking water standards developed by the Environmental Protection Agency and the workplace guidelines published by the American Council of Governmental Industrial Hygienists. The effect of the risk levels is demonstrated by comparing the standards for 55 chemicals common to both groups. In order to make a direct comparison of the two groups of standards, the doses permitted under each standard were calculated. The doses permitted under the workplace guidelines were 0.6 to 20,000 (median value 50) times greater than those permitted under the water standard. The doses permitted also differed according to chemical type, with there being a relatively close agreement between some pesticides and maximum disagreement between halogenated solvents. These differences are largely due to the different margins of safety used by the standard-setting bodies. The methodologies for setting the margins of safety are, in turn, determined by the acceptable level of risk. Based on this study, different levels of acceptable risk may affect chemical standards by up to four orders of magnitude.

KEY WORDS: Standards, Acceptable Risks, Occupational Risks, Drinking Water Risks, Policy Analysis, Risk Assessment Methodologies

INTRODUCTION

The risk from a toxic chemical is a function of both the dose received and the toxicological properties or potency of the chemical. By extension, a toxic chemical standard is a function of both the level of risk acceptable to the standard setting body and the chemical's potency. This presentation reviews the role of the acceptable level of risk in setting chemical standards by examining the doses permitted for the same chemicals regulated under two different types of standards. The two types of standards reviewed are the occupational standards published by the American Conference of Governmental Industrial Hygienists (ACGIH) and the water standards developed by the water programs of the Environmental Protection Agency (EPA).[1]

STANDARD SELECTION

The two types of standards used in this study were selected for several reasons. First, these types are the two largest bodies of published chemical standards. The large size of the types provides a large number of chemicals (greater than 100) with both types of standards. Second, the standards were developed independently, one by a Federal agency and one by an independent organization. Third, both types have been developed under different levels of acceptable risk.

Acceptable Levels of Risk

The level of risk accepted under each type of standard differs greatly, with lower levels of risk tolerated under the water standards. Summarizing the level of acceptable risk under which the water standards were established is somewhat complicated by the fact that the EPA water standards are composed of four groups, each developed under different circumstances and for different purposes.[1-5] However, all four groups have similar statements on the acceptable level of risk (Table I). The level of risk under the water standards is essentially a zero risk--no adverse effect is considered acceptable under any circumstances.[2] Because of the absolute nature of this goal and the inherent uncertainty of any scientific finding, statements on the acceptable levels of risk contain an emphasis on the certainty of the data; stating, for example, that there is "a practical certainty" or "adverse health effects are not anticipated."

Occupational standards accept a higher level of risk than water standards (Table I). The goal of the occupational standards is not zero risk but manageable risk;[6,7] that is sufficiently small as to be manageable by industrial hygienists. A chemical's risk may be considered to be small if either the severity of its effects or the probability of their occurring is low.

TABLE I. STATEMENTS OF ACCEPTABLE LEVELS OF RISK

Table I. Statements of Acceptable Levels of Risk	
Water	
...a practical certainty that no harm will result	National Academy of Sciences(1,2)
...adverse health effects are not anticipated...to protect the most sensitive members of the present population	EPA Water Advisories(3)
...to prevent adverse health effects in humans	EPA Water Quality Criteria(4)
...to protect (human health) to the maximum extent possible.	EPA Maximum Containment Levels(5)
Occupational	
...nearly all workers may be repeatedly exposed day after day without adverse effect. Because of wide variation in individual susceptibility, however, a small percentage of workers may experience discomfort from some substances at concentrations at or below the threshold limit; a smaller percentage may be affected more seriously by aggravation of a pre-existing condition or by development of an occupational illness.	(ACGIH Threshold Limit Values)[6]

Problems in Comparing Standards

There are several issues which must be considered carefully when comparing toxic chemical standards. First, many of the chemicals have standards which are not established upon the same toxicological basis. For example, a chemical may have a drinking water standard based upon taste and an occupational standard based upon eye irritation. To avoid this, all chemicals used in this study are based upon chronic systemic effects. To avoid differences in standards which are due to inconsistencies in standard setting methodology, water standards which were set on the basis of non-threshold models were not included in this study. A second consideration is that occupational exposures occur by inhalation and water exposure by ingestion. To place the standards on an equivalent basis, the doses absorbed by the body under each standard were estimated. These doses may be viewed as allowable daily intakes adjusted for absorption. Tables II and III give the assumptions used to estimate the absorbed doses. The third problem in using occupational and water standards is the difference in the exposed populations. Drinking water standards are meant to protect the entire population; occupational standards only relatively healthy adults. The implication of this difference will be discussed later in this presentation.

TABLE II. DOSE ESTIMATE ASSUMPTIONS

TABLE II Dose Estimate Assumptions

Weight - 70 kg
(Unless specified otherwise)

Occupational Exposure - 8 hr/day
5 days/week

Drinking water exposure - 7 days/week

Food contamination exposure - 7 days/week

Inhalation rate - 1/2 m^3/hr

Water consumption rate - 2 l/day

RESULTS

A total of 53 chemicals met the criteria given in the previous section. These chemicals fell into four classes: pesticides, metals, halogenated aliphatics, and others (Table IV). Because some of the 53 chemicals have more than one water standard, a total of 70 water standards and 53 occupational standards are included in this study.

As one would expect from the acceptable levels of risk, for the majority of chemicals, occupational standards permit higher doses than drinking water standards. Plotting the occupational permitted dose versus the drinking water permitted dose is shown in Figure 1. Because some chemicals have more than one drinking water standard, there are more than 53 points on the graph. Only three of the drinking water standards permit higher doses than occupational standards. Based upon the scatter in the plot, no simple relationship exists between the two permitted doses. The complexity of the relationship is also shown by the distribution of the ratios between the two groups of doses (Figure 2). The ratios between the doses ranged from 0.6 to 20,000, with a median value of 50. The distribution is roughly lognormal with a geometric mean of 56.

The distribution of ratios also varies greatly with the class of chemicals. The median value varies from 24 for pesticides to 290 for

TABLE III. ABSORPTION ASSUMPTIONS

Table III. Absorption Assumptions

Inhalation		Ingestion	
Type	Absorption	Type	Absorption
Polar vapors	100%	Inorganic	Case by case
Nonpolar vapors	50%	Organic	100%
Particulates	100%		

TABLE IV. CHEMICAL STANDARDS

Table IV. Chemical Standards

Pesticides		
2,4 - D	Phorate	Atrazine
2,4,5 - TP	Carbaryl	Carbofuran
Paraquat	Captan	Endrin
Methoxychlor	p - Dichlorobenzene	Lindane
Toxaphene	Parathion	Endosulfan
Azinphos-methyl	Malathion	Isopherone
Diazinon	Thiram	Chlordane
Pentachlorophenol		
Metals		
Arsenic	Chromium III	Selenium
Antimony	Lead	Silver
Barium	Mercury (alkyl)	Thallium
Cadmium	Nickel	
Halogenated Solvents		
Methyl chloride	Methyl bromide	Trichloroethylene
Methylene chloride	Methyl chloroform	Perchloroethylene
Dichlorofluoromethane	Vinylidene chloride	
Others		
Toluene	Dinitro-o-phenol	Cyanide
Xylenes	Chlorobenzene	Fluoride
Ethylbenzene	o-Dichlorobenzene	Ethylene glycol
Styrene	Phenol	Methyl methacrylate

halogenated aliphatics, with metals and others falling in between (Table V).

In summary, the occupational standards permit higher doses than water standards. The differences between the two range from a factor of less than 1 to 20,000. The differences are smallest for pesticides and greatest for halogenated aliphatics.

DISCUSSION

Background

The mechanism by which the acceptable level of risk influences the permitted dose is the safety factor. All the standards used in this survey are set in relation to thresholds.[1,8] The safety factor is defined as the ratio of the permitted dose to the threshold or, to

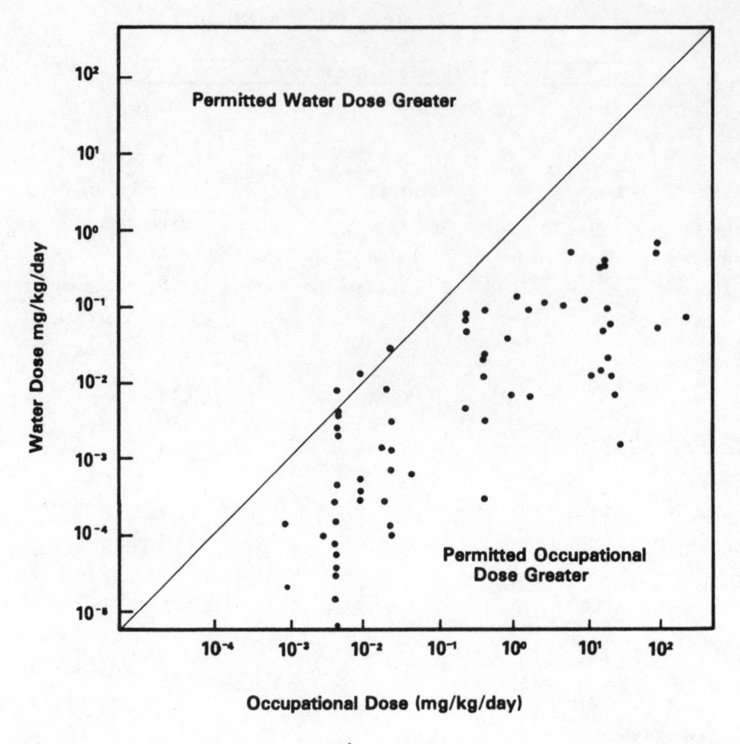

Figure 1

Permitted Doses: Occupational Standards vs Water Standards

be exact, the toxicological endpoint representing the threshold. Assuming that for a given chemical similar estimates of the threshold are used in both types of standards, the ratio of permitted doses is equal to the inverse of the ratio of the safety factors.

Occupational and Water Safety Factors

Safety factors used in setting occupational and drinking water standards differ according to their range and selection. Occupational standards use safety factors ranging from less than one to 10;[8] water standards from 10 to 10,000.[4,9] The higher level of acceptable risk in the occupational standards allows for a lower range of safety factors.

The criteria for selecting safety factors for a specific chemical for water standards reflect the standard's goal of zero risk and are based upon the certainty of the toxicological evidence (Table VI). Where the data is sufficient to accurately predict the threshold dose level a small safety factor is used. Where the data is less certain, larger safety factors are used. Because no effect is tolerated, the

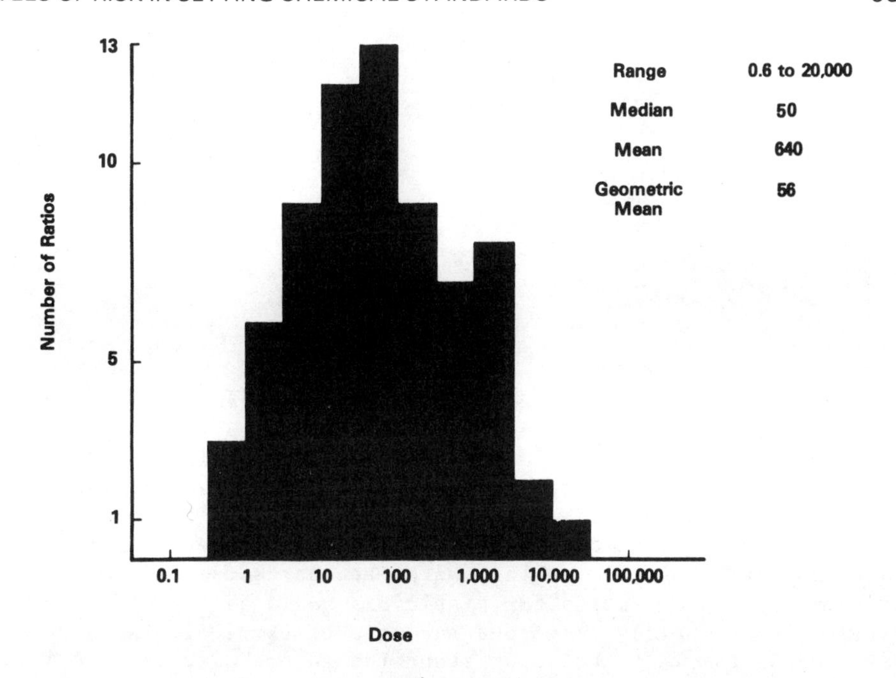

Figure 2

Distribution of Occupational/Water Dose Ratios

size of the safety factor is independent of the severity of the effect.

In contrast to safety factors used in water standards, occupational safety factors are selected on the basis of the severity of the effect (Table VI). Because occupational standards do not seek a zero risk there is a much lower emphasis on the sufficience of the data. Thresholds are therefore established on the basis of either animal or human studies and subchronic studies may be used.

The Effect of Safety Factor Criteria on Dose Ratios

These differences in criteria appear to be the reason for the wide range in dose ratios. As illustrated in the two hypothetical cases given in Table VII, the ratio of the doses permitted for the same chemical under the two types of standards may range from less than one to greater than 1,000 as the result of the different criteria. Further support for the role of the criteria is shown in the variation of dose ratio by class of chemicals. The halogenated aliphatics lowest observed effects are behavioral or central nervous system depression effects. They are reversible and non life-threatening.

TABLE V. MEDIAN VALUES OF RATIOS BY CLASS

Table V. Median Values of Ratios by Class

Class	Median Value
Pesticides	24
Metals	60
Halogenated aliphatics	290
Others	150
All chemicals	50

Therefore, they have small occupational safety factors. However, because these effects are established in short-term studies they have large water safety factors. The dose ratios of halogenated compounds would be expected to be large. Pesticides' toxic effects are generally more severe than the halogenated aliphatics and the occupational safety factors should be larger. Because pesticides generally have thorough toxicological testing, water standards use smaller safety factors. The dose ratios for pesticides would be expected to be small. Unfortunately, rigorous analysis of chemical standards is hampered by the fact that many standards do not clearly identify the threshold and safety factors on which they are based.

The Effect of Different Populations

As stated earlier, the two types of standards are intended to protect two different populations--workers and the general population. While this difference in population may justify some of the variation between the standards, it does not appear to be sufficient to justify all. Stokinger and Woodward[11] suggested that an additional margin of safety of two would be appropriate to allow for the difference in population. Calabrese,[12] based upon a review of the maximum contaminant levels, suggested a factor of 10. These adjustments in the margin of safety for population do not explain 75 percent of the standard ratios calculated.

SUMMARY

This presentation has shown that occupational standards permit larger doses than those permitted under corresponding water standards. The differences between the doses appears in large part to be due to the size of the safety factors used in setting each standard. The range of the safety factors used in setting each type of standard corresponds to the level of the acceptable risk under which the standard was established. The two types of standards use different criteria for the selection of specific safety factors. The differences in criteria are a result of the standard's different levels of acceptable risk. The differences in criteria are partly responsible for the wide range of variation between the permitted doses.

TABLE VI. SELECTION OF SAFETY FACTORS

Table VI. Selection of Safety Factors

Basis	Size
Water	
Chemical's threshold is established in long-term human studies	10[1]
Chemical's threshold is established in lifetime animal studies	100[1]
Chemical's threshold is established in sub-chronic animal studies	1,000[1]
Chemical's threshold is not clearly established in animal study	1,000-10,000[9]
Occupational	
Chemical's effects are non-life-threatening, reversible, or self-limiting	1[7,8]
Chemical's effects are life-threatening, irreversible, little or no warning	10[7,8]

TABLE VII. EFFECT OF SAFETY FACTOR CRITERIA ON CHEMICAL STANDARDS

Table VII. Effect of Safety Factor Criteria on Chemical Standards

	Safety Factors		Ratio
	Water	Occupational	
Case 1: Chemical has --Non-life-threatening effects --Reversible effect --Threshold established in short-term animal studies	1,000	1	1000
Case 2: Chemical has --Life-threatening effects --Threshold established in long-term human studies	10	10	1

REFERENCES

1. National Research Council, Drinking Water and Health Volume 1 (National Academy Press, Washington, DC, 1977).
2. National Research Council, Drinking Water and Health Volume 3 (National Academy Press, Washington, DC, 1980).
3. United States Environmental Protection Agency, Office of Drinking Water, "SNARL for Tetrachloroethylene," (February 6, 1980).
4. "Water Quality Criteria Documents Availability," Federal Register 45 FR 79318 (November 28, 1980).
5. Safe Drinking Water Act, Report No. 93-1185 (U.S. Government Printing Office, Washington, DC, 1975).
6. American Conference of Governmental Industrial Hygienists, Threshold Limit Values for Chemical Substances and Physical Agents in the Workroom Environment with Intended Changes for 1981 (ACGIH, Cincinnati, OH, 1981).
7. J. A. Zapp, An acceptable level of exposure, American Industrial Hygiene Association Journal 38, 425 (1977).
8. H. E. Stokinger, Concepts of thresholds in standards setting, Archives of Environmental Health, 25, 153 (1972).
9. M. L. Dourson and J. F. Stara, Regulatory history and experimental support of uncertainty (safety) factors, Regulatory Toxicology and Pharmacology, in press.
10. W. J. Hayes, Toxicology of Pesticides (Williams and Wilkins, Baltimore, MD, 1975).
11. H.E. Stokinger and R.L. Woodward, Toxicological methods for establishing drinking water standards, Journal of the American Water Works Association, 50, 515 (1958).
12. E.J. Calabrese, Methodological Approaches to Deriving Environmental and Occupational Health Standards (John Wiley and Sons, New York, NY, 1978).

*This article was written by the author in his private capacity. No official support or endorsement by the Environmental Protection Agency is intended or should be inferred.

FOOTNOTES

1. It should be made clear at this time that this presentation uses the term water standard to cover suggested or recommended levels, or criteria as well as legally enforceable standards.
2. The one exception to this goal is the Maximum Contamination Levels, which also consider technical and economic feasibility.

INTEGRATED RISK INDEX SYSTEM

G. R. Rosenblum
W. S. Effron
J. L. Siva
G. R. Mancini
R. N. Roth

Atlantic Richfield Company
Health, Safety and Environmental
Protection Department

ABSTRACT

Although large numbers of chemicals need to be assessed for potential risk, financial and human resources for evaluation are often limited. Therefore, prioritization of materials for assessment is a necessity. The Integrated Risk Index System prioritizes materials through use of a rapid and simple scoring procedure. Some advantages of this system are: (1) materials are assessed and ranked with efficient use of health professional time; (2) materials can be ranked with reasonable accuracy using readily available data; and (3) certain types of health, safety, and environmental data concerning carcinogenicity and teratogenicity are weighted in relation to societal concerns. The Integrated Risk Index function is described by the mathematical expression RI = P(2PH + 2HH + EH). The function "RI" is an expression of relative risk since P is derived from potential exposure factors, and (2PH + 2HH + EH) is the summation of the inherent hazards associated with a material. The potential exposure factor, (P) incorporates annual production and the number of potentially exposed populations. The physical hazard factor (PH) is based on flammability and reactivity. The health hazard factor (HH) is derived from acute and subchronic toxicity, carcinogenicity, mutagenicity, teratogenicity, and reproductive effects. The environmental hazard factor (EH) quantifies the potential for ecological damage under hypothetical spill conditions. This system can quickly and efficiently rank in an appropriate and meaningful way a large number of materials according to their relative risk.

KEY WORDS: Risk, Prioritization, Hazard, Warning.

INTRODUCTION

This paper presents an Integrated Risk Index System which prioritizes materials for the timely development of hazard warnings. This system is "integrated" because it combines an assessment of the three basic elements of hazard, which are health hazard, fire/explosive hazard (physical hazard), and environmental hazard with an assessment of the exposure potential. By assigning numbers to the degrees of the various hazards as well as exposure potential, a numerical ranking of materials can be performed. This system has been developed to prioritize materials for hazard warning development. However, since the system is flexible, it can be used for prioritizing materials for toxicity testing with only minor modifications.

Presently, with thousands of materials in use, and thousands of materials developed for use each year, there is an extremely large number of materials to assess. The present regulatory climate, with burgeoning right-to-know laws, has created a situation where detailed hazard warnings must be developed for many materials that may not have been carefully scrutinized before. This increased review effort must sometimes be performed with limited personnel and financial resources.

A systematic approach to this situation has numerous advantages. The optimal systematic approach is one that can identify the materials that present the greatest risk and allow the development of timely hazard warnings for those materials first. The system must be efficient, consistent, and defensible. An efficient system is one which avoids the effect that occurs when the time spent prioritizing the materials approaches infinity: the time spent for health assessment and hazard warning development approaches infinity: the time spent for health assessment and hazard warning development approaches zero. A consistent system allows different persons to reach similar conclusions. This allows the workload to be distributed. In addition, a systematic approach to prioritization such as the Integrated Risk Index System is defensible in a courtroom where, depending on the circumstances, it might be necessary to explain why hazard warnings for certain materials are completed, and others are in progress.

THE INTEGRATED RISK INDEX SYSTEM CRITERIA

When the criteria for developing a ranking system were considered, one important consideration was that the system be effective with limited data. This is significant because available health hazard and environmental hazard data are sometimes very limited. Professional judgments must then be made concerning suspicion of hazard without hard data. The system must be able to incorporate not only this type of judgment, but also the judgment that there is insufficient data and a particular hazard cannot really be assessed. How the system ranks a material when there is insufficient health hazard data is one difference between prioritization for timely hazard warning development, and prioritization for toxicity testing. This will be discussed later.

The well recognized primary principle for assessing risk is illustrated in Figure 1. This concept was expanded for the development of the Integrated Risk Index System in the manner outlined in Figure 2.

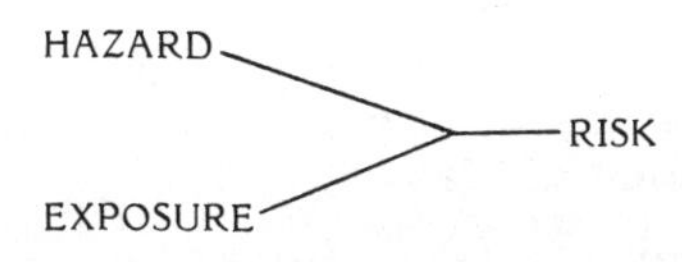

Figure 1

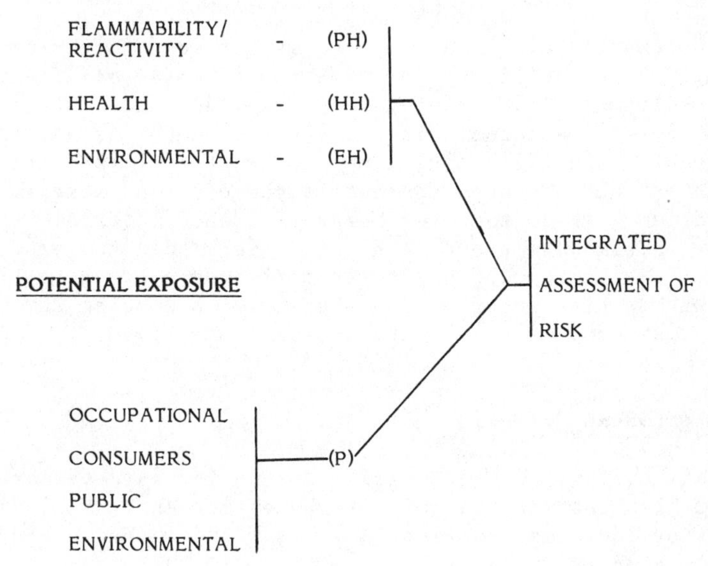

Figure 2

Potential hazards were defined by three elements. Each element is assessed and has a number assigned based on the level of hazard inherent in each material as determined by specific criteria.

The potential for exposure to the hazards is assessed by considering the three groups that can possibly be exposed. These are the occupationally exposed, the consumers of raw and finished materials, and the general public. Environmental discharge, a factor in whether a material can reach the general public, is considered as well. All these elements of risk are considered in the Integrated Risk Index System.

The function that determines the Risk Index (RI), the number that the materials are ranked by, is: RI = P(2PH + 2HH + EH). The terms in this function are defined as follows:

P is the Potential Exposure index
PH is the Physical Hazard index
HH is the Health Hazard index
EH is the Environmental Hazard index

The final product of this function is a number with meaning only within the group of materials ranked by this system. This number indicates which materials in the group have higher or lower relative risk and consequently which materials should have hazard warnings developed first. Although the relative risk is determined by a number, this is not to be confused with quantitative risk assessment where the potential for harm is predicted by probabalistic interpretation of toxicity data.

The Integrated Risk Index System is not used to perform risk assessment; it is used for prioritization of materials for hazard warning development. Therefore, the Risk Index does not in any way suggest or determine actual risk, i.e., the number of adverse effects per number of a specific population exposed to a particular dose level. It can only be used to shuffle the order of materials on a prioritization list so that the hazards of each material can be assessed in a timely manner. Also, the Risk Index number does not allow quantitative comparisons to be performed between materials. A material with a Risk Index of 120 therefore, cannot be said to present twice the risk of a material with a Risk Index of 60.

POTENTIAL EXPOSURE CRITERIA

In this system, the potential exposure (P) is determined from two major variables that can be quickly and easily estimated. These are the weight of annual production (N_{ap}), and the number of populations potentially exposed (N_{pe}). (Table I)

TABLE 1. POTENTIAL EXPOSURE INDEX (P)

ANNUAL PRODUCTION (N_{ap})

SCORE	CRITERIA
1	Less than 10,000 lbs.
2	10,000 to 1 million lbs.
3	Greater than 1 million lbs.

POPULATIONS POTENTIALLY EXPOSED (N_{pe})

A) OCCUPATIONAL, B) CONSUMER, OR C) GENERAL PUBLIC:

SCORE	CRITERIA
1	One Population **OR** Limited Environmental
2	Two Populations **OR** Moderate Environmental
3	Three Populations **OR** Widespread Environmental

The criteria for exposed populations are that a point is given for each potentially exposed population type: occupational, consumer, or public. Environmental discharge may be included in the decision, and without supporting population exposure data, may be used alone to estimate the score given in this category.

The Potential Exposure index (P) is derived from the preceding ratings in an additive fashion: $P = N_{ap} + N_{pe}$.

HAZARD CRITERIA

The three major elements of hazard are Physical Hazard (PH), Health Hazard (HH), and Environmental Hazard (EH). These are broken into various factors which have specific criteria for assigning ranking numbers.

Physical Hazard

The Physical Hazard index (PH) results from an assessment of a material's flammable and explosive properties. The criteria here are based on flash points and reactivity criteria developed by the National Fire Protection Association (NFPA).[1] These criteria are applied to a material in normal conditions of use and storage, not during fire conditions.

These criteria lead to an assignment of a zero through four (0-4) rating for flammability (N_f) and reactivity (N_x). (Tables 2,3) The Physical Hazard index rating for each material can then be derived from a physical hazard matrix[2] and can range from 1.00 - 10.00. (Table 4a) Materials with flammability characteristics that are not defined by Table 2 may be rated from an alternate matrix. (Tables 4b,4c)

TABLE 2. PHYSICAL HAZARD CRITERIA

FLAMMABILITY (N_f)

SCORE	CRITERIA
0	Materials that will not burn in air when exposed to a temperature of 1500°F for a period of five (5) minutes.
1	Materials that must be preheated before ignition can occur (F.P. >140°F.)
2	Materials that must be moderately heated or exposed to relatively high ambient temperatures before ignition can occur (F.P. >100°F <140°F.)
3	Liquids and solids that can be ignited under almost all ambient temperature conditions (F.P. < 100°F; B.P. >100°F.)
4	Materials which will rapidly or completely vaporize at atmospheric pressure and normal ambient temperature, or which are readily dispersed in air and which will burn readily (F.P. < 100°F; B.P. < 100°F.)

TABLE 3. PHYSICAL HAZARD CRITERIA

REACTIVITY (N_x)

SCORE	CRITERIA
0	Materials which are normally stable, even under fire exposure conditions, and which are not reactive with water.
1	Materials which are normally stable, but which can become unstable at elevated temperatures and pressures or which may react with water with some release of energy but not violently.
2	Materials which are normally unstable and readily undergo violent chemical change but do not detonate. Also materials which may react violently with water or which may form potentially explosive mixtures with water.
3	Materials which are capable of detonation or explosive reaction but require a strong initiating source or which must be heated under confinement before initiation or which react explosively with water.
4	Materials which are readily capable of detonation or of explosive decomposition or reaction at normal temperatures and pressures.

TABLE 4. PHYSICAL HAZARD MATRIX

A) FLAMMABLE LIQUIDS AND GASES

	N_x=0	N_x=1	N_x=2	N_x=3	N_x=4
N_f=0	1.00	3.50	6.00	7.25	10.00
N_f=1	1.00	3.50	6.00	7.25	10.00
N_f=2	2.50	3.50	6.75	7.75	10.00
N_f=3	4.00	4.00	7.50	8.50	10.00
N_f=4	5.25	5.25	8.00	9.00	10.00
B) COMBUSTIBLE DUST OR MIST					
Coarse Particles	4.00	4.00	6.00	7.25	10.00
Fine Particles	6.00	6.00	7.25	8.50	10.00
C) COMBUSTIBLE SOLIDS					
Dense Solids, eg., wood, metals	1.00	3.50	6.00	7.25	10.00
Open Solids, eg., pellets, rolls, bags	2.50	5.00	8.00	8.50	10.00
Rubber Goods	5.25				

Environmental Hazard

The Environmental Hazard index (EH) is assessed in a manner similar to physical hazard. The two factors used to estimate the environmental hazard are: persistence and bioaccumulation or bioconcentration (N_b), and adverse effects on exposed animals and plants (N_e). (Tables 5,6)

The criteria, based on hypothetical spill conditions into a running stream, delineate seven levels of hazard, rated from 0-6. Any significant toxicity endpoint may be used to define the adverse effects considered here. The environmental hazard matrix gives a hazard rating of 1.00 - 10.00. (Table 7)

Health Hazard

The health hazard assessment is more complex than the other hazard assessments in this system because of the wide variety of toxicity endpoints that are considered for each material. (Table 8)

TABLE 5. ENVIRONMENTAL HAZARD CRITERIA

CHEMICAL PERSISTENCE/BIOACCUMULATION POTENTIAL (N_b)

SCORE	CRITERIA
0	Experimental evidence showing nonpersistence and nonaccumulation.
1	No data; bioaccumulation not expected.
2	Persistence and bioaccumulation is low.
3	Testing needed; bioaccumulation judged to be appreciable.
4	Persistence and bioaccumulation is appreciable.
5	Testing needed; bioaccumulation judged to be high or no known information.
6	Persistence and bioaccumulation is high.

The criteria were based on a hypothetical accidental spill of 100 lbs of product into a flowing stream.

TABLE 6. ENVIRONMENTAL HAZARD CRITERIA

ECOLOGICAL EFFECTS (N_e)

SCORE	CRITERIA
0	Experimental evidence with negative results.
1	No test data; low probability for adverse effect.
2	Adverse effects at high concentrations.
3	Testing needed; probability of minor or local adverse effects at moderate comcentrations.
4	Adverse effects at moderate concentrations.
5	Testing needed; probability of major or widespread adverse effects or no known information.
6	Adverse effects at low concentrations.

The criteria were based on a hypothetical accidental spill of 100 lbs of product into a flowing stream.

TABLE 7. ENVIRONMENTAL HAZARD MATRIX (EH)

	N_b=0	N_b=1	N_b=2	N_b=3	N_b=4	N_b=5	N_b=6
N_e=0	1.00	2.25	3.50	6.00	6.25	7.50	10.00
N_e=1	1.00	2.25	3.75	6.00	6.25	7.50	10.00
N_e=2	1.00	2.25	4.00	6.25	6.75	7.75	10.00
N_e=3	2.50	3.50	4.25	6.50	7.00	8.00	10.00
N_e=4	2.50	3.50	4.25	6.50	7.00	8.00	10.00
N_e=5	4.00	4.25	4.75	6.75	7.75	8.50	10.00
N_e=6	5.25	5.50	5.75	7.00	8.00	9.00	10.00

The rating scale for each health factor varies, since we did not start with a formula and then try to fit the criteria into it. It was first decided what were appropriate criteria to define the levels of each hazard, and judge the conclusiveness or uncertainty of the toxicology data. The formula was then fitted to these criteria.

Tables 9-14 are the criteria for rating health hazards. A variety of toxicity endpoints can be considered within the areas of acute or subchronic/chronic adverse effects.

The acute and subchronic/chronic criteria are fairly straightforward. (Tables 9,10) In the carcinogenicity assessment criteria (Table 11), the rating increases as the knowledge of the material's carcinogenic effects increases, as well as with an increasing suspicion of carcinogenicity. If there is insufficient data for a definitive conclusion, a professional toxicological judgment that a material

TABLE 8. HEALTH HAZARD (HH)

FACTOR	RATING
Acute Toxicity (N_a)	0-7
Subchronic Toxicity (N_s)	0-7
Carcinogenicity (N_c)	0-6
Mutagenicity (N_m)	0-5
Teratogenicity (N_t)	0-5
Reproductive Effects (N_r)	0-5

TABLE 9. ACUTE TOXICITY (N_a)

SCORE	CRITERIA
0	No test data, suspected to be minimally toxic, non-irritating or non-sensitizing.
1	Minimally toxic, irritating or sensitizing Acute oral LD_{50}: > 5g/kg Acute dermal LD_{50}: >500 mg/kg Acute inhalation LC_{50}: >500 ppm Skin irritation Draize: < 0.9 Skin sensitization: minimal Eye irritation Draize: < 24.9
2	No test data, suspected to be slightly toxic, irritating or sensitizing.
3	Slightly toxic, irritating or sensitizing Acute oral LD_{50}: 0.5-5 g/kg Acute dermal LD_{50}: 50-500 mg/kg Acute inhalation LC_{50}: 50-500 ppm Skin irritation Draize: 1.0-1.9 Skin sensitization: slight Eye Irritation Draize: 25-44.9
4	No test data, suspected to moderately toxic, irritating or sensitizing.
5	Moderately toxic, irritating or sensitizing Acute oral LD_{50}: 50-499 mg/kg Acute dermal LD_{50}: 1-49 mg/kg Acute inhalation: 5-49 ppm Skin irritation Draize: 2.0-5.9 Skin sensitization: moderate Eye Irritation Draize: 45-64.9
6	No test data, suspected to be extremely toxic, irritating or sensitizing or cannot judge probable toxicity.
7	Extremely toxic, irritating or sensitizing Acute oral LD_{50}: < 50 mg/kg Acute dermal LD_{50}: < 1 mg/kg Acute inhalation LC_{50}: < 5 ppm Skin Irritation Draize: 6.0-8.0 Skin sensitization: extreme Eye Irritation Draize: > 65

TABLE 10. SUBCHRONIC/CHRONIC TOXICITY (N_s)

SCORE	CRITERIA
0	No test data, suspected to be minimally toxic.
1	Minimally toxic NOEL oral: >100 mg/kg-day NOEL dermal: >10mg/kg-day NOEL inhalation: >10 ppm-day
2	No test data, suspected to be slightly toxic.
3	Slightly toxic NOEL oral: 10-100 mg/kg-day NOEL dermal: 1-10 mg/kg-day NOEL inhalation: 1-10 ppm-day
4	No test data, suspected to be moderately toxic.
5	Moderately toxic NOEL oral: 1-9 mg/kg-day NOEL dermal: 0.02-0.9 mg/kg-day NOEL inhalation: 0.1-0.9 ppm-day
6	No test data, suspected to be extremely toxic or cannot judge probable toxicity.
7	Extremely toxic NOEL oral: < 1 mg/kg-day NOEL dermal: < 0.02 mg/kg-day NOEL inhalation: < 0.1 ppm-day

NOEL: No Observable Effect Level

Data used may be subchronic or chronic depending on availability.

TABLE 11. CARCINOGENICITY (N_c)

SCORE	CRITERIA
0	Adequately tested, with negative results in two animal species.
1	Insufficient test data; no suspicion.
2	Insufficient test data; equivocal as to positive or negative based on structure or biological activity.
3	Insufficient animal test data; positive mutagenicity tests (3 or higher on Mutagenicity Score).
4	Positive in one animal species.
5	Insufficient test data; strong suspicion as human carcinogen or cannot judge probable carcinogenicity.
6	Known human carcinogen or positive in two animal tests.

is suspect results in a higher rating. Short-term mutagenicity data of a positive nature can increase the suspicion of carcinogenicity and thus result in a higher rating. A positive carcinogenicity finding in only one animal species should not be considered completely conclusive, but is rated still higher. Professional judgment, in the face of strongly suggestive but inconclusive evidence, can rate the material as a strongly suspected carcinogen. The dividing line, the "one positive" rating and the "strongly suspected" rating, is flexible.

TABLE 12. MUTAGENICITY (N_m)

SCORE	CRITERIA
0	Adequately tested with negative results.
1	Inadequately tested; no suspicion of mutagenicity.
2	Inadequately tested; suspicion of mutagenicity.
3	Tested in multiple systems with mixed (positive and negative) results.
4	Positive in one mutagen system.
5	Tested in multiple systems with all positive results.

TABLE 13. TERATOGENICITY (N_t)

SCORE	CRITERIA
0	Adequately tested, with negative results in at least two animal species.
1	No test data; no suspicion.
2	No test data; suspected or cannot judge probable teratogenicity.
3	Confirmed teratogen in one animal species.
4	Confirmed teratogen in two animal species.
5	Confirmed or strongly suspected human teratogen.

This allows for the impact of a positive finding through an inappropriate species or route of exposure to be slightly reduced, or conversely if the evidence is very relevant, the rating may be increased by considering the one positive finding as "strongly suspected."

The "cannot judge possible carcinogenicity" criterion is for a material that should not get a low ranking because it has not undergone testing, and has no database. The end result of the use of the Integrated Risk Index System is important to consider here. By ranking a material with no health effects data relatively high, it will

TABLE 14. REPRODUCTIVE EFFECTS (N)

SCORE	CRITERIA
0	Adequately tested with negative results in at least two animal species.
1	No test data; no suspicion.
2	No test data; suspected or cannot judge reproductive effects.
3	Positive for reproductive effects in one animal species.
4	Positive for reproductive effects in two animal species.
5	Confirmed or strongly suspected human reproductive effect.

have a higher priority for a detailed health assessment, which is an integral part of the hazard warning development process. The material will therefore be more likely to be tested when the data gaps are evaluated.

The Health Hazard index is determined by the equation in Table 15. The health hazard factors are first weighted according to their severity of effects and societal concerns. This is one area of the system which incorporates values that are not strictly within the scientific realm. Carcinogenicity, teratogenicity, and reproductive effects were considered to carry slightly more weight for prioritizing materials for hazard warnings than acute or subchronic toxicity. Thus, a material with little or no toxic effects in five factors, but in which there is a known carcinogen, will maintain a high Health Hazard index, and will still be ranked as a relatively high priority for hazard warning.

Mutagenicity, which has predictive value for further testing priorities, was not weighted strongly for hazard warning prioritization because the direct significance of a positive mutagenicity finding for human health is still uncertain. The mutagenicity weighting would be increased if the system were altered to perform testing prioritization.

The 35 in the denominator of the Health Hazard index is the sum of the maximum number of all the health hazard factors. The constant 4.32 is used in order for the Health Hazard index to be within the 1.0 - 10.00 scale which was used for the other two hazard elements.

TABLE 15. HEALTH HAZARD (HH) INDEX CALCULATION

FACTOR	WEIGHTING	SCORE
Acute Toxicity	2x	= (N_a)
Subchronic Toxicity	2x	= (N_s)
Carcinogenicity	3x	= (N_c)
Mutagenicity	1x	= (N_m)
Teratogenicity	3x	= (N_t)
Reproductive Effects	3x	= (N_r)

$$\text{HH Index} = \frac{4.32\,(N_a + N_s + N_c + N_m + N_t + N_r)}{35}$$

After P, PH, EH, and HH are determined, the Risk Index can then be calculated. (Figure 3) The group of assessed materials can then be ranked according to their Risk Index number. Of the materials that have had a Risk Index number calculated, three examples have been chosen.

ETHYLENE GLYCOL

(P) POTENTIAL EXPOSURE INDEX

$N_{ap} = 3$
$N_{pc} = 3$
$P = 6$

(PH) PHYSICAL HAZARD

$N_f = 1$
$N_x = 0$
$PH = 1.00$

(HH) HEALTH HAZARD

$N_a \quad 1 \times 2 = 2$
$N_s \quad 1 \times 2 = 2$
$N_c \quad 1 \times 3 = 3$
$N_m \quad 1 \times 1 = 1$
$N_t \quad 1 \times 3 = 3$
$N_r \quad 1 \times 3 = \underline{3}$

$14 \times 4.32/35 = 1.73$

$HH = 1.73$

(EH) ENVIRONMENTAL HAZARD

$N_b = 1$
$N_e = 4$
$EH = 3.5$

(RI) RISK INDEX

$RI = 6(2.00 + 3.45 + 3.50)$
$RI = 53.7$

This material rated fairly low in each hazard category. It is a fairly high volume production material, but its inherent hazards are relatively low, and therefore a relatively low RI of 53.7 is the result.

FLUORINE

(P) POTENTIAL EXPOSURE INDEX

$N_{ap} = 2$
$N_{pc} = 3$
$P = 5.00$

(PH) PHYSICAL HAZARD INDEX

$N_f = 0$
$N_x = 3$
$PH = 7.25$

(HH) HEALTH HAZARD INDEX

$N_a \quad 3 \times 2 = 6$
$N_s \quad 5 \times 2 = 10$
$N_c \quad 1 \times 3 = 3$
$N_m \quad 2 \times 1 = 2$
$N_t \quad 3 \times 3 = 9$
$N_r \quad 2 \times 3 = \underline{6}$

$36 \times 4.32/35 = 4.4$

$HH = 4.4$

(EH) ENVIRONMENTAL INDEX

$N_b = 1$
$N_e = 4$
$EH = 3.5$

(RI) RISK INDEX

$RI = 5.00(14.50 + 8.8 + 3.5)$
$RI = 134.0$

Fluorine does not rate as high in potential exposure as ethylene glycol, but it is a significantly more hazardous material. It poses both a physical hazard and health hazard, and its 134.0 Risk Index indicates that this material should be a higher priority for hazard warning development than ethylene glycol.

MERCURY

(P) POTENTIAL EXPOSURE INDEX

$N_{ap} = 3$
$N_{pe} = 3$
$P = 6$

(PH) PHYSICAL HAZARD

$N_f = 0$
$N_x = 0$
$PH = 1.00$

(HH) HEALTH HAZARD

$N_a \quad 7 \times 2 = 14$
$N_s \quad 6 \times 2 = 12$
$N_c \quad 1 \times 3 = 3$
$N_m \quad 1 \times 1 = 1$
$N_t \quad 1 \times 3 = 3$
$N_r \quad 1 \times 3 = \underline{3}$

$36 \times 4.32/35 = 4.4$

$HH = 4.4$

(EH) ENVIRONMENTAL HAZARD

$N_b = 6$
$N^e = 6$
$EH = 10.00$

(RI) RISK INDEX

RI = 6(2.00 + 8.80 + 10.00)
RI = 124.8

While mercury has a far lower physical hazard than fluorine, its environmental hazard is a maximum rating. With the same health hazard rating as fluorine, this demonstrates that a significant environmental hazard can produce a Risk Index rating of relatively high priority.

The significance of the 10-point difference between fluorine and mercury can vary depending on the group of materials they are ranked with. For example, if they are part of a group of 50 materials, all of which range from 120-140, then the 10-point difference may lead to a significant difference in priority for hazard warning development. If, on the other hand, the group of 50 materials were evenly distributed between 50-150, then fluorine and mercury could be similarly prioritized for hazard warning development.

CONCLUSION

The systematic prioritization of large numbers of materials for hazard warning development can be accomplished by use of the Integrated Risk Index System. By evaluating potential exposure as well as the inherent hazards of a material, and assigning numerical ratings to these factors based on specific criteria, a Risk Index number can be derived. Materials can then be ranked and prioritized for hazard warning development based on this ranking scheme.

TABLE 16. SUMMARY OF RI FOR 3 MATERIALS

MATERIAL	P	PH	HH	EH	RI
Ethylene Glycol	6	1.00	1.73	3.5	53.7
Fluorine	5	7.25	4.4	3.5	134.0
Mercury	6	1.00	4.4	10.0	124.8

RISK INDEX CALCULATION SHEET

MATERIAL: ______________________ RISK INDEX: __________

CAS NO.: ______________________

COMPILED BY: ______________________ DATE: __________

PRODUCTION (P)	RATING	NOTES
Annual Production (N_{ap})		
Population Exposed (N_{pe})		
P Index =		

PHYSICAL HAZARD (PH)	RATING	NOTES
Flammability (N_f)		
Reactivity (N_x)		
PH Index =		

HEALTH HAZARD (HH)	RATING	SCORE
Acute Toxicity (N_a)	2x	=
Subchronic Toxicity (N_s)	2x	=
Carcinogenicity (N_c)	3x	=
Mutagenicity (N_m)	1x	=
Teratogenicity (N_t)	3x	=
Reproductive Effects (N_r)	3x	=

HH Index = $\frac{4.32\ (N_a + N_s + N_c + N_m + N_t + N_r)}{35}$

ENVIRONMENTAL HAZARD (EH)	RATING	NOTES
Bioaccumulation (N_b)		
Ecological Effects (N_e)		
EH Index =		

RISK INDEX CALCULATION (RI) (RI) = P(2PH + 2HH + EH)

Risk Index =

Figure 3

The rating criteria outlined in this system should not be considered definitive. Additions, deletions, and modifications to these criteria are necessary if the system is to be effective for a variety of purposes other than strictly hazard warning prioritization. Therefore, this system can also be considered a flexible template for developing similar systems where efficient and effective assessments of relative risks are needed for risk management activities.

REFERENCES

American Institute of Chemical Engineers: Fire and Explosion Index Hazard Classification Guide, 5th Edition. American Institute of Chemical Engineers, New York, NY (1981).

Environmental Protection Agency: Pesticide Chemical Active Ingredients; Proposed Registration Standards Ranking Scheme, Federal Register, Friday, November 14, 1980, pages 75486 - 75497.

National Fire Protection Association: Fire Protection Guide on Hazardous Materials, 7th Edition. NFPA, Boston, MA (1978).

FOOTNOTES

1. National Fire Protection Association; Fire Protection Guide on Hazardous Materials, 7th Edition. NFPA, Boston, MA (1978).
2. American Institute of Chemical Engineers: Fire and Explosion Index Hazard Classification Guide, 5th Edition. American Institute of Chemical Engineers, New York, NY (1981).

*The System described within is not an officially adopted Atlantic Richfield Company procedure and represents the views of the authors.

SECTION 2: OCCUPATIONAL RISK ANALYSIS

OCCUPATIONAL RISK ANALYSIS: INTRODUCTION

P. F. Deisler, Jr.

Shell Oil Company
One Shell Plaza
P.O. Box 2463
Houston, TX 77001

The five papers presented during this session represent a considerable diversity of types of risk analysis and of the uses of risk analysis, and they offer a similar diversity of perspectives on the subject. That so few papers could be so diverse should not surprise anyone familiar with the field; by its nature, the field engenders diversity and, more than that, controversy.

Gratt,[1] in the first paper, employs a method which describes in quantitative terms the occupational risks that may be run in an industry yet to become a significant part of the industrial scene, the oil shale industry. He does this using overall risk factors (and their ranges) which compare risks in segments of the new industry with comparable segments of selected, existing, similar industries. His factors are based on both quantitative, measured data, and subjective judgments, and they include consideration of the numbers of workers potentially exposed. The objective of the work is to aid in recommending research areas which, if undertaken, might result in reducing the uncertainties in the estimated risks.

Clearly, opportunities for detailed controversy abound at every point in Gratt's work. They range from questions of the real comparability and relative quality of the input data, through questions about the assumptions made along the way, to questions of the general validity of the subjective inputs. How the many considerations are combined to yield the numerical risk factors and their ranges offers still further opportunity for discussion. Certainly his method is not one designed to illuminate small differences between segments of the future shale oil industry and current, comparable industrial segments. However, if used carefully and thoughtfully by research decisionmakers, this method should be an aid to highlighting major areas needing work, assessing their priorities, and so accomplishing the stated purpose of the method. Used along with other considerations (recognizing that the quantitative factors should be taken as condensed descriptions of relative risks rather than measurements of them, and with a thorough exposition of the way the methodology was applied in mind) this technique should give valuable additional insights in cases where large and costly programs are at stake. The ordering of thought required to bring all factors together into the risk factors will have value by itself.

The second paper by Yurachek,[1] Brauer, and Stein, deals with occupational risks to health from inorganic arsenic in current industrial settings. It summarizes and explains the work published recently by the Occupational Safety and Health Administration or OSHA (1) and is unique in several respects. It is an historic event since it is the first quantitative risk assessment made by OSHA in support of a regulation; it utilizes OSHA's four-step process; it addresses the significance of risk; and it is based on a relatively large sample of exposed workers and not on animal data as is more often the case in risk assessment. In this case, an animal model does not exist at present. Indeed, if the existing animal data alone were available, inorganic arsenic might not today be considered a human carcinogen. The unfortunate fact of the uncertainty of animals as simple predictors of carcinogenic hazards to humans (2) is once again demonstrated here, though the utility of animal testing, provided a program is broad enough in its research, cannot be totally discounted and is rapidly improving.(3)

The four-step process (1, p. 1,865) is an important innovation. Though OSHA does not have a formal cancer policy in operation at present (4), this process can be viewed as a possible element that such a policy might contain. The first step in the process, in which "risk assessments are performed where possible and considered with other relevant factors to determine whether the substance to be regulated poses a significant risk to workers," includes (in principle) in one step the several steps of Hazard Identification, Dose-Response Assessment, Exposure Assessment and Risk Characterization proposed by the National Research Council or NRC.(5) The way in which the risk assessment for inorganic arsenic is carried out, however, follows these latter steps to a fair degree, implicitly if not explicitly. Moreover, OSHA's four-step process distinguishes between what the NRC report identifies as the scientifically-based risk assessment portion of the regulatory process and the "risk management" portion. The risk management step that the NRC report proposes consists of the development of regulatory options, the evaluation of public health, economic, social and political options, and, combined with the characterization of the risk, agency decisions and actions. OSHA's last three steps correspond to the NRC's risk management activity. They are the second step, in which "OSHA considers which, if any, of the proposed standards being considered for that substance will substantially reduce risk;" the third step, in which "OSHA looks at the best available data to set the most protective exposure limit necessary to reduce significant risk that is both technologically and economically feasible;" and the fourth step in which "OSHA considers the most cost-effective way to achieve the objective." In taking decisions and actions, OSHA combines the results of its last three steps with its characterization of risk to the extent of concluding whether a significant reduction of a significant risk will result. Here, too, in its own specific way, there is a parallel between OSHA's four-step process and the NRC's proposal. In characterizing the risk, OSHA, in the case of inorganic arsenic as explained by Yurachek, et al., describes the basis on which the risk is estimated, gives the ranges of risk found, and explains the reasons for selecting a most likely estimate of the risk. OSHA does not, however, parallel the types of formal peer review mechanisms described by the NRC.

Lamm[1] and Lederer, in the third paper, illustrate the point that scientific examination of the data and what it can tell us is a most difficult art. Moreover, how the data are interpreted can have a profound effect on the resulting assessment of risk.

The data which Lamm and Lederer discuss are from a partial, randomly selected set of the data considered in OSHA's risk assessment and which have been the subject, themselves, of a separate substudy. Because of the low statistical power of the subset, OSHA did not give these results weight in their final assessment using instead, the full study. The point made by Lamm and Lederer is that when the exposures in the substudy are recategorized using a larger number of categories, recategorizing some jobs to their correct exposure groups, a very much lower risk is shown at the lower exposure levels. Indeed, when workers are classified according to their highest 30-day exposures, this effect is very marked and predicted risks at the level of the new standard drop from a preferred estimate of eight lung cancer deaths per thousand workers per 45-year working life to a minuscule or even a zero value. The data in the substudy indicate the possibility of a threshold under the new interpretation.

Yurachek, et al., comment on the fact that the eight per thousand risk at the new standard appears still to be significant compared to other risks, noting that the Supreme Court has opined that a risk of one in a thousand in a working lifetime is a risk that should be reduced or eliminated. Indeed, a risk of eight in a thousand is about twice as high as the risk of death from industrial accidents in the manufacturing and service industries, (6) extrapolated to a forty-five-year working life. Should the further work Lamm and Lederer indicate is in progress to study the full set of subjects bear out the conclusions of the partial study, the resulting drastic reduction in the risk at the new standard should be most welcome news. The understanding of the mode of action, if it were possible to achieve, would also help to distinguish which correlation is the more plausible in such a case, and the search for an animal model would take on new significance.

One further point emphasized by this entire consideration of the risks associated with inorganic arsenic is the potential sensitivity of the risk assessment not only to the uncertainty in the exposure data but also to the interpretation of the exposure data. This has implications for animal studies in cases where animal models are available. Such studies are most often done only using steady, regular exposures and regular, pulsed exposures are rarely studied. It also has implications for epidemiologic studies and may add greatly to the already large uncertainties which normally exist about interpreting exposures in such studies.

Cobler and Hoerger,[1] in the fourth paper, undertake a very timely look backward to offer an answer to the question of to what levels risks have actually been reduced through agency action. For obvious reasons, decisions by agencies in addition to OSHA are included, and the general conclusion is that the true risks probably lie about two orders of magnitude above the estimates of risk implied in the agencies' decisions. They go on to conclude that target risk levels might better be in the vicinity of 10^{-4} and not 10^{-7} to 10^{-5}. Moreover, quantitative risk assessment based on animal studies should utilize as much ancillary information as possible and should be used with care and in fairly narrowly defined situations.

The possible utility of defining an upper level of risk above which risks are considered "high" or even, in a societal sense, "unacceptable," as opposed to relying solely on levels of "*de minimis*" or "target" risks for deciding on action versus no action has been proposed. (7) In this case, the upper level of risk would apply to the aggregate risk posed by all carcinogens found in a specific exposure situation. The thought is a simple one and it involves setting up a yardstick against which risks would be judged "high," "low," or "insignificant." When risks are found to be high, action would be taken to lower those risks, first, into the low risk zone at least. When found to be low, they would be lowered to the extent feasible and as effectively as possible. When found to be insignificant, no significant action would be required and resources could be applied instead to lowering other high or low risks. If a risk is judged to be significant, but there is insufficient information to assess whether it is high or low, it is to be treated as a low risk, as above, while further information is developed. The process is visualized as stepwise and iterative, the idea being to make some progress in reducing the high risks significantly and the low risks to some degree over a broad front and to revisit such decisions at a later time when more information is available. The upper level of risk on the yardstick is the boundary between the high and low regions and, for cancer, it has been quantified in terms of probability alone using a number of criteria, including the goal of containing total risk to *below* a level corresponding to a very low contribution of industrial sources to total cancer. Other criteria include comparative risk. The range within which the probability boundary for cancer might lie, and within which it may be chosen, is not above one chance in two hundred in a lifetime nor, roughly, below one in a thousand in a lifetime in the occupational setting. The boundary between the low and insignificant risk regions is the more usually considered de minimis risk level.

This three-risk-region yardstick can be useful within a private concern in setting its own internal standard when none exists or when data indicate an existing standard may allow exposures to be too high. Such a concept and process is currently used within the author's company.

It is hoped that the subject of the fourth paper will be re-examined in several years' time, when further examples are available and, perhaps, when additional data improve risk estimation. Considering such results should help, in time, to put occupational risks into perspective and even to help define what acceptable risk means. A definition of truly unacceptable risk for specific exposure situations would also be helpful in characterizing and reducing the most serious risks first. At the present time, as Hohenemser and Kasperson have said, "if there is one word that summarizes the general questions of determining risk acceptability, it is 'unsolved'."(8, p. 201)

The final paper by Compton[1] deals with risk analysis and assessment from an entirely different perspective, that of the risk taker. In this paper the author describes ways to inform workers and to train them in using the information so they can attain a personal assessment and understanding of the risks they may run from the materials they handle in their work. This concept goes well beyond the usual concepts of safety training; in the author's own words, it is aimed "to create a frame of reference for non-scientific persons to use in assessing hazard and risk." Given the difficulties of risk assessment by scientifically trained persons, this is a difficult goal to achieve and Compton's contribution, based on her own practical professional experience, is a most welcome one. Achieving an understanding of risk by the worker should also lead to intelligent risk-reducing behavior--an important adjunct to the usual means of complying with whatever standards may exist. Clearly, the bottom line of all occupational risk analysis and assessment (however quantitative or qualitative it may be, and whether it is highly complex or relatively simple) is to make real reductions in occupational risk.

REFERENCES

1. Occupational Safety and Health Administration, Occupational Exposure to Inorganic Arsenic, Federal Register, 48 (1983), 1,864-1, 903.

2. G. B. Gori, Regulation of Cancer-Causing Substances: Utopia or Reality, Chemical and Engineering News 60:25 (Sept. 6, 1982).
3. E. J. Calabrese, Principles of Animal Extrapolation, John Wiley & Sons, New York (1983).
4. S. G. Campbell, OSHA Cancer Policy: Nearing an End or a New Beginning?, Occupational Hazards 88 (April, 1983).
5. National Research Council, Risk Assessment in the Federal Government: Managing the Process, National Academy Press, Washington, DC (1983).
6. R. Wilson, Direct Testimony Before the United States Department of Labor Assistant Secretary of Labor for Occupational Safety and Health Administration, OSHA Docket No. H-090, Washington (1980).
7. P. F. Deisler, Jr., A Goal-Oriented Approach to Reducing Industrially Related Carcinogenic Risks, Drug Metabolism Reviews, 13(5), 875 (1982).
8. C. Hohenemser and J. X. Kasperson, Risk in the Technological Society, AAAS Selected Symposium 65, Westview Press, Inc., Boulder, Colorado (1982).

FOOTNOTES

[1] Authors whose names are given the superscript, 1, are those who presented the paper.

OCCUPATIONAL HEALTH AND SAFETY RISK ANALYSIS FOR OIL SHALE*

Lawrence B. Gratt

IWG, Corp.
975 Hornblend Street, Suite C
San Diego, CA 92109

ABSTRACT

The potential human health and environmental risks of a hypothetical one-million-barrels-per-day oil shale industry have been analyzed to serve as an aid in the formulation and management of a program of environmental research. Occupational safety and illness have been analyzed for the oil shale fuel cycle from extraction to delivery of products for end use based on existing and surrogate industries. A methodology was developed to estimate risks for this hypothetical industry. The surrogate methodology is based on using modifications of the corresponding exposure and toxicological potency of oil shale materials to conventional petroleum and underground mining risks. The methodology includes a treatment of uncertainty. Pneumoconiosis from the dust environment is the worker disease resulting in the greatest number of fatalities, followed by chronic bronchitis, internal cancer, and skin cancers, respectively. The results have been used to formulate recommendations for research for reducing the uncertainties.

KEYWORDS: Oil Shale Risk Analysis, Ocupational Safety, Occupational Illness

INTRODUCTION

The objective of the 1982 Oil Shale Risk Analysis (OSRA) was to estimate the potential human health and environmental risks of an oil shale industry in order to establish important research needed to reduce the uncertainties in the estimated risks. The results are reported and disseminated in the Health and Environmental Effects

Document (HEED) for oil shale.[1] The information contained herein is not intended for and should not be used for regulatory purposes. The methodology developed for occupational safety and illness will be presented for potential application to other synfuel technologies.

A commercial domestic oil shale industry does not yet exist, thus a hypothetical industry at a production level of one million barrels-per-day (BPD) shale oil was analyzed as representative of a size supplying a significant fraction of the United States' energy demand. As of this writing, the first commercial U.S. oil shale facility is scheduled to begin production at 10,000 BPD (Union Oil's Parachute Creek Project). The resource of western oil shale is of sufficient magnitude that such an industry may exist for several hundreds of years at this production level. To reach this level, an installation phase of up to 30 years is possible. As the resource exploitation ceases after decommissioning, a phase of hundreds of years may be necessary before potential environmental impacts occur.

This paper will consider the human health concerns for the occupational workforce for the steady-state production scenario. The risk analysis formulation is probabilistic and is presented in a simplified form. In general, the risk is the product of factors: the source term times the exposure-dose function times a health effect or damage function times the population at risk. A risk analysis involves the construction of scenarios for industry development, production, and distribution; pollutant source terms; media exposure for populations at risk; and workforce requirements. The derivation of research recommendations is based on both the magnitude of the risk estimate, R, and the uncertainty factors, (u_R), used to generate the uncertainty range. The risks and uncertainties generated in this analysis have been based on both objective and subjective techniques. The resulting range on the risk estimate is computed by R/u_R to $R \cdot u_R$. The research recommendations are based on reducing the magnitude of the uncertainty range.

SCENARIO

The risk analysis scenario shown in Figure 1 has 14 production sites, projected to produce a total of one million BPD, distributed along five of the six major oil shale region creek systems feeding the Colorado River. The production level for sites 1, 5, 7, 8, 12, and 13 is 100,000 BPD. All other sites are projected to produce 50,000 BPD. The production is based on underground room-and-pillar mining with above-ground retorting (AGR) for all but one site. The remaining site (13) is a MIS (modified-in-situ) operation with 50 percent of the production coming from the MIS and 50 percent from AGR. The year when the hypothetical scenario occurs is 2010.

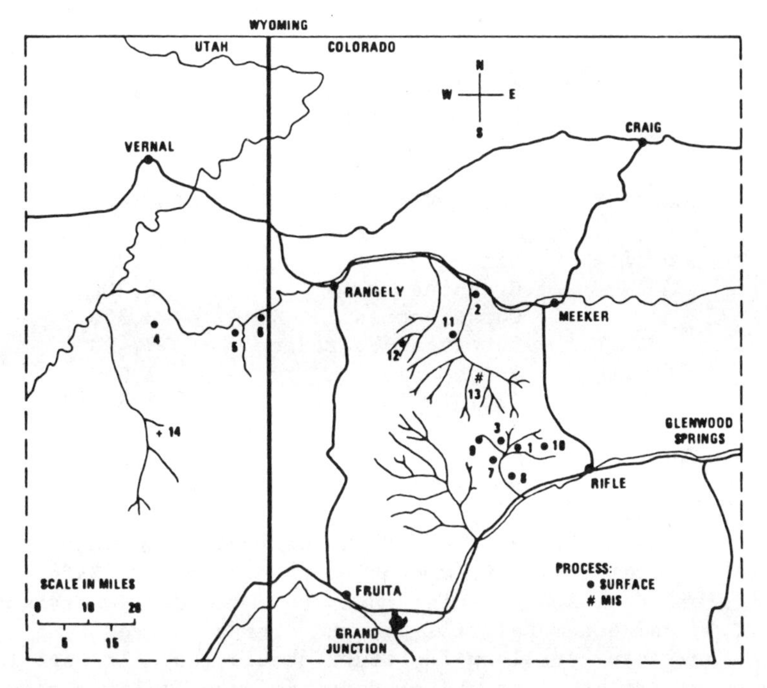

Figure 1

Oil Shale Region Used in the Risk Analysis with Site Locations

OCCUPATIONAL SAFETY

The occupational workforce considers an oil shale fuel cycle ending when products are delivered to the consumers (e.g., gas stations or fuel distributors). The production site occupational workforce estimates were based on total on-site workforce necessary for the proposed technologies (2, 3, 4, 5, 6, 7) and an allocation of the workers to the different job classifications. The result was an average of 0.032 on-site workers per BPD produced. A key assumption made during this analysis was the underground mining productivity of 150 tons per man-shift (with a range of 100 to 170). The refining and transportation workforce estimates were based on statistics for 1978, 1979, and 1980 normalized to one million BPD. Table I summarizes the workforce estimates for the occupational safety and health analysis.

Accident and injury occurrences were estimated for each step of the oil shale fuel cycle using the incidence rates (normally specified in occurrences per 100 workers or 200,000 man-hours per year) from surrogate or actual industries. In some cases the surrogate industry was assumed to have a similar risk environment to that which would be encountered in the oil shale industry, such as refining. Three year

TABLE I

WORKFORCE ESTIMATES FOR A ONE MILLION BPD OIL SHALE INDUSTRY

LOCATION	WORKFORCE COMPONENT	ESTIMATE
On-site	Mining	14,200
	Crushing	6,200
	Construction	3,300
	Retorting-Upgrading	9,400
Off-site	Transportation	2,200
	Refining	5,600
Total		41,000

averages (1978-1980) were used for the estimated incidence rate. Mining rates were derived from statistics of the Mine Safety and Health Administration. Crushing rates were found from statistics on coal, metal, and non-metal milling. Retorting and refining rates are from the American Petroleum Institute statistics. Transportation incidence rates are based on API statistics for pipeline transport and on National Safety Council statistics for rail transport. Construction incidence rates are also based on National Safety Council statistics. Uncertainty ranges were determined by the maximum and minimum values for the years averaged. The mining and crushing rates were also averaged across different types of mined material but restricted to underground mining. A reduction factor based on available statistics was calculated for large mines and applied to the oil shale scenario.

Larger mining operations were reputed to be safer than others. Commercial oil shale mines are expected to be the largest mines in the U.S. Mining safety statistics were analyzed to establish a mine size factor, a number to multiply the mining incidence rates by to adjust for the size of the mining operation. The mine size data available were based on the size of the workforce broken into various classifications. The highest classification in available data was 250 or more employees for coal, metallic mineral, and non-metallic mineral mining. The largest employment size group for stone mining was 150 to 249 employees. For example, in 1978 the fatality incidence rate for all underground coal mining was .07 deaths per 100 workers per year. The mine size factor for this incidence rate is .04 divided by .07, or .57. In this way, an adjustment can be made which compares large--scale mining statistics to average mining operation statistics. An average mine size factor for each accident category for each material was determined using an hour's worked weighted average of the data for the three years (1978-1980). The resulting mine size factor was .59 for fatalities and approximately one for the other incidence rates.

The transportation scenario was the most complex, using different combinations of rail and pipeline utilization to produce the uncertainty range for accident and injury rates. This range varied from 100 percent pipeline to 50 percent rail for transport of upgraded shale oil into the existing pipeline system. The assumed level of rail transport of shale oil is a key assumption for the transportation sector of the fuel cycle.

Table II is a summary of the estimated workforce accidents and injuries along with associated uncertainty ranges for each step of the oil shale fuel cycle. The estimates indicate that from 10 to 19 worker accident-related deaths will occur per year. This corresponds to an "industry" fatality incidence rate of 0.03 to 0.05 deaths per 100 workers per year based on 41,000 total workers. Similarly, the expected non-fatal accidents with days lost from work range from 1800 to 3600 (incidence rate of 4.5 to 8.8) and accidents with no days lost from 1200 to 2000 (incidence rate of 2.9 to 4.9).

The sources of the greatest uncertainty in the accident and injury analysis are the incidence rates in the extraction (mining and crushing) phase. The extraction workforce is roughly half of the total workforce, and fatality incidence rates are .050 and .035 for mining and crushing, respectively. These rates are substantially higher than the fatality rates of .018 to .023 for the other components of the fuel cycle. Out of a total of 13 deaths per year, the extraction phase accounts for 9, or almost 70 percent of the total. The mining fatality rate was reduced by 41 percent from the rate for all mining to adjust for large mine sizes. Without this reduction, the rate would be .085, mining fatalities would increase from 7 to 12. Research is recommended in the areas of mining and crushing accident and injury rates to make the greatest reduction in total occurrence uncertainties. Analysis of large mines safety statistics as applicable to oil shale mines is also needed.

OCCUPATIONAL HEALTH

There are a number of potential occupational health hazards involved in the extraction, retorting, upgrading, transporting, and refining of shale oil including exposure to dusts, toxic gases, heat, noise, and the oil. Human health risks are best estimated from epidemiologic data with the support of in vivo and in vitro test results. Estimation of the occupational health effects of a million BPD oil shale industry is constrained by the paucity of epidemiologic data from the oil shale industry itself. This analysis is based on a surrogate approach where data from industries of similar or analogous exposures and practices is extrapolated to the estimated oil shale exposure and the resulting risk is modified according to other relevant knowledge.

TABLE II

ACCIDENT AND INJURY OCCURRENCES FOR A ONE MILLION BPD OIL SHALE INDUSTRY

Fuel Cycle Component	Fatalities	Occurrences (Range) NFDL	NDL
Mining	7.1 (4.6 - 11)	1600 (1100 - 2400)	440 (290 - 660)
Crushing	2.2 (1.1 - 4.4)	310 (210 - 460)	170 (110 - 280)
Retorting & Upgrading	1.7 (.61 - 4.8)	130 (51 - 340)	300 (120 - 770)
Refining	1.00 (.67 - 1.5)	78 (74 - 82)	180 (170 - 190)
Transportation	0.5 (.10 - 2.5)	140 (18 - 1100)	81 (27 - 240)
Construction	0.77 (.54 - 1.1)	140 (100 - 190)	280 (200 - 390)
Total	13.00 (10 - 19)	2400 (1800 - 3600)	1500 (1200 - 2000)

NFDL is non-fatal occurrences with workdays lost.
NDL is occurrences with no workdays lost.

Methodology

The health risks were analyzed using the simplified equation:

$$R = H \times E \times T \times P$$

R is the estimated number of illnesses in a million BPD oil shale industry workforce. H is the excess rate of a health effect in a similar "surrogate" industry. The excess (or attributable) rate is the rate of the disease above the background or baseline level. This rate is derived from (1) relative risk estimates resulting from epidemiological studies, and (2) the baseline incidence rate of the disease in the general population. The total rate for workers in the surrogate industry is found by multiplying the baseline rate by the relative risk, and the excess rate is found by subtracting the baseline rate from the total. E is the ratio of the exposure in the oil shale industry to the exposure in the surrogate industry. T is the biologic or toxicologic modifier which adjusts the risk result to account for the relative biologic potency of oil shale exposure as compared with the surrogate exposure. The data for this factor come from in vivo systems toxicologic tests. P is the estimated size of the exposed worker population.

The health risk methodology requires the selection of the proper occupational health effects and an appropriate surrogate. The criteria for selection of a surrogate is based on three considerations: (1) data availability, (2) similarities with the new industry and the degree of differences, and (3) the analogy for the population at risk. For each occupational health effect a surrogate is selected.

Hydrocarbon Cancer Risk

The selection of a surrogate can be illustrated by the example of considering the risk of hydrocarbon-induced occupational cancers. The oil refining industry was chosen as the optimal surrogate for oil shale retorting for the following reasons. First, the materials (petroleum and shale oil) are similar in the sense that both are liquid hydrocarbons. Second, both are handled with closed systems where the primary mechanism of exposure is through leaks, accidents, or in maintenance procedures. The products are both potentially toxic through inhalation or direct skin contact. Third, the involvement of essentially the same corporations in both oil refining and oil shale development suggests that industrial hygiene approaches will be similar. The primary alternative to oil refineries as the surrogate industry is coke ovens. The strength of coke oven workers as a population from which to extrapolate retorting risks[8] is the uniquely thorough epidemiologic research on this population. The reason this population was not used is the absence of exposure data for retorting operations. Without scaling the risks, it would be very difficult to determine which subgroup of oven workers best represented retort workers. Thus, to select risks from a surrogate industry, oil refining was thought to be more directly applicable to shale retorting counterbalancing the admittedly inferior epidemiologic data.

Selection of the surrogate is followed by a review of the surrogate data to derive the excess rate of the health effect. Continuing the hydrocarbon example, nine epidemiologic studies of oil industry employees were available.[9-17] While there is some overlap of the study cohorts, all of the studies were considered as independent rather than as follow-ups of the same cohort. In the effort to derive risk estimates, consideration of the shortcomings of each study is important and should be used to determine the relative use of the reported results. The nine investigations were judged to have useful estimates of carcinogenic risks for lung, stomach, kidney, and brain cancer sites. An approach to using this data set to estimate low, average, and high relative risk estimates was derived. The lower bound was established on conceptual grounds as 1.0 (the absence of any adverse occupational effect). All relative risk estimates below 1.0 were considered to be 1.0. Each cancer site had one or more studies which contained no indication of an occupational risk.

The high relative risk was an upper bound chosen as the highest observed relative risk estimate. An average risk estimate was calculated as the arithmetic mean for the reported risk estimates for refinery workers. The H-term in the methodology is in the form of an excess risk above background attributable to the occupational exposure. The relative risk multiplied by the baseline incidence cancer rate produces the estimated incidence rate. Subtracting the baseline value results in the estimates of the occupationally attributable risks. The uncertainty factor for H was derived based on the upper bound divided by the average.

The occupational health methodology includes two terms to modify the risk estimate according to information on the exposure and the toxicity differences between the surrogate and the oil shale industry. Continuing this example, there is a paucity of data on the exposure in both settings and a complete lack of data to quantify any differences. In both cases, a complex hydrocarbon mixture, having an equally complex polycyclic aromatic hydrocarbon component, is present in the workplace setting at varying concentrations for varying periods of time. Based on expected similar industrial hygiene practices, the estimate of the exposure ratio chosen was to be unity, i.e., E=1. Given the lack of data, it was estimated that the aggregate exposure is the same in both settings but the uncertainty about this estimate could be a factor of 25.

The toxicologic modifier scales the risk estimate according to differences in the biological potency of the materials exposing the workers. This requires comparative toxicity data between petroleum and shale oils, including many raw and intermediate products and process stream materials. For the analysis, no such comparative data were available for inhalation exposures. Skin painting test data indicated a toxicologic modifier ratio of one as an initial estimate. For every process stream in the oil shale process, there is a process stream of similar toxicity in the refinery surrogate, but the actual exposures to those fractions is unquantifiable in either industry at the present time. Based on the limited ratios available, the uncertainty range about this estimate was chosen to be 9 (or the ratio of some oil shale substances may exhibit from 1/3 to 3 times the potency of petroleum).

The portion of the work force applicable to these risk estimates are those workers in the retort and upgrading facility. This group will be heterogeneous, but application of the H-term derived here as

Happy Holidays!
Manou Kirk & Olivia

an average for the group appears reasonable. As more precise workforce estimates and relative exposure and toxicologic risk levels are derived for subsets of these workers, the method can be adapted to make use of such data.

Additional uncertainties in deriving the hydrocarbon health risk estimates fall into two distinct categories: (1) risk calculations in the petroleum industry and (2) extrapolation of those risks to oil shale retort workers. The major problems in estimating risks among refinery workers are poor characterization of exposure and imprecision in regard to latency of effect. The virtual absence of quantitative exposure data precludes any dose-effect estimation other than approximate categorizations. This lack of detail probably operates to dilute the observed risks since the cohorts studied would include a fairly large proportion of unexposed workers. Similarly, inadequate consideration of latency can only result in false negatives (missed risks) rather than false positives.[18] Extrapolation of these estimates to oil shale retort workers is another major source of uncertainty. The model includes E and T terms to adjust directly for the different level of exposure and carcinogenicity, but the inability to obtain precise measurements of either constitutes a significant unknown. An ideal E term would include a person- and time-weighted exposure estimate for both refining and retorting, for example, in the form of a total hydrocarbon or polycyclic aromatic hydrocarbon exposure. Similarly, the T-term ideally represents a person-, time-weighted indicator of the carcinogenicity of the measured hydrocarbons. The discrepancy between these ideals and the limitations of the proposed methodology using available data should be noted.

Dust-in-the-Lungs

In applying the surrogate methodology to other health effects, the individual nature of the data and the interrelationships of each term in the formula were considered. Inhalation of dust as a result of mining, crushing, and retorting operations raises the possibility of adverse physiological reaction to that dust in the oil shale workforce. The ability for self-protection and repair of injury can be exceeded. Three non-neoplastic lung responses to respirable dust particles potentially occurring in the oil shale industrial environment were analyzed: pneumoconiosis, chronic bronchitis, and chronic airway obstruction. Silicosis, the pneumoconiosis caused by the inhalation of silica, was analyzed independently. Because these diseases can occur simultaneously in a given individual, the summation of the separate risk estimates will overestimate the total incidence. The estimate of pneumoconiosis was based on British coal worker data and total respirable dust exposures. Silicosis risk was estimated from data on Vermont granite shed workers and Peruvian metal miners with respirable silica as the exposure measurement. The only available oil shale mine dust measurements come from studies of pilot operations and may not adequately represent exposures in a commercial operation. The current

TABLE III

Summary of Oil Shale Occupational Health: Incidence, Mortality, and Uncertainty

Disease	Excess Incidence per 1000 per year	Uncertainty Factor* u_H	u_E	u_T	u_P	u_R	Number of Workers at Risk (1000's)	Case Fatality (%)	Cases Per Year Estimate	Cases Per Year Uncertainty Range	Deaths Per Year Estimate	Deaths Per Year Uncertainty Range
Internal Cancers												
Lung (hydrocarbons)	0.115	3.16	5	3	1.5	10.0	15.0	91	1.73	0.17 - 17.3	1.57	0.16 - 16.00
(radioactivity)	$8.7x10^{-4}$	2.00	3	1	1.5	3.9	20.4	91	$1.77x10^{-2}$	$5x10^{-3}$ - $7x10^{-2}$	$1.62x10^{-2}$	$4.1x10^{-3}$ - $6.3x10^{-2}$
(arsenic)	0 012	2.20	2	1	1.5	3.1	20.4	91	0.25	0.08 - 0.78	0.23	0.07 - 0.71
Stomach	0.039	3.16	5	3	1.5	10.0	15.0	88	0.59	0.06 - 5.9	0.52	0.052 - 5.2
Kidney	0.022	3.16	5	3	1.5	10.0	15.0	56	0.33	0.03 - 3.3	0.18	0.018 - 1.8
Brain	0.069	3.16	5	3	1.5	10.0	15.0	82	1.04	0.10 - 10.4	0.85	0.085 - 8.5
Skin Cancers												
Melanoma	0.029	3.16	5	3	1.5	9.96	15.0	38	0.44	0.04 - 4.4	0.17	0.017 - 1.7
Basal Cell	1.080	5	5	3	1.5	12.93	15.0	1	16.20	1.25 - 210	0.16	0.012 - 2.1
Squamous Cell	0.290	5	5	3	1.5	12.93	15.0	1	4.35	0.33 - 57	0.04	0.003 - 0.52
Pneumoconiosis												
Method 1: Coal Dust Surrogate												
0/1+	2.9	2	2	2	1.5	3.55	20.4	7	59.2	16.70 - 210.	4.03	1.1 - 14.3
2/1+	0.652	2	2	2	1.5	3.55	20.4	7	13.3	3.75 - 47	0.90	0.25 - 3.2
PMF	0.95	2	2	2	1.5	3.55	20.4	35	19.4	5.46 - 69	6.69	1.9 - 23.7
Combined (0/1+ and PMF)									78.6	22.2 - 279	10.7	3. - 38.
Method 2: Silica	2.5	2.5	2	1	1.5	3.38	20.4	43	51.0	15.10 - 172	21.90	6.5 - 74.
Chronic Bronchitis	4.5	2	2	2	1.5	3.55	20.4	4	91.8	25.9 - 326	4.04	1.1 - 14.
Airway Obstruction	0.18	2	2	2	1.5	3.55	20.4	3	3.67	1.03 - 13	0.11	0.031 - 0.39
High Frequency Hearing Loss	1.4	2	2	1	1.5	2.88	20.4	-	28.6	9.9 - 82.3	-	- -

$*\log u_R = \sqrt{(\log u_H)^2 + (\log u_E)^2 + (\log u_T)^2 + (\log u_P)^2}$

Note: Pneumoconiosis, Chronic Bronchitis, and Airway Obstruction estimates may overlap.

nuisance dust standard in oil shale mines of 5 mg/m^3 respirable dust was selected as the estimate for dust exposure.

Health Estimates

The excess incidence rates, the number of workers at risk, the annual new cases, the case fatality rate, and the annual deaths for each disease as a result of applying the methodology are presented in Table III. Fifteen thousand workers involved in the retorting, upgrading, and refining of shale oil are at risk of the cancers. There are 20,400 workers exposed to dusts which can cause non-neoplastic respiratory diseases: the miners and the crusher operators. The population at risk of high frequency hearing loss are the 20,400 workers in the mines and around the crushing facility. The result is the number of excess cases of each disease which could be seen annually in a million BPD industry. The case fatality rates are the estimated percentage of all cases who eventually die as a result of that disease.

SUMMARY

The methodology and results for an occupational health and safety risk analysis for oil shale have been presented. The methodology can be used for other synfuels technologies and new technologies where appropriate surrogate data are available. Based on the magnitude of the risk estimates and the importance of the disease in terms of worker productivity and health, non-neoplastic lung diseases stand out among the diseases considered as very important concerns for future consideration of mine dust control. The results have been used to recommend research areas to reduce the uncertainty in the risk estimates.

REFERENCES

1. IWG Corp., Health and Environmental Effects Document for Oil Shale - 1982. San Diego, California, December 1982.
2. U.S. Department of Interior, Proposed Development of Oil Shale Resources by the Colony Development Operation. Final Environmental Impact Statement, 1975.
3. Rotariu, G.J., Western Oil Shale Development: A Technology Assessment. Volume 1: Main Report. Pacific Northwest Laboratory, Richland, Washington, November 1981.
4. Colorado Energy Research Institute, Oil Shale 1982: A Technology and Policy Primer. CERI and Colorado School of Mines Res. Inst., November 1981.
5. Energy Development Consultants, Inc., Colorado Oil Shale: The Current Status. EDC, Inc., for U.S. DOE, 1979.
6. Nazaryk, P., Colorado Department of Health, memorandum: "Cumulative Environmental Impact Study - Production Scenarios and Workforce Projections." February 1, 1982.

7. Long, R., TOSCO, Denver, Colorado, Personal communication via telephone with L. Gratt on May 24, 1982.
8. Redmond, C.K., H.S. Wieand, H.E. Rockette, R.S. Sass, and G. Weinberg, Long-Term Mortality Experience of Steelworkers. U.S. Deprtment of Health and Human Services (NIOSH) Publ. No. 81-120. U.S. DHHS, PHS, CDC, NIOSH, Cincinnati, Ohio, 1981.
9. Hanis, N.M., K.M. Stavraky, and J.L. Fowler, "Cancer mortality in oil refinery workers." J. Occup. Med., 21: 167-174, 1979.
10. Hanis, N.M., T.M. Holmes, L.G. Shallenberger, and K.E. Jones, "Epidemiologic study of refinery and chemical plant workers." J. Occup. Med. 24: 203-212, 1982.
11. Rushton, L., and M.R. Alderson, "An epidemiological survey of eight oil refineries in Britain." Brit. J. Indust. Med. 38: 225-234, 1981.
12. Schottenfeld, D., M.E. Warshauer, A.G. Zauber, J.G. Meikle, and B.R. Hart, "A prospective study of morbidity and mortality in petroleum industry employees in the United States--A preliminary report." In: Peto, R. and M. Schneiderman, eds., Quantification of Occupational Cancer. Cold Spring Harbor Laboratory, Cold Spring Harbor, New York, pp. 247-265, 1981.
13. Tabershaw/Cooper Associates, "A mortality study of petroleum refinery workers." Project OH-1, Prepared for the American Petroleum Institute, September 15, 1974.
14. Theriault,, G. and L. Goulet, "A mortality study of oil refinery workers." J. Occup. Med. 21: 367-370, 1979.
15. Thomas, T.L., P. Decoufle, and R. Moure-Eraso, "Mortality among workers employed in petroleum refining and petrochemical plants." J. Occup. Med. 22: 97-103, 1980.
16. Thomas, T.L., R.J. Waxweiler, R. Moure-Eraso, S. Italya, and J.F. Fraumeni, Jr., "Mortality patterns among workers in three Texas oil refineries." J. Occup. Med. 24: 135-141, 1982.
17. Wen, C.P., S.P. Tsai, N.S. Weiss, W.A. McClellan, and R.L. Gibson, "A population-based cohort study of brain tumor mortality among oil refinery workers with a discussion of methodological issues of SMR and PMR." In: Petro, R. and M. Schneiderman, eds., Quantification of Occupational Cancer. Cold Spring Harbor Laboratory, Cold Spring Harbor, New York, pp. 413-432, 1981.
18. Rothman, K.J., "Induction and latent periods." Am. J. Epidemiol. 114: 253-259, 1981.

*This analysis was performed under U.S. Department of Energy (DOE) contract DE-AC02-82ER60087 by the University of Colorado, Center for Environmental Sciences and the IWG Corp. The assistance and direction of Dr. Paul Cho (project officer) and Dr. Nat Barr (program manager) of DOE's Health and Environmental Risk Analysis Program, Office of Energy Research were appreciated. Dr. W. Chappell, Mr. B. Perry, Dr. W. Marine, Dr. D. Savitz, and Mr. J. Feerer were all key contributors to this effort.

QUANTITATIVE RISK ASSESSMENT FOR INORGANIC ARSENIC:

DETERMINING A SIGNIFICANT RISK

Mary M. Yurachek
Elisa R. Braver
Edward Stein

Directorate of Health Standards Programs
The Occupational Safety and Health Administration
Room N-3718
200 Constitution Avenue, N.W.
Washington, DC 20210

ABSTRACT

On January 14, 1983, the Occupational Safety and Health Administration published its first formal quantitative risk assessment as part of the rulemaking for the inorganic arsenic standard. In addition to making determinations concerning the scientific issues surrounding the carcinogenicity of arsenic, the Agency had to make initial determinations about general risk assessment methodology and policy issues such as the definition of "significant." This paper will discuss the approach OSHA took in its first risk assessment endeavor.

KEY WORDS: Quantitative Risk Assessment, Inorganic Arsenic, Lung Cancer, The Occupational Safety and Health Administration, Regulation.

INTRODUCTION

The Occupational Safety and Health Administration (OSHA) published its first formal quantitative risk assessment on January 14, 1983, as part of the rulemaking proceedings for inorganic arsenic. This risk assessment was contained in a 40-page document in the Federal Register (1) devoted to the quantitative assessment of risk from exposure to inorganic arsenic and a determination of the significance of that risk, based on substantial epidemiologic evidence associating excess lung cancer with exposure to inorganic arsenic. This paper offers a description of a few of the many complex, scientific issues before the Agency with regard to inorganic arsenic in particular and risk assessment in general.

Section 2 will describe some of the legal and judicial constraints placed on OSHA in promulgating a standard and how these constraints may influence the risk assessment. Section 3 will present the major issues of the risk assessment and a brief explanation of the resolution of some of these issues, and Sections 4 and 5 will offer OSHA's conclusions and justification for these conclusions.

This paper is meant only to summarize the approach taken by OSHA in its first formal quantitative assessment of risk and will address only a limited number of the issues involved in promulgating a standard for inorganic arsenic. For a highly detailed and technical exposition of all the data before the Agency when it made its final evaluation of risk, the reader is directed to the _Federal Register_ (1).

BACKGROUND

The Occupational Safety and Health Administration was established in 1970 as a result of the Occupational Safety and Health Act (OSH Act, 84 Stat. 1590). The OSH Act outlined specific guidelines to be followed in the promulgation of standards for the toxic substances. It stated, in particular, that the Secretary shall be empowered to set standards that ensure that an "employee shall not suffer material impairment of health or functional capacity even if such employee has _regular exposure_ to the hazard dealt with by such standard for the _period of his working life_" [6(b)(5), emphasis added].

For the purposes of risk assessments, these phrases were used to establish basic assumptions for the determination of risk (e.g., risk was calculated for exposure over an eight-hour day and a working lifetime of 45 years). The OSH Act further states that standards must be set "to the extent feasible" (both economically and technologically) and that determinations are to be made on the basis of the "best available evidence."

In July, 1980, in its interpretation of the OSH Act in the case of _Industrial Union Department_ v. _American Petroleum Institute_ [commonly referred to as the "benzene decision," 448 U.S. 607 (1980)], the Supreme Court held that in the promulgation of a standard "the Secretary is required to make a threshold finding that a place of employment is unsafe--in the sense that significant risks are present and can be eliminated or lessened by a change in practices." That is, the Secretary must make a determination that a significant risk exists at the current permissible exposure limit (PEL) and that by lowering the PEL, one can lower the risk. Until the ruling in the benzene case, OSHA had always regulated carcinogens on the assumption that there was no demonstrated safe level for any human carcinogen, and therefore the PEL was set at the lowest feasible level.

In the case of inorganic arsenic, the Court decision was interpreted to mean that the Agency should quantify the risk, if possible, though the Court noted that this determination should not be a "mathematical straitjacket" and that the Secretary may "use conservative assumptions in interpreting the data with respect to carcinogens, risking error on the side of over-protection rather than under-protection." (448 U.S. 655, 656). The Court continued stating that "OSHA is not required to support its finding that a significant risk exists with anything approaching scientific certainty." The Court did not use the term "significant" to refer to statistical significance; rather, the Court used "significant" to refer to important substantial health risks posed by exposure to a specific level of an agent.

The final standard regulating inorganic arsenic as a carcinogen was published on May 5, 1978, reducing the PEL from 500 $\mu g/m^3$ to 10 $\mu g/m^3$ and establishing requirements for monitoring, control strategy, medical surveillance, and other provisions. The risk assessment discussed below summarizes the Agency's "post-benzene" efforts to establish the significance of the risk at the former PEL of 500 $\mu g/m^3$ in order to justify lowering the PEL to 10 $\mu g/m^3$.

MATERIALS AND METHODS

The Data

In writing a final rule, OSHA relies on both published and unpublished data submitted to the Agency during the rulemaking proceedings, including data submitted in testimony during informal public hearings and pre- and post-hearing comment periods. (Some of these documents will be referred to in the reference list by exhibit numbers and are available for public inspection.)

The 1978 final standard was promulgated on the basis of a number of high quality epidemiologic studies[2] and the qualitative determination that occupational exposure to inorganic arsenic was associated with excess respiratory cancer, skin cancer, and other diseases, such as tuberculosis.

The quantitative assessment of risk (1983) relied on six epidemiologic studies in two industrial settings: smelting[3] (where workers are exposed to inorganic arsenic as a by-product of the smelting of sulfide ores of copper, lead, and zinc) and arsenical pesticide manufacturing.[4]

These six studies were chosen for the risk assessment because, when coupled with data in the rulemaking record, each contained sufficient information to establish a dose-response relationship of increasing risk of respiratory cancer mortality with increasing arsenic exposure.

Another characteristic of many of these studies which strengthened the confidence that could be placed in mathematical extrapolation of the lung cancer risk was that excess risk was actually observed as distinguished from predicted, at exposures estimated to be at or below 500 $\mu g/m^3$. Furthermore, all of these epidemiologic studies were well-designed and well-conducted. These data are of exceptionally high quality; in fact, they are considered to be substantially more evidence than is required for a determination of significant risk under the guidelines set forth in the benzene decision.

In addition to these data, there were several other epidemiologic studies which were important in OSHA's final determination of significance of risk.[5] The data contained in several of the studies were of a quantitative nature, but were not necessarily suitable for risk assessment because of the paucity of exposure measurements. The data in these studies, however, contributed to the formation of a conclusion about the quantitative nature of the risk from exposure to inorganic arsenic and strengthened the causal association between inorganic arsenic exposure and excess lung cancer risk.

The Model

Initially three models were considered for the purpose of the risk assessment:(20,21)

- The linear non-threshold model expressed as

  $$R = Bd$$

  where R is a measure of risk, d is a measure of dose and B is a measure of the carcinogenic potency. This model assumes that the risk is directly proportional to dose.

- The quadratic non-threshold model expressed as

  $$R = Bd^2.$$

  Here risk is assumed to be proportional to the square of the dose.

- A threshold "model" proposed to OSHA on the basis of the Higgins et al. data.

Before a model was fit to the (risk, dose) data, components of the model were considered separately. The first determination to be made was the choice of the most appropriate measure of risk to be employed. The first measure of risk considered was the relative risk as measured by the standardized mortality ratio (SMR). The SMR is usually computed as (Observed/Expected x 100) and is a common measure of risk employed in epidemiologic studies, where the excess risk in the exposed population is computed relative to an appropriate control population. An example of the SMRs for various exposure categories in the Lee and Fraumeni study can be found in Table I. As can be seen in this table, there is a clear relationship of increasing risk with increasing dose characterization, as well as with increasing duration of exposure.

In contrast, the data of Table II (Enterline and Marsh [13]) do not show as clear a dose-response relationship--that is, the SMRs do not clearly rise with increasing dose. This was puzzling, since in an earlier study of retirees from this same smelter (3,4) a clear dose-response relationship was observed. Upon further investigation it was discovered that the measure of dose in the later cohort study (of all workers) was accumulated in a manner such that dose was confounded with age. That is, older workers had higher cumulative doses. Hence, the SMRs tended to level off at higher doses (older workers) because arsenic-induced cancer rates were not rising as rapidly as lung cancer rates in the background population. This phenomenon was not observed in the study of the retiree population because all of the individuals in the population were approximately of the same age.

Age-confounding of the comparison of SMRs among different groups of workers is a potential problem for many studies, in that the SMR is a risk measure that is directly dependent on the age distribution of the study population (i.e., the SMR is a multiple of the expected number of deaths, which itself is age-dependent).

For such studies, OSHA adopted a measure of absolute risk for use in the risk assessment model. Absolute risk was defined as (Observed - Expected/Person-Years) and is a measure of the excess risk attributable to inorganic arsenic in each dose category. Absolute risk is not a multiple of the expected value, which makes it a risk measure less age-dependent than the SMR. As can be seen from Table II, the use of absolute risk for the Enterline and Marsh data provides a much clearer dose-response relationship, a relationship more suitable to risk assessment. Similar determinations on appropriate risk measures were made for each of the six studies used in the risk assessment.

The second component of the model to be considered was the appropriate measure of dose. Three such measures were prominent during the rulemaking: (1) cumulative dose, a measure which incorporates both the intensity and the duration of dose, (2) simple intensity of dose, and (3) "ceiling" dose or short-term "peaks" of intensity.

TABLE I

Mortality from Respiratory Cancer by Length of Employment and Exposure Category
Lee and Fraumeni, 1969

Cohort	Respiratory cancer mortality		Maximum exposure to arsenic (12 or more months) (ug/m^3)			Avg. Length of employment (yrs.)
		Overall	11,270[1]	580	290	
All cohorts combined	Obs.	147	18	44	45	
	Exp.	44.7	2.7	9.2	18.8	
	SMR	329**[2]	667**[3]	478*	239*	
Cohort 1	Obs.	61	8	22	14	31.7[4]
	Exp.	13	1	3.3	5.6	
	SMR	469**	800*	667*	250*	
Cohort 2	Obs.	37	6	12	9	18.1[5]
	Exp.	10	0.9	2.2	2.9	
	SMR	370**	667*	545*	310*	
Cohort 3-5 Combining Cohorts 3,4,5	Obs.	49	4	10	22	5.4[6]
	Exp.	21.7	0.9	3.8	10.3	
	SMR	226*	444**	263**	214*	
Number of persons in each category		8,047	402	1,526	3,257	

1 Morris [22]
2** Significant at the 5% level
3* Significant at the 1% level
4 15 (given 15 years or more of employment prior to 1938) + 15799/947 (Person-years/no. of persons in cohort 1) Computed by Chu
5 1/2 (15) (assumes 1/2 exposed years occurred prior to 1938) + 17460/1644 (Person-years/no. of persons in cohort 2); Computed by Chu
6 Weighted average of no. of years of exposure in cohorts 3-5; Computed by Chu

TABLE II

Respiratory Cancer Deaths and SMR's by Cumulative Arsenic Exposure Lagged 10 Years
Enterline and Marsh, 1982

Cumulative Exposure (ug/As/l) (urine-years)	Person-Years at risk	Observed Deaths	Expected Deaths	SMR	Absolute Risk x 10^5
<500 (302)[1]	27802	10	6.44	155.4	12.8
500-1500 (866)	16453	22	12.46	176.6*[2]	58.0
1500-3000 (2173)	11213	26	11.48	226.4*	129.5
3000-6000 (4543)	9571	22	12.39	177.6*	100.4
7000+ (13457)	5423	24	9.75	246.2**[3]	262.7

[1] () = Mean of class interval
[2] * $p < .05$
[3] ** $p < .01$

Cumulative dose was the measure employed by most of the risk assessments submitted to the Agency for consideration. For the Lee and Fraumeni data, for example, cumulative dose was computed by multiplying average length of employment by the maximum exposure. On the other hand, data from the Tacoma smelter was characterized by total cumulative micrograms of arsenic per liter of urine. There was evidence to support a correlation between urine arsenic levels and air arsenic levels (23). The intensity levels from both smelters were based on direct industrial hygiene measurements made in the plants under study, measurements often going back to the 1940s.

It was suggested that cumulative dose may not be the appropriate dose measure to be used for the assessment of risk posed by inorganic arsenic and that intensity alone was the major determinant of risk level. After examining several studies, (11, 14, 15, 17) it was concluded that duration of exposure did indeed contribute to risk; in fact, Brown and Chu concluded that duration of exposure was observed to be the most important single factor in the excess lung cancer risk.

Relying on the Higgins et al. data for a 22 percent sample of the Anaconda cohort, several participants in the rulemaking hypothesized a threshold for carcinogenicity at 500 $\mu g/m^3$, the former PEL. This hypothesis was based on the Higgins et al. "ceiling" analysis where workers were classified not only by their average or cumulative exposure, but also by the highest level to which they had 30 or more days of exposure. An example of this data is given in Table III. This data may be considered striking upon first inspection, since it actually implies a deficit of risk at lower levels. Upon analysis, however, it can be shown that the study lacked adequate statistical power to detect a risk in lung cancer mortality among workers receiving lower exposures; for example, in the ceiling category "less than 500 $\mu g/m^3$," the power to detect a 1.5-fold increase in risk (50 percent excess) in the lowest cumulative exposure category (less than 500 $\mu g/m^3$-years) is only 25 percent and the power in the next higher cumulative exposure category is 15 percent.[6]

Most epidemiologic investigators, when initiating a study, attempt to choose a study cohort of sufficient size to have at least 80 percent power to detect a true difference of a specific magnitude

TABLE III

Respiratory Cancer Mortality by Cumulative Lifetime Exposure to Arsenic and Lifetime Ceiling Dose[1,2]

Higgins et al., 1982
(Method V)

Ceiling Exposure (ug/m^3)		Cumulative Lifetime Exposure (ug/m^3 - years)			
		500	500-2000	2000-12000	12000+
Total	Obs.	4	9	27	40
	SMR	69	157	400*[3]	550*
<500	Obs.	3	1	0	1
	SMR	67	79		940
≥500	Obs.	1	8	27	39
	SMR	77	180	422*	542

1 Ceiling is defined in Higgins et al. as the maximum exposure category in which a worker spent 30 or more days.
2 CEOH [24]
3 *Significant at .05 level

in the variable of interest. Given that lung cancer is a relatively common cause of death, it seems reasonable that studies of lung cancer risk should have at least 80 percent power to detect a 50 percent increase in risk.

The lack of power in this study can be attributed to a "small numbers" problem, which is exacerbated when examining the power of individual cells. Further evidence of the low statistical sensitivity of the study by Higgins et al. was their failure to observe significant excess mortality from tuberculosis and digestive cancer, both of which had been observed in Lee-Feldstein's study of the entire cohort.

As noted before, we have only touched on a small number of the statistical and biologic factors which influenced the determination of significance of risk. Other issues, such as the effects of smoking in the population, the influence of concomitant exposures, mode of carcinogenic action, an animal model for the carcinogenicity of inorganic arsenic, mutagenicity and the role of arsenic as a trace metal, are addressed in the Federal Register (1) and the reader is directed there for information on these subjects.

CONCLUSIONS

After the close of the rulemaking record, OSHA analyzed all the data submitted to make a final evaluation of the level and significance of the risk posed by occupational exposure to inorganic arsenic. This decision was based on nine individual risk assessments performed by OSHA experts, other government experts, the World Health Organization-Arsenic Working Group and private participants in the rulemaking. A summary of the methods employed and results from each of these quantitative risk assessments is given in Table IV and summarized in graphic form in Figure 1.

OSHA concluded that the prediction of the risk from inorganic arsenic should be based on a linear non-threshold model based on a measure of cumulative dose. The linear model was chosen because (1) it uniformly showed better "fit" to the data than the quadratic model, (2) it was consistent with biological evidence indicating linear dose-response relationships for both carcinogenesis and mutagenesis, (3) it was a conservative position reflecting uncertainty about the effects of low-dose exposures, and (4) it was the model recommended by virtually all of the experts participating in the rulemaking. Due to the methodological limitations of the Higgins et al. data, and the weight of evidence from the other studies, a threshold model for the carcinogenicity of inorganic arsenic has been rejected at this time.

The Federal Register document explains the rationale used by the Agency in arriving at its preferred estimate of risk. However, an exploratory data analysis technique (EDA) introduced by Tukey (33) can

TABLE IV

Summary of Risk Assessments
Excess Risk of Lung Cancer per 1000 Workers
Risk at 10/50/500 UG/Cubic Meter

	Model Used	Lee and Fraumeni	Pinto et al.	Ott et al.	Lee-Feldstein	Enterline and Marsh		Higgins et al.
CHU [20]	Linear RR[1]	10/51/393	25/125/713	29/146/767				
	Quadratic RR	0.3/7/-	2/49/-	2.1/52/-				
CLEMENT ASSOCIATES[26]	Linear RR	7.7/39/394	9.1/46/465					
EPA-CAG[27]	Linear RR	8/38/375[2]						
CRUMP [28]	Linear RR	8.7/43/321	19/92/518	25/117/578	8.3/40.6/310	-		9.4/45.8/342[3]
	Linear AR[4]				3.2/16.0/148	7.8/38.6/315	LAG 0	5.2/25.9/228
						7.6/37.3/303	LAG 10	
RADFORD [29]	Linear RR			18.8-37.5/-/-[5]				
ENTERLINE [30]	Linear RR					2.7/13.4/-	LAG 10	
						2.2/11.2/-	Table 12	
BLEJER AND WAGNER [31]		Recommends 2 $\mu g/m^3$						
WORLD HEALTH ORGANIZATION [32]	Linear RR		61/305/-[6]					
CEOH [24]								No excess risk if exposure <500 $\mu g/m^3$

1 Relative risk (Observed/Expected)
2 Combined estimates from Lee and Fraumeni, Pinto et al., and Ott et al. studies.
3 Cumulative exposure data. Ex. 202-8, Appendix D.
4 Absolute risk (Observed-Expected)/Person-Years.
5 Range of estimates from Ott et al., Lee-Feldstein, and Enterline and Marsh studies.
6 Calculated by OSHA using WHO's methodology for risk analysis.

a. 10 ug/m^3 Risk, Cut[2]

0	2 2 2 3 5 7 7 7 8 8 8 9 9
1	0 9
2	5 5 9
3	

Five-value summary N=18

.	2 (2.2)[3]
H	5 (5.2)
M	8
H	10
.	29

b. 500 ug/m^3 Risk x 10^{-1}, Cut

1	4
2	2
3	0 1 1 2 4 7 9 9
4	6
5	1 7
6	
7	1 6

Five-value summary N=15

.	14 (148)
H	31 (310)
M	37 (375)
H	51 (513)
.	76 (767)

[1] From Table IV

[2] Cutting refers to the process of dropping decimal places, rather than rounding

[3] ( ) = Result corresponding to this entry in the diagram

Figure 1

Combined Results of Risk Assessments
"Stem and Leaf" Diagrams at 10 and 500 ug/m^3 [1]
Excess risk per 1000 workers

be used in evaluating the data from the many risk assessments. Figure 1 illustrates the application of the EDA. The "stem and leaf" diagram shown in the figure can be thought of as a type of histogram of the data. The "stem" is the "decade" that is graphed; the "leaves" are the individual values within the decade. The five value summary gives the range of the data, the 25 percent and 75 percent quartiles (designated H), and the median (designated M).

As can be seen in Figure 1a, the lifetime estimates of risk (from 45 years' exposure) at 10 $\mu g/m^3$ range from 2.2 to 29 excess deaths per 1,000 workers. (The WHO estimate of 61 excess deaths per 1,000 was eliminated as an outlier.) In examining the diagram it can be seen that first, the range of estimates is only of one order of magnitude; given the uncertainties in risk assessment, this is considered a very narrow range. Second, there is a distinct mode in the estimates of risk less than 10 excess deaths per 1,000. Third, the quartile range is small (5 to 10) showing a tight cluster around the median, which is eight excess deaths per 1,000 workers (exact value from the EPA-CAG risk assessment). Consequently, OSHA chose a lifetime risk of eight excess deaths per 1,000 workers as its best estimate of risk at the 10 $\mu g/m^3$ level. This estimate was within the range proposed by OSHA in April, 1981. In addition, it did not represent the extremes of risk presented to the Agency.

Based on this determination, and the assumption of a linear model, OSHA calculated the best estimate of lifetime risk at the 500 $\mu g/m^3$ as 400 excess deaths per 1,000 workers. This estimate agrees well with the stem and leaf diagram in Figure 1b and observed risk at approximately the 500 $\mu g/m^3$ level. Again, the range is less than one order of magnitude (148 to 767), the median is 375 excess deaths per thousand workers and the quartile range (310 to 518) shows a tight cluster around the median. Slight variations in the estimates at higher levels can be explained by variability in dose data (past exposures) and rounding error, as well as general statistical variability inherent in studying diverse occupational populations.

SIGNIFICANCE OF THE RISK

The qualitative determination that OSHA's preferred estimate represents a significant risk was based on five criteria:

- The high quality of the data on which the risk assessment was based--The final estimates of risk were the consensus of nine risk assessments based on combinations of six epidemiologic studies. These epidemiologic studies were of very high quality conducted in occupational environments similar to those regulated by the standard; four of them contained exposure measurements of the populations studied and two had exposure data supplied in the original rule-making hearings. The studies clearly associated inorganic arsenic exposure with substantial excess risk of lung cancer, had adequate follow-up, and indicated that risk is proportional to the degree of arsenic exposure. In addition, there were a number of other studies associating inorganic arsenic with lung cancer, but most did not contain exposure data that could be used in establishing a dose-response relationship.

- The serious nature of the risk--Lung cancer is usually a fatal disease. It evades early detection and, according to the American Cancer Society, only about 9 percent of lung cancer patients live five or more years after diagnosis.

- The reasonableness and consistency of the estimates of risk.

- The statistical significance of the risk demonstrated in the epidemiologic studies--Exposure to inorganic arsenic is associated with as high as a seven-fold excess risk in some categories, which was a highly significant statistic. The statistical significance of this risk was the motivation for the initial qualitative determination that inorganic arsenic is a human carcinogen. In addition, the linear dose-response curves showed highly significant positive slope and good "fit" to the data.

- An unfavorable comparison with other occupational risks--The prediction of lifetime risk at the former PEL of 500 $\mu g/m^3$ of 400 excess deaths per 1,000 workers clearly exceeds risk rates in even some of the most dangerous occupations, such as mining and firefighting. Moreover, the Agency concluded that the risk posed at 10 $\mu g/m^3$, eight excess deaths per 1,000 should still be considered significant because it was greater than the risks of occupations of average risk, and because the Supreme Court gave the example that a "reasonable person" might well take steps to reduce a risk of one in a thousand.

REFERENCES

1. The Occupational Safety and Health Administration, Occupational Exposure to Inorganic Arsenic; Supplemental Statement of Reasons for Final Rule, Federal Register, January 14, 1983, Vol. 48, No. 10, 1864-1903.
2. A. M. Lee and J. F. Fraumeni, Jr., Arsenic and respiratory cancer in man--an occupational study, J. National Cancer Institute 42, 1045-52 (1969).
3. S. S. Pinto and P. E. Enterline, Mortality study of ASARCO Tacoma smelter workers, unpublished (1975).
4. S. Pinto, P. Enterline, V. Henderson, and M. Varner, Mortality experience in relation to a measured arsenic trioxide exposure, Environmental Health Perspectives 19, 127-130 (1977).
5. M. Ott, B. Holder, and H. Gordon, Respiratory cancer and occupational exposure to arsenicals, Archives of Environmental Health 29, 250-55 (1974).
6. S. Tukodome and M. Kuratsune, A cohort study on mortality from cancer and other causes among workers at a metal refinery, Int. J. Cancer 17, 310-17 (1976).
7. A. Baetjer, M. Levin, and A. Lillienfeld, Analysis of mortality experience of Allied chemical plant, unpublished (1974).
8. A. B. Hill and E. L. Faning, Studies in the incidence of cancer in a factory handling inorganic compounds of arsenic-I. Mortality experience in the factory, Br. J. Ind. Med. 5, 1-6 (1948).
9. R. Denk, H. Holzmann, H. Lange, and D. Greve, [Concerning the long-term effects of arsenic as seen from autopsies on Moselle vineyard workers], Mediz. Welt 11, 557-67 (1967).
10. F. Roth, [Concerning the long-term effects of chronic arsenical poisoning of Moselle vineyard workers], Dtsch. Med. Wschr. 6, 211-16 (1957).
11. A. Lee-Feldstein, Arsenic and respiratory cancer in man: follow-up of an occupational study, prepublication version from Arsenic-Industrial, Biomedical, Environmental Perspectives, W. H. Lederer, R. J. Fensterheim, eds. (Van Nostrand Reinhold Company, New York, 1983).
12. I. T. T. Higgins, K. B. Welch, M. S. Oh, and C. Burchfiel, Mortality of Anaconda smelter workers in relation to arsenic and other exposures, April 12, 1982.
13. P. E. Enterline and G. M. Marsh, Mortality among workers exposed to arsenic and other substances in a copper smelter, prepublication version from Arsenic-Industrial, Biomedical, Environmental Perspectives, W. H. Lederer, R. J. Fensterheim, eds. (Van Nostrand Reinhold Company, New York, 1983).

14. C. C. Brown and K. C. Chu, Approaches to epidemiologic analysis of prospective and retrospective studies: example of lung cancer and exposure to arsenic. Presented at the SIAM Institute for Mathematics and Society (SIMS) 8th Research Application Conference on Environmental Epidemiology: Risk Assessment, June, 1982.
15. C. C. Brown and K. C. Chu, Implications of the multistage theory of carcinogenesis applied to occupational arsenic exposure. Submitted for publication for the J. Natl. Cancer Institute (1982).
16. J. Lubin, L. Pottern, W. Bolt, S. Tokudome, B. Stone, and J. Fraumeni, Respiratory cancer among copper smelter workers: recent mortality statistics, J. Occupational Medicine 23, 779-784 (1981).
17 K. Mabuchi, A. Lillienfeld, and L. Snell, Cancer and occupational exposure to arsenic: a study of pesticide workers, Preventive Medicine 9, 51-77 (1980).
18. O. Axelson, E. Dahlgram, C. D. Janoson, and S. O. Rehnlund, Arsenic exposure and mortality: a case referent study from a Swedish copper smelter, British J. Indust. Med. 35, 8-15 (1978).
19. S. Wall, Survival and mortality patterns among Swedish smelter workers, Int. J. Epidemiology 9, 73-87 (1980).
20. K. C. Chu, Addendum B: Risk assessment for inorganic arsenic, Exhibit 201-2B (1980), as corrected, Exhibit 201-2D.
21. Consultants in Epidemiology and Occupational Health, Arsenic risk assessment, critique and alternative (1982).
22. H. F. Morris, Testimony submitted to inorganic arsenic hearings, Exhibit 28B (1975).
23. S. S. Pinto, M. O. Varner, K. Nelson, A. L. Labbe, L. D. White, Arsenic trioxide absorption and excretion in industry, J. Occupational Medicine 18, 677-680 (1976).
24. Consultants in Epidemiology and Occupational Health, Inc., Supplementary tables and figures for oral presentation before the Occupational Safety and Health Administration hearings on arsenic, Washington, DC, Exhibit 219 (1982).
25. Occupational Safety and Health Administration staff submission to the record, Power to detect a statistically significant increase in risk of 1.5-fold (SMR=150). No. 2. Higgins et al. (1982), Exhibit 237B (1982).
26. Clement Associates, Assessment of human respiratory cancer risk due to arsenic exposure in the workplace, Exhibit 201-4, September 5, 1980.
27. The Carcinogen Assessment Group, U.S. Environmental Protection Agency, Final risk assessment on arsenic, Exhibit 201-5, May 2, 1980.

28. K. S. Crump, Testimony submitted to the record, Exhibits 206, 212 (1982).
29. E. P. Radford, Testimony submitted to the record, Exhibit 207 (1982)
30. P. E. Enterline, Testimony submitted to the record, Exhibit 202-8 (1982)
31. H. P. Blejer and W. Wagner, Case study 4: inorganic arsenic--ambient level approach to the control of occupational carcinogenic exposures, N.Y. Acad. Sci. 271, 179-186 (1976).
32. World Health Organization, Environmental health criteria 18--arsenic, Geneva, Switzerland, (1981).
33. J. Tukey, Exploratory Data Analysis, (Addison-Weseley Publishing Co., Inc., Reading, Massachusetts, 1977).

FOOTNOTES

1. This article summarizes the approach of the Occupational Safety and Health Administration to the assessment of risk from exposure to inorganic arsenic. However, the official position of the Agency is as stated in the January 14, 1983, Federal Register document.
2. Lee and Fraumeni (1969) [2], Pinto et al. (1977) [3,4], Ott et al. (1974) [5], Tukodome and Kuratsune (1976) [6], Baetjer et al. (1974) [7], Hill and Faning (1948) [8], Denk et al. (1967) [9], and Roth (1957) [10].
3. Lee and Fraumeni (1969) [2], Lee-Feldstein (1982) [11] and Higgins et al. (1982) [12] studied workers at the Anaconda copper smelter; Pinto et al. (1977) [3,4], and Enterline and Marsh (1982) [13] studied workers at the Tacoma copper smelter, Tacoma, Washington.
4. Ott et al. (1974) [5].
5. Brown and Chu (1982) [14,15], Lubin et al. (1981) [16], Mabuchi et al. (1980) [17], Axelson et al. (1978) [18], Wall (1980) [19].
6. Power calculations were based on a one-tailed Poisson distribution [25].

ANALYSIS OF AGENCY ESTIMATES OF RISK FOR CARCINOGENIC AGENTS

John G. Cobler
Fred D. Hoerger

Health and Environmental Sciences
2020 Dow Center
Midland, Michigan 48640

ABSTRACT

Agency decisions involving 14 "carcinogens" were examined by an interdisciplinary group with experience in risk assessment. Analysis of these agency decisions led to observations and conclusions of broad applicability.

The scope of the data base available to the agencies was highly variable. This factor has contributed to a tendency to focus on bioassay results for risk assessment rather than the total data base.

Reliance upon quantitative risk assessment by agencies has been varied. OSHA has not utilized quantitative risk assessment but based regulatory action on feasibility. EPA has relied largely upon a standardized logic pattern for carcinogen classification and a conservative policy which standardizes and confines judgmental consideration of data. FDA uses quantitative risk assessment with "prudent" scientific flexibility, but is limited in areas of applicability by statutory policy restraints.

Recent court decisions, such as the Supreme Court decision on benzene, increase the likelihood of use of risk assessment by the agencies, especially OSHA. There is variation in the level of risk control achieved by the agencies. For example, 16 of 21 agency instances of risk control were in the range of 1 in 10,000 to 1 in 1 million lifetime risks. Considering all the other agency decisions displayed in this report, the majority of decisions cluster around 10^{-4} or 10^{-5}.

Credible judgmental estimates of risk are two orders of magnitude lower than agency estimates. (Comparisons were possible with 6 of 14 substances.) Strong evidence suggests credible judgmental bases for adjusting agency estimates downward for an additional 5 of the 14 substances.

KEYWORDS: Carcinogens; Carcinogenic Policies; Regulation of Carcinogens; Risk Assessment

During the past several years, the federal regulatory agencies have regulated, or proposed for regulation, a wide variety of substances classified as potential human carcinogens. Selected literature relating to agency discussions on 14 of these substances was collected and reviewed by an ad hoc group of scientists convened by the American Industrial Health Council (Appendix II). The group's purpose was to evaluate the basis for agency regulatory action.

A number of questions were addressed in this review with particular attention being given to the role of quantitative risk assessment by EPA, OSHA, and FDA and the estimated risk protection achieved.

The substances considered in this paper are:

Acrylonitrile	Lead acetate
Aflatoxin	Lindane
Benzene	4-Methoxy-m-phenylenediamine
Chlorobenzilate	Nitrilotriacetic acid
Chloroform	Saccharin
bis-Chloromethyl ether	Tetrachloroethylene
Formaldehyde	Vinyl chloride

These were selected because there was a reasonable amount of literature relating to the agency decisionmaking and covering a spectrum of agency activities. The majority of the agency deliberations occurred during the period of 1975 to 1981. In several cases toxicological data has become available since the agency decisions so that one can gain a greater perspective on risk than was possible at the time of agency decisions.

It should be noted that the many health statutes administered by the four principal regulating agencies differ in their statutory direction for risk assessment. Risk assessments by OSHA are permitted under the Act and now appear to be required as a result of the Supreme Court decision on benzene. OSHA controls to the extent feasible, but must demonstrate that a regulation results in a significant reduction of risk.

The Federal Hazardous Substances Act does not explicitly mention risk assessment but has been interpreted as allowing it, and the Consumer Product Safety Commission does use such analyses. Risk assessment is explicitly required by the Consumer Product Safety Act.

The Clean Air Act (except for Sec. 112), Water Pollution Control Act, Safe Drinking Water Act, and the Federal Insecticide, Fungicide, and Rodenticide Act permit the use of risk assessment, while the Toxic Substances Control Act explicitly requires that risk assessments be considered.

The Food and Drug Administration uses quantitative risk assessment with "prudent" scientific flexibility, but is limited in areas of applicability by statutory restraints. For example, the Delaney Clause which prescribes banning of carcinogens is a zero-risk criteria; the utility of risk assessment is limited to establishing de minimis levels.

In general, the various statutes suggest that there should be a safe level or low risk to the public or the worker. This suggestion should then be modified with such caveats as reasonable certainty or technological feasibility.

During the 1975 to 1979 period, two significant policies toward risk assessment emerged. First, EPA evolved a policy that relied upon "weight of the evidence" as a basis for identifying a substance as a carcinogen.[1] A second part of the policy related to the extrapolation of the dose-response curve from high dose to low dose. The policy tended toward reliance on animal bioassay data and the one-hit model, later modified to rely upon a multi-stage model.

The second significant policy was the Interagency Regulatory Liaison Group (IRLG) document of 1979.[2] The IRLG policy was more generic and even more conservative than the EPA policy. Its thrust was toward almost complete reliance upon animal bioassay positive results and linear extrapolation. For the purposes of this paper, the IRLG policy of utilizing the most sensitive animal species and linear extrapolation is utilized in several instances when an agency did not make its own risk assessment.

In evaluating the basis for the regulatory decisions, six basic topics were evaluated for each substance as follows:

- o The scope of the data base
- o Agency reliance upon risk assessment
- o Methodology for risk assessment
- o The level of risk control achieved
- o The consistency of risk level between agencies
- o A comparison of risk level from models vs. judgmental weighting of biological variables

In examining the scope of the data base relied upon by the agencies, particular attention was given both to the quality and extent of information relating to:

- o Epidemiology
- o Animal bioassays--species, target organ, route of exposure
- o Illucidating data--metabolism, pharmacokinetics, DNA studies, mutagenicity tests
- o Ambient and population exposure profiles

For 21 decisions or proposals relating to the 14 substances, risk assessment was utilized by EPA in all instances and by FDA in selected instances as shown in Table I.

Four of the 14 substances reviewed have been regulated as carcinogens by OSHA--bis-chloromethyl ether[3], vinyl chloride[4], acrylonitrile[5] and benzene.[6] OSHA did not utilize risk assessment in support of any of these regulations. Failure to make a "threshold finding of significant risk" led to the Supreme Court rejection of the 1 ppm proposal for benzene.[7] The 10 ppm exposure level established in 1971 by OSHA is, therefore, still effective.

The level of control established by the four OSHA regulations are listed in decreasing order of stringency:

o	VCM	1 ppm
o	bis-CME	work practice controls
o	AN	2 ppm
o	Benzene	10 ppm

TABLE I. AGENCY RELIANCE ON RISK ASSESSMENT

OSHA	Pre-benzene		No
	Current (arsenic, EO, EDB)		Yes
EPA	-		Yes
FDA	Saccharin		No
	Aflatoxin		Yes
	Acrylonitrile:	past	No
		current	Yes
	Lead acetate		Yes
	Vinyl chloride:	past	No
		current	Yes

Judgmental estimates of potency probably rank the four substances in a differing order:

- o bis-CME
- o AN
- o VCM
- o Benzene

EPA, of the four agencies, has most extensively relied on quantitative risk assessment. Weight of the evidence (one or two animal experiments, or epidemiological data) classifies a material as a carcinogen for regulatory appraisal. Generally, a multi-stage mathematical model is applied to an animal bioassay model for risk extrapolation. In a few instances, adjustments are made to accommodate the estimate to negative epidemiological data. Metabolism, pharmacokinetic, or comparative DNA data are not utilized in analysis of species differences.

Despite EPA's emphasis on quantitative risk assessment policies, William Ruckelshaus recently stated that EPA now is using 10 different risk assessment methods under the 10 environmental laws it administers.[8]

FDA tends to utilize risk assessment in a variety of ways. However, their approach is primarily to fine-tune decisions which are largely directed by legislated policies. A few examples include saccharin, aflatoxin, acrylonitrile, and lead acetate.

FDA did not rely on risk assessment in addressing saccharin.[9] Strict interpretation of the Delaney Clause is a disincentive to its use. The Congressionally-mandated NAS Study Panel on Saccharin calculated risks by all available models, but was unable to select a model to represent the risks. The panel did agree to characterize saccharin as a carcinogen of low potency.

FDA has more discretion in regulating natural or unavoidable food contaminants than is permitted for intentional additives under the Delaney Clause. In the case of aflatoxin, FDA developed a risk assessment document which is perhaps the best state-of-the-art assessment.[10] Species differences were considered, along with national and regional epidemiology. Metabolism studies were available, but in this instance did not clarify extrapolation considerations.

It can be said that the aflatoxin risk assessment considered a wide data base; considered various mathematical models as a tool in a judgmental risk assessment; and that the final agency decision balanced the results of risk assessment (the benefits of regulatory alternatives) against qualitative cost and feasibility considerations of food processing.

In the case of the acrylonitrile bottle issue, FDA largely ignored risk estimation and thus was directed by the courts to ignore "de minimus" quantities.[11] FDA has also inferred the concept of "virtually safe doses" from this court decision.

In the recent case of lead acetate use in hair dyes, FDA relied heavily on the fact that lead acetate is not significantly absorbed through the skin.[12] Use of a "product absorption" factor and a conservative animal model extrapolation gave an estimated low risk of one in five million lifetimes. It should be noted that FDA established that a condition of normal use presented an acceptable risk; the agency did not establish a standard that divides an acceptable risk from an unacceptable risk.

The data in Appendix II, and illustrated in Figure 1, show the risk levels achieved by various agency decisions. The data indicate that agency actions have provided calculated risk protection levels which range from 1 in 10,000 to 1 in 1 million. Two caveats are important:

- The histogram includes only decisions where the agency made an estimate of risk (or their risk assessment logic was clear so that the risk level could be derived).

- Secondly, in almost every case of risk estimation, the agency estimates give little consideration to epidemiology, metabolism, pharmacokinetics, or other species difference factors.

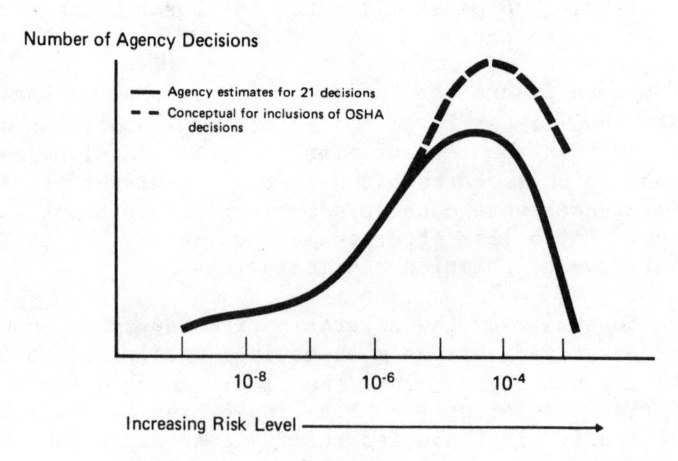

Figure 1

Conceptual Histogram of Risk Levels Achieved by Agency Decisions (Agency or IRLG Method of Estimating Risk)

Disregarding the above two caveats on uncertainty for the moment, it seems important to note that the majority of agency risk decisions fall in the calculated range of 10^{-4} to 10^{-6} (16 of 21).

Two considerations appear to warrant further fact finding and discussion:

- Estimates for the omitted decisions, if based on policy and methodology recommended by the Interagency Regulatory Liaison Group (IRLG) in 1979, would probably spread the distribution, but probably skew the distribution toward the higher risk levels (10^{-4}). (See Figure 2).

- Although there is still lingering controversy about many of the agency decisions on the 14 substances, collectively there is at least some degree of societal acceptance of the risk levels represented in Figure 1. On this acceptance premise, along with the anticipated skewing projected in Figure 2, it seems that debates on the target level for regulatory controls might focus on risks of 10^{-4}, rather than 10^{-5} to 10^{-7} as often debated today. In short, while certain current thrusts are to achieve levels of 10^{-6} risks, experience suggests 10^{-4} levels of control. Again, note that this analysis focuses on

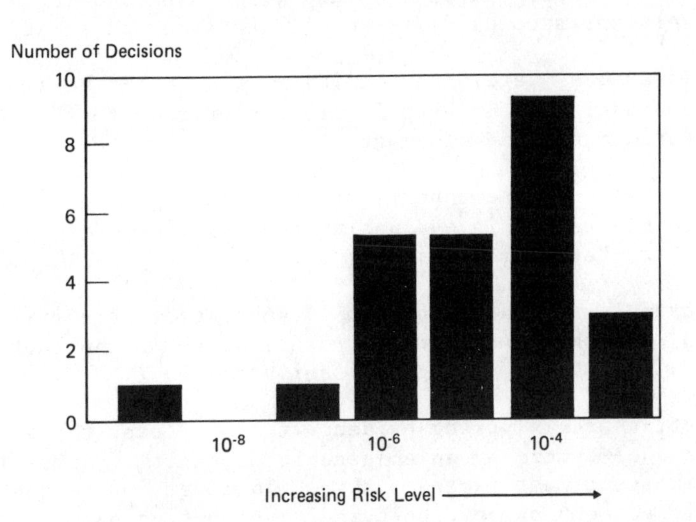

Figure 2
Histogram of Risk Levels Achieved by 21 Agency Decisions
(Agency Estimates of Risk)

ultra-conservative IRLG methodology and further, that these are calculated values.

The standardized generic process of data treatment and statistical modeling does not account for differences in tumor site, adequacy and completeness of the chronic animal study, sex, strain, or differences in species sensitivity.

There appears to be emerging consensus that the data base for comprehensive scientific risk assessment should include the following:

- o Animal carcinogenicity studies
- o Comparative organ toxicity
- o Pharmocokinetics
  - --Absorption
  - --Distribution
  - --Metabolism
  - --Excretion
- o Routes and degree of exposure
- o Consideration of short-term tests
- o Genotoxicity
- o Animal to human extrapolation
- o Epidemiology

Quantitative risk assessments based on the standardized generic process of data treatment and model extrapolation should therefore be limited to:

- o Those instances where all the data suggest a direct acting carcinogen and where data showing relevant species difference factors do not exist

- o Instances where upper limits of risk might be a consideration in establishing priorities for further data development

- o Instances where upper limit of risk might be a consideration in evaluating the likelihood of short-term harm or injury

A careful examination of the 14 substances identified overestimates of risk and led to characterization of the factors omitted (given low weight) in the agency estimates.

First, there is strong evidence that at least three of the substances induce tumors by an epigenetic mechanism. Simplified, an epigenetic agent must exceed a threshold level for tissue injury (cytotoxic action) or physicological dysfunction before normal DNA synthesis rates are accelerated. Only after DNA synthesis is accelerated does the probability of tumor formation above background levels increase.

The three nongenetic substances are chloroform[13], NTA[14], and tetrachloroethylene.[15] Traditional quantitative risk models extrapolated to zero are not applicable to these materials.

In the case of vinyl chloride, differential pharmacokinetic data for rates of excretion versus rates of metabolism forces an adjustment in conventional risk extrapolation.[16]

Evidence suggests or forces conventional risk extrapolation models to be modified in order to be consistent with epidemiological data in six instances:

acrylonitrile[17]
aflatoxin[10]
benzene[18]
formaldehyde[19]
saccharin[9]
vinyl chloride[20]

Species differences are of significance to risk assessment in at least nine instances:

acrylonitrile
aflatoxin
benzene
chlorobenzilate
chloroform
formaldehyde
saccharin
tetrachloroethylene
vinyl chloride

Based upon the judgmental considerations of mechanism, epidemiology, and species differentiation factors, the agency calculated risks were adjusted downward as shown in Table III. In many cases, the level of risk articulated by the agencies contrasts markedly with credible judgments of risk based upon the broader data base.

TABLE III. STANDARDIZED METHODOLOGY FOR RISK ESTIMATION COMPARED TO JUDGMENTAL ESTIMATION

Substance	Decision	Risk Level: Agency Estimate	Risk Level: Judgmental Estimate
Lead Acetate	Hair Dye	2×10^{-7} (12)	5×10^{-8}
NTA	Water	4×10^{-6} (21)	4×10^{-7}
VCM	OSHA 1 ppm	3×10^{-4}*	10^{-8}
$CHCl_3$	Water	3×10^{-4} (22)	8×10^{-6}
Tetrachloroethylene	Air	8×10^{-6} (23)	4×10^{-8}
Lindane	Food	2×10^{-5} (24)	10^{-8}

*Calculated by IRLG-CAG methodology

This tabulation shows that the six judgmental estimates were approximately two orders of magnitude lower than the agency estimates.

In summary, credible judgments show six instances out of 14 where agency or IRLG methodology overstates risks by about two orders of magnitude. Credible judgments also indicate that modification of the standardized IRLG methodology should be applied to five additional substances. Standardized IRLG methodology applies to no more than three of the 14 substances.

REFERENCES

1. Environmental Protection Agency, Health Risk and Economic Impact Assessments of Suspected Carcinogens, Fed. Reg. 41:21402 (1976).
2. Interagency Regulatory Liaison Group; Work Group on Risk Assessment, Scientific Bases for Identification of Potential Carcinogens and Estimation of Risks, Fed. Reg. 44:39858 (1979).
3. Occupational Safety and Health Administration, bis-Chloromethylether, Code of Federal Regulations, Title 29, Section 1910.1008 (1981).
4. Occupational Safety and Health Administration, Vinyl Chloride, Code of Federal Regulations, Title 29, Section 1910.1017 (1981).
5. Occupational Safety and Health Administration, Acrylonitrile, Code of Federal Regulations, Title 29, Section 1910.1045 (1981).
6. Occupational Safety and Health Administration, Benzene, Code of Federal Regulations, Title 29, Section 1910.1028 (1981).
7. Industrial Union Department, AFL-CIO v. American Petroleum Institute, 448 U.S. 607, 659-62 (1980).
8. Chemical Regulation Reporter, 7, 451, 1983.
9. National Academy of Sciences, Committee for a Study on Saccharin and Food Safety Policy, Food Safety Policy: Scientific and Social Considerations, NAS-NRC, March 1979.
10. Food and Drug Administration, Assessment of Estimated Risk Resulting from Aflatoxins in Consumer Products Peanut Products and Other Contaminants, Food and Drug Administration (1979).
11. Monsanto Company, et al. v. Donald Kennedy, 613F. 2d 947 (D.C. Cir 1979).
12. Food and Drug Administration, Lead Acetate: Listing as a Color Additive in Cosmetics That Color the Hair on the Scalp, Fed. Reg. 45:72112 (1980).
13. R. H. Reitz, T. R. Fox, and J. F. Quast, Mechanistic Considerations for Carcinogenic Risk Estimation: Chloroform, Environmental Health Perspective, 46, 163 (1982).
14. National Cancer Institute, Bioassays of Nitrilotriacetic Acid (NTA) and Nitriolotriacetic Acid, Trisodium Salt Monohydrate (Na_3 NTA H_2O) for Possible Carcinogenicity, NCI Tech. Ser. No. 6, January 1977, DHEW Publication No. NIH77-806.
15. A. M. Schumann, J. F. Quast, and P. G. Watanabe, The Pharmacokinetics and Macromolecular Interactions of Perchloroethyl-

ene in Mice and Rats as Related to Oncogenicity, Toxicol. Appl. Pharmacol. 55, 207 (1980).

16. P. J. Gehring, P. G. Watanabe, and C. N. Park, Resolution of Dose-Response Toxicity Data for Chemicals Requiring Metabolic Activation: Example-Vinyl Chloride, Toxicol. Appl. Pharmacol. 44, 581 (1978).
17. M. O'Berg, Epidemiologic Studies of Workers Exposed to Acrylonitrile: Preliminary Results, E. I. DuPont de Nemours and Company, 1977.
18. CMA-NIOSH Meeting On Industry Concern About Study Used For Hazard Assessment, Chemical Regulation Reporter 6(26), 763 (1982).
19. B. W. Karrh, Epidemiologic Studies of Workers Exposed to Formaldehyde: Preliminary Results. E. I. Dupont de Nemours and Company, May 20, 1982.
20. J. J. Beaumont and N. E. Breslow, Power Considerations in Epidemiologic Studies of Vinyl Chloride Workers, American Journal of Epidemiology 114 (5), 725 (1982).
21. Environmental Protection Agency, Draft Final Report: NTA Assessment, April 22, 1980.
22. Environmental Protection Agency, Part III National Interim Primary Drinking Water Regulations: Control of Trihalomethanes in Drinking Water; Final Rule, Fed. Reg. 44:68697 (1979).
23. Environmental Protection Agency, The Carcinogen Assessment Group's Carcinogenic Assessment of Tetrachloroethylene (Perchloroethylene), July 25, 1980.
24. Environmental Protection Agency, Notice of Rebuttable Presumption Against Registration and Continued Registration (RPAR) of Pesticide Products Containing Lindane, Fed. Reg. 42:9816 (1977).
25. R. H. Reitz, P. J. Gehring, and C. N. Park, Carcinogenic Risk Estimation for Chloroform: An Alternative to EPA's Procedure, Food Cosmetic Toxicol. 16, 511 (1978).
26. D. D. McCollister, W. H. Beamer, G. J. Atchison, and H. C. Spencer. The Absorption, Distribution, and Elimination of Radioactive Carbon Tetrachloride by Monkeys Upon Exposure to Low Vapor Concentrations, J. Pharm. Exp. Tox. 102 (2), 112 (1951).
27. Occupational Safety and Health Administration, Occupational Exposure to Acrylonitrile (Vinyl Cyanide): Final Standard, Fed. Reg. 43:45762 (1978).
28. Environmental Protection Agency, The Carcinogen Assessment Group's Carcinogenic Assessment of Acrylonitrile, July 25, 1980.
29. Environmental Protection Agency, Water Quality Criteria, Availability, Fed. Reg. 44:56628 (1979).
30. Food and Drug Administration, Cancer Assessment Committee, Acrylonitrile Risk Assessment, November 24, 1981.
31. Donald A. Olson, Monsanto Chemical Intermediate Company, personal communication.

32. National Toxicology Program, Technical Bulletin (1980): Vol 1 (2), April 1980.

33. N. Cerna and H. Kypenova, Mutagenic Activity of Chloroethylenes Analyzed by Screening System Tests, Mut. Res. 46, 214 (1977).

34. L. W. Rampy, J. E. Quast, M. F. Leong, and P. G. Gehring, Results of a Long-Term Inhalation Toxicity Study: Perchloroethylene in Rats, Toxicology Research Laboratory, The Dow Chemical Company, October 1978.

ACKNOWLEDGMENT

The initial analysis of agency regulatory decisions was made by an ad hoc group of scientists in March, 1981. The authors gratefully acknowledge the contributions of John Barr, Air Products & Chemicals; Robert Barnard, Cleary, Gottleib, Steen, and Hamilton; Clyde Burnett, Clairol; Ann Norberg, Monsanto; Malcolm Smook, DuPont; and Carol Weil, Union Carbide.

APPENDIX I

Examples of Judgmental Considerations

1. Chloroform. EPA's National Interim Primary Drinking Water Regulations (22) established a maximum contaminant level (MCL) for total trichloromethanes in drinking water of 0.1 mg/L. The interim MCL was based on the balancing of public health considerations and the feasibility of achieving such levels. Based on the NCI animal data and using a multistage model with 95 percent UCL, the environmental lifetime cancer risk was estimated at 3.4×10^{-4}.

The toxicity of $CHCl_3$ is due not to the parent compound, but to the production of a reactive metabolite(s) which binds covalently to tissue protein. Thus, species variations in the capacity to metabolize $CHCl_3$ must be considered. The relative carcinogenicity of $CHCl_3$ decreases rapidly with decreasing dose suggesting that there may be a detoxification mechanism which protects the animal until the mechanism is overwhelmed by massive doses.

Application of an appropriate species correction factor to account for metabolism rate differences reduces the estimated risk to man by at least an order of magnitude.[25] Thus, the best judgmental value for incremental lifetime cancer risk at 0.1 mg/L is approximately 8×10^{-6}.

Contrasted to EPA's MCL, NIOSH recommended that occupational exposure not exceed 2 ppm (10 mg/m^3) as the TWA (time-weighted average) for a 10-hour day. No risk estimates were provided. Man is estimated to inhale 10 m^3 per 8-hours or 12.5 m^3 per 10-hour working day. During breathing about 30 percent of the $CHCl_3$ contained in the air could be absorbed.[26] The permissible exposure of humans to $CHCl_3$ under the NIOSH recommended TWA would therefore be 37.5 mg/day.

2. Acrylonitrile. OSHA[27] established an 8-hour TWA of 2 ppm acrylonitrile (4.5 mg/m^3). No estimate of past or future risk was made by OSHA. OSHA concluded that "based on the best available scientific evidence, no safe level for exposure to a carcinogen can be established at present." "The establishment of a so called 'safe level' of human exposure to AN is not possible when serious effects are being reported at increasingly lower dose levels." The 2 ppm worker exposure limit was, in the absence of a risk assessment, set as the lowest feasible for the AN industry as a whole.

The EPA Cancer Assessment Group, the EPA Water Quality Assessment, and the FDA used data from a drinking water animal bioassay to estimate potential risk. The agencies either did not have available at the time of assessment, or largely ignored, later animal bioassays which established a no-observed effect level for acrylonitrile. The carcinogen Assessment Group (EPA) estimated a lifetime probability of cancer of 1.58×10^{-4} (based on the drinking water bioassay of acrylonitrile) and a risk of 8.5×10^{-5} (based on the 1980 O'Berg epidemiological study) for continuous exposure to 1 μg AN/m^3.[28] CAG used a multistage model to estimate the 95 percent upper limit for the extra risk for humans.[29] Using interim results from the drinking water bioassay study, EPA[29] considered setting criteria for water quality at an interim target risk level in the range of 10^{-5}, 10^{-6}, or 10^{-7} with corresponding criteria of 0.08 μg/l, 0.008 μg/l, and 0.0008 μg/L respectively (based on the daily consumption of 2 L water plus 18.7 g fish and shellfish). EPA used a modified one-hit extrapolation procedure with rat to man conversion based on the 1/3 power rule to calculate the risk. In the case of the acrylonitrile copolymer bottle issue, FDA largely ignored risk assessment. Recently, however, the Cancer Assessment Committee of the Food and Drug Administration completed a risk assessment of acrylonitrile, using the results of the two bioassay studies and a strictly linear proportional extrapolation.[30] An "overall" calculated safe dose at the 10^{-6} level was estimated to be about 0.5 to 1.0 g/day.

OSHA - Worker Exposure	20.7 mg/day Risk not provided
CAG - Ambient Air	11.5 μg/day Risk of 10^{-4}
EPA - Water Qualtiy	0.016 μg/day Risk of 10^{-6}
FDA - Food	0.5 - 1.0 μg/day Risk of 10^{-6}

The agency risk assessments appear to be ultraconservative. No evidence exists to show AN binding to DNA in any metabolism study with intact animals. AN was found to bind evenly to calf thymus DNA in vitro (1000 $<$ than VCM and 100 $<$ than trichloroethylene). AN is considered to be a weak mutagen; weak mutagenic activity is characteristic of a nongenotoxic mechanism. A threshold below which AN will not cause tumors is consistent with such a mechanism. A no-observed-effect-level (NOEL) of 0.1 to 0.3 mg/kg/day was observed in three bioassay studies.[31] Further, the result of epidemiological studies on workers have not indicated a statistically significant effect.

<u>3. Tetrachloroethylene (TCE).</u> The Cancer Assessment Group estimated that the cancer risk from continuous lifetime exposure to 1 $\mu g/m^3$ tetrachloroethylene from ambient air would be 7.6×10^{-6}.[23] The risk was calculated from the results of the NCI bioassay (hepatocellular carcinomas in mice which metastasized to lung and kidney) using a multistage model with a 95 percent upper limit.

The CAG unit risk assessment for TCE is severely flawed. The CAG document concluded that: "The available information on the mutagenicity of TCE is limited and inconclusive. Thus TCE cannot be classified as either mutagenic or nonmutagenic." The negative Ames' test result by NTP[32] and the negative cytogenic assay by Cerna and Kypernova[33] were not cited by CAG. Overall, TCE has been found to be positive in only one of six bacterial mutagenicity assays. Additionally, TCE was negative in two cytogenetic studies. These results strongly support the nonmutagenicity of TCE. Recurrent tissue damage produced by TCE precedes tumor formation indicating that the mechanism of tumor formation in the mouse was nongenetic. The nongenetic mechanism is in agreement with the preponderance of the data from mutagenetic studies. There was no evidence of carcinogenicity in rats in an inhalation study involving exposure to 300 to 600 ppm daily for one year followed by observation of the animals for an additional year.[34]

Despite this, if one still assumes the mouse as the animal model for estimating human risk, the CAG estimate should be modified. The use of an interspecies conversion factor of

$$\frac{(0.03)^{1/3}}{70}$$

by CAG assumes that TCE is directly acting. However, TCE is biotransformed to a reactive intermediate(s). Thus, since mice metabolize TCE to a greater extent than rats and rats greater than man,[15] the interspecies conversion should be used in reverse. The correct use of the interspecies conversion factor lowers the estimated risk by over two orders of magnitude from 7.59×10^{-6} to 4.36×10^{-8}.

APPENDIX II

TABLE II. RISK ASSESSMENT SUMMARY

Substance	Agency	Exposure		Lifetime Risk Estimation		Notes
		Route	Level	Agency	Best Judgement	
Acrylonitrile	CAG	Air	1 $\mu g/m^3$	1.58×10^{-4}	Current risk levels based on linear extrapolation should be accepted pending generation of metabolism and pharmacokinetic data.	Insufficient data to attempt interspecies conversion. Epidemiology does not indicate potent carcinogen.
	OSHA	Air	2 ppm TWA	No Estimate		
	EPA	Water	0.16 $\mu g/l$	10^{-5}		Based on interim data from animal bioassay. Final data indicate possible threshold.
		Water	0.08 $\mu g/l$ 0.008 $\mu g/l$ 0.0008 $\mu g/l$	10^{-5} 10^{-6} 10^{-7}		
Aflatoxin	FDA	Food (Corn, Peanuts)	20 ppb Action Level	$20\text{-}67 \times 10^{-5}$ (Epidemiology) $17\text{-}2500 \times 10^{-5}$ (animal bioassay)	$20\text{-}67 \times 10^{-5}$	Agency decision based on epidemiology data which, although still conservative, represents a better judgment of risk as compared to the animal bioassay risk data.
Benzene	OSHA	Air	10 ppm (1971 TWA)	Not Estimated	No detectable human risk at current exposure level	Human epidemiology shows leukemia and aplastic anemia at high concentrations. No evidence of elevated risk at 10 ppm level.
			1 ppm (proposed-remanded)	Not Estimated		
	EPA	Air-vent gas	5 ppmv (EB/Sty)	1.7×10^{-6} to 1.2×10^{-5}		Risk estimated from human exposure using CAG methodology. No animal models available.
		Air-vent gas	BAT (benzene storage)	2.7×10^{-5} to 1.9×10^{-4}		

APPENDIX II

TABLE II. (CON'T) RISK ASSESSMENT SUMMARY

Substance	Agency	Exposure		Lifetime Risk Estimation		Notes
		Route	Level	Agency	Best Judgement	
Chloro-benzilate	CAG/ EPA	Food (US)	0.0038 mg/day (0.002 ppm)	2.14×10^{-6}		Calculation based on Innes-mouse study rather than NCI study. Oncogenic response per unit dose in Innes study was five times greater. No significant carcinogenic response was obtained in rats.
		Food (Florida)	0.0095 mg/day (0.006 ppm)	6.53×10^{-6}		
		Food - Citrus Applicators	1.39 ppm	1.38×10^{-3}		
Chloroform	EPA	Water	0.1 mg/l	3.4×10^{-4}	8×10^{-6}	Use of appropriate inter-species conversion factor lowers estimated risk.
	FDA		Proposed ban on direct addition	No Estimate		Possible threshold which, if documented, will also lower risk level.
bis-Chloro-methyl Ether	OSHA	Air	Exposure level not established. Mixtures exempt if below 0.1%.	Not Estimated		Exposure estimated at somewhat lower than 5-10 ppm resulted in 40% tumor incidence in humans. Medium and low exposure groups showed no elevation of incidence. Complete carcinogen - does not need activation.
		Air	0.1 ppb		8×10^{-6}	Linear extrapolation from epi-demiology data.
	EPA	Water	0.038 ng/l 0.0038 ng/l 0.0038 ng/l	10^{-5} 10^{-6} 10^{-7}		BCME not detected in water-hydrolysis and photolysis. Any action would be superfluous.
Formaldehyde	CPSC	Air	1 ppm	$3 \times 3 \times 10^{-2}$	No direct evidence to support carcin-ogenicity in humans	Nasal tumors in rats but not mice. No human cancers from formaldehyde exposure have been reported. Eye and upper respiratory tract irritant. Risk of 290×10^{-6} appears to be significantly exaggerated.
			30 ppt	1×10^{-6}		
			8×10^{-3} ppm	290×10^{-6} (yearly)		

APPENDIX II

TABLE II. (CON'T) RISK ASSESSMENT SUMMARY

Substance	Agency	Exposure: Route	Exposure: Level	Lifetime Risk Estimation: Agency	Lifetime Risk Estimation: Best Judgement	Notes
Lead[acetate]	FDA	Percutaneous/ Absorption	0.4 μg/ day	$1x10^{-6}$	No direct evidence to support carcinogenicity in humans	Agency considered bioavailability from gastrointestional tract.
			1.0 μg absorbed/ application	$2x10^{-7}$		Risk associated with use of hair dye. Number of applications per week, years of application and number of users considered.
					$5.4x10^{-8}$	No direct evidence to suggest that exposure to lead salts cause cancer of any site in man.
Lindane (γ-1,2,3,4,5,6-hexachloro-cyclohexane)	EPA	Food	0.001 ppm in diet	$1.7x10^{-5}$		Metabolite may be direct acting carcinogen. Appears to have threshold level in mice. No oncogenic changes in rats.
			0.006 ppm in diet		10^{-8}	Log-probit extrapolation.
4-MMPD (4-methoxy-m-phenylenediamine)	FDA	Skin Absorption (hair dye)		10^{-4} to 10^{-7} (depends on shade)	No detectable risk under conditions of use	Existing data do not indicate that hair dyes cause cancer in humans. MMPD does not bind to DNA.
Nitrilotri-acetic acid	EPA	Water	0.2-24.5μg/1 (mean 2.82 μg/1)	$3.58x10^{-6}$	Human cancer risk relatively low.	Carcinogenic at high dose levels, no observed effect at low dose levels.

APPENDIX II

TABLE II. (CON'T) RISK ASSESSMENT SUMMARY

Substance	Agency	Exposure Route	Exposure Level	Lifetime Risk Estimation Agency	Lifetime Risk Estimation Best Judgement	Notes
		Water/ Workplace	100-200 g/day	2.5×10^{-4}	Best-fit modeling indicate a risk level several orders of magnitude less than results of linear extrapolation.	Epigenetic. Epidemiology not used in calculations.
Saccharin	FDA	Food		None		Epidemiology studies show no association between saccharin and liver cancer in humans.
	NAS	Food	0.12 g/day	5200×10^{-6} to 0.001×10^{-6} $<1 \times 10^{-9}$		NAS concluded estimation of human risk from animal data not feasible owing to variations in extrapolation.
					Best-fit modeling indicate a risk level several orders of magnitude less than results of linear extrapolation.	
Tetrachloro-ethylene	CAG	Air	$\mu g/m^3$	7.6×10^{-6}	4.36×10^{-8}	Use of appropriate inter-species conversion factor lowers estimated risk.
	NAS	Water	$1 \mu g/l$	6.2×10^{-8} (1.4×10^{-7} at 95% CL)		No human data exist to suggest a carcinogenic risk to humans.
						Carcinogensis in mice results from epigenetic mechanism. No significant carcinogenic response in rats.

APPENDIX II

TABLE II. (CON'T) RISK ASSESSMENT SUMMARY

Substance	Agency	Exposure Route	Exposure Level	Lifetime Risk Estimation: Agency	Lifetime Risk Estimation: Best Judgement	Notes
Vinyl Chloride	OSHA	Air	1 ppm TWA (BAT)	Not Estimated		Procarcinogen becoming active upon metabolism. Human carcinogen at high (ppm?) levels. NOEL at 1-2 ppm. Negative epidemiology at low exposure (10 ppm). Best judgment risk based on animal/human difference in metabolic rate (biotransformation). Strong evidence for threshold.
			1 ppm	2.72×10^{-4}	10^{-8}	
	EPA	Air	1 ppm	71×10^{-6} yearly		
	EPA	Water-Draft (+ fish)	517 μg/l	10^{-5}		
			51.7 μg/l	10^{-6}		
			5.17 μg/l	10^{-7}		
		- Final	20 μg/l	10^{-5}		Rationale for difference between draft and final documents not given.
			2 μg/l	10^{-6}		
			0.2 μg/l	10^{-7}		
	FDA	Food	6.7 ppb	1×10^{-6}	10^{-9}	No final regulation.

A TRAINING PROGRAM ON HAZARDOUS MATERIALS FOR NON-SCIENTISTS

Aileen T. Compton

Smith Kline & French Laboratories
1500 Spring Garden Street
Philadelphia, PA 19101

ABSTRACT

It is imperative to educate and to train personnel at all levels in the organization on hazardous materials. Special concern is raised when the person to be trained has minimum education and experience in the basic physical and chemical sciences. These persons are indirectly involved in the use of chemicals through housekeeping, glassware cleaning, receiving and distribution, maintenance, building services, and intrasite transportation of hazardous materials.

The two obstacles to overcome in the training are the trainee's perception of hazard and risk, and the trainer's anxiety to implement company policy, as well as meet regulatory compliance.

Effective training and education can be accomplished by establishing the following objectives for the outcome of the training:

- To create a frame of reference for non-scientific persons to use in assessing hazard and risk
- To establish a baseline for the training group members with respect to their perception of hazard and risk
- To inform personnel on potential hazards, internal policies, procedures, programs, and available resources

Training time is one and a half to two hours. It requires active participation by the trainees. The training aides are household products that contain chemicals and texts on toxicology and hazardous materials.

The four basic components of the training session are:

- Structured participation by the class
- Discussion and preparation of a hazards sign and label
- Slides on policies, programs, and regulations
- Questions, answers, and open discussion on the work environment, home, or other topics of interest

This program is being used to make personnel aware of their work environment. It can also be expanded to meet public education needs for the recent Right-to-Know legislation.

KEY WORDS: Education/Training Hazardous Materials; Training Toxic Materials Hazards; Education Chemical Safety; Right-to-Know Public Education; Hazards Training Non-Scientist

INTRODUCTION

Hardly a day goes by that the layman doesn't pick up a newspaper and read an article on dioxin, formaldehyde, the dangers of Three Mile Island, or other incidents involving potentially hazardous materials. A special concern is raised when the person to be trained has minimal education and experience in the basic physical and chemical sciences. This is not to say that if one were educated, there would be a different viewpoint on potentially hazardous materials. Anyone who has worked with chemicals, radioactive materials, or other physical hazards is aware that a certain amount of panic exists, even among scientists, when there is overstimulation from sensationalism in the news media.

All of the above factors must be considered when a training program is designed.

IDENTIFICATION OF THE POPULATION AT RISK

In a facility where chemicals and hazardous materials are used, there is a group of support personnel who have the potential for exposure to these materials. The initial step prior to training is to identify the population at risk. This task can only be accomplished by making several tours through the facility to follow the flow of materials--all points of internal transportation, consumption (intended/unintentional spills), and disposal. Job descriptions give very little, if any, indication of how the job is actually accomplished. A camera is used in this tour to take a picture of each individual observed with a chemical, entering or leaving the laboratory or performing services in the laboratory during regular and off-shift hours. It takes about a month to get a physical, as well as a mental, picture of the described activities.

CREATING A FRAME OF REFERENCE ON HAZARD AND RISK

In order to create a frame of reference, a common source closely associated with the personnel must be identified. This source is the hazardous substances and chemicals used at home. Potentially hazardous materials and chemicals are stored and handled in the home on a routine basis. These materials are packaged according to the Poison Prevention Act of 1970, and they also contain warning labels. Substances used in the home that are subject to this act include:

- Over-the-counter drugs (aspirin, and its substitutes)
- Furniture polish
- Lye, oven cleaners, drain cleaners
- Controlled and prescription drugs
- Anti-freeze
- Sulfuric acid
- Lighter fluids

Most of the products listed above are in the home. Other common products found in the home are bleach and ammonia. Employees are familiar with products used in the home for cleaning, and it is relatively easy to develop a frame of reference using these familiar items.

A baseline determines the level at which the training will proceed. It is a subjective judgment or decision made by the trainer. Factors that enter into the judgment are:

- Observations of the group during participation in class discussions and small group work assignments
- The past history of the group with respect to present and past experiences with hazardous materials

- Recent incidents report by the news media

All of the above factors affect the overall attitude of the trainees and influence certain mind sets that I categorize as:

- A closed mind
- An open mind
- An over-stimulated mind

A closed mind operates at the extreme where everything is bad and serves as the prophet of doom during the session. The open-minded individual has come to class to obtain new knowledge and will make a decision on the facts presented. The over-stimulated mind is operating at a high level of sensitivity and concern based on events that may have occurred currently or in the recent past.

STRUCTURED PARTICIPATION BY THE CLASS

A variety of household products (examples of the labels are shown in Figure 1) are displayed in the classroom. The activities centered around the products are as follows:

- The class is asked to divide into small groups (no more than three persons per group)
- Each group chooses a product(s) to work with
- Standard toxicology and text books on toxic materials are made available
- Each group member is instructed to select a chemical from the list of ingredients on the label and record the chemical properties and toxicity data information from the textbooks
- Where time permits, each member is asked to read aloud the warning on the label and the hazardous properties of the chemical
- When there is a shortage of time, the group is asked to select a chemical(s), and a leader/recorder is selected to present the warning label and chemical hazard's information

Following the classroom exercise, the hazard and warning signs posted in the laboratories are shown to the class. These signs/symbols are matched with the chemical hazard information gathered by

DANGER: Harmful or fatal if swallowed. Keep out of reach of children. If swallowed, do not induce vomiting. Call a physician immediately. Eye irritant. In case of eye contact, flush with water. Call a physician. If skin contact occurs, wash thoroughly.

Keep from heat and flame. Keep away from small children Harmful if taken internally. In case of accidental ingestion, induce vomiting and consult a physician. Harmful to synthetic fabrics, wood finishes and plastics.

Warning: This product contains sodium hydroxide (lye). You must follow directions carefully to avoid skin and scalp burns, hair loss and eye injury.

CAUTION: Keep out of the reach of children. Replace cap firmly.
DANGER: Extremely flammable! Keep away from fire, sparks and heated surfaces.

Extremely Flammable
Eye Irritant
Vapors May Cause Flash Fire
Harmful or Fatal if Swallowed

Figure 1

Examples of Labels Used to Train Workers

the students; e.g., carcinogen, suspect carcinogen, flammable, etc. This information is also compared to labels on drums or bulk materials used in the laboratories.

At this point the class is asked to prepare a data sheet on the chemical. The data provided is to inform another worker on the potential hazards of the compound, using terminology that is precise and clear; e.g., "Do not get into eyes;" "Do not get on skin." The next component is introduced by informing the class that the company has internal policies and responsibilities to inform all workers on potential hazards. The policies are based on Government regulations as well as the intent of the company to protect the health and safety of the worker and the environment.

PRESENTATION OF REGULATIONS, POLICIES, AND PROGRAMS

Audiovisual aides such as slides, video tapes, and films are used to familiarize employees with the applicable federal, state, and local regulations. Company policy on the regulations, along with information on where to locate the written policies, is a part of the instruction. Internal training programs on safety, handling material, respirators, and emergency procedures are outlined in detail. Where feasible, a copy of the regulations and policies is handed out; otherwise, the employee is informed where the documents are available for inspection.

Employees are advised on the type of training that is needed based on their job and level of interest. If the job or level of interest is such that participation in an emergency team is desirable, the employee is directed toward the training resources for the emergency response team. Only the basic emergency response actions such as location of fire alarms and emergency numbers are covered in this session. The goal is to give instructions on where to go or whom to contact in case of an emergency.

OPEN DISCUSSION AND QUESTIONS

By conducting a participatory training class, a high level of interest and concern is raised in the session. Employees usually have many questions and anxieties on topics of personal interest that are not job related. This session allows the employees to get information on personal concerns, thus reinforcing the benefit of the program to the employee, as well as the employer. The trainer may use this session to his/her benefit by deferring issues that could side track the original intent of the training to this session.

SUMMARY AND CONCLUSION

This is the era of the informed worker. The employer is obligated to inform employees on regulations as well as the potential hazards of chemicals in the work place. The recent Right-to-Know legislation passed at state and local levels has provisions for informing the general public as well as the worker.

The support personnel in a research facility are being trained on hazardous materials using participatory training. Participatory training includes establishing a frame of reference based on a common familiar source, setting a baseline for the group based on the group's perception of hazard and risk. The baseline is a subjective judgment by the trainer using the criteria of open minded, closed minded, or over-stimulated. A trainer must have experience and make careful observations during the small group sessions to make this judgment. Common household products are used to establish a frame of reference for the training. Employees become educated on the household products through participation in small groups using basic textbooks on toxicology and hazardous materials. The small group sessions are structured such that the trainer readily moves into an education process directed towards company policies, procedures, and regulations. The open discussion at the end of the session allows employees to surface issues that may be of a personal nature either on or off the job.

This method of training is time-consuming, and it requires a high degree of commitment on the part of management. Supervisors, as well as their employees, must attend sessions in order for the organization to obtain the maximum benefits from the training. Reinforcement of the training should be carried out on a periodic basis by handing out pamphlets and reminders at regularly scheduled safety meetings.

BIBLIOGRAPHY

1. N. Estrin (et al.). Cosmetic Ingredient Dictionary (Cosmetic and Toiletry and Fragrance Association, Washington, DC, 1982).
2. National Academy of Sciences. Principles and Procedures of Evaluating the Toxicity of Household Substances. (National Academy of Sciences. Washington, DC, 1977).
3. T. A. Loomis. Essentials of Toxicology, pp. 1-12. (Lea & Febiger, Philadelphia, 1976).
4. F. A. Patty (et al.). Industrial Hygiene and Toxicology. V. 2. (Interscience Publishers, New York, 1963).
5. N. I. Sax. Dangerous Properties of Industrial Materials. (Van Nostrand Reinhold, New York, 1979).
6. M. Windholz (et al.). The Merck Index. (Merck & Co., Rahway, NJ, 1976).
7. U.S. Department of Health and Human Services. Registry of Toxic Effects of Chemical Substance. (NIOSH, Rockville, MD, 1980).
8. U.S. Department of Labor. OSHA Standard 1910, Subpart 7, Toxic and Hazardous Substances. (U.S. Department of Labor. OHSA, 1972).

INORGANIC ARSENIC: IMPORTANCE OF ACCURATE EXPOSURE CHARACTERIZATION FOR RISK ASSESSMENT

Steven H. Lamm*
William H. Lederer**

*Consultants in Epidemiology and Occupational Health, Inc.
2423 Tunlaw Road, N.W.
Washington, DC 20007

**Koppers Company, Inc.
Pittsburgh, Pennsylvania

ABSTRACT

Workers at the Anaconda Smelter in Montana were studied to investigate the patterns of association between exposure to airborne inorganic arsenic and respiratory cancer risk. This group has been previously studied using as exposure categories the qualitative exposure intensity groupings developed by the National Cancer Institute (NCI): "Heavy," "Medium," and "Light." Significant excess risk had been observed in each of these qualitative exposure categories. The exposure data upon which this classification was based have subsequently been lost. In recent years, several quantitative analyses have been attempted to relate the risks in these broad qualitative groupings to available summary exposure data. Analysis of all known exposure data indicates wide variation in the actual exposure levels of the workers within each group, particularly for the "Light" category, and gross overlap in exposure ranges between groups.

Recently, Higgins et al., of the University of Michigan, have developed a rather well defined quantitative exposure classification scheme for the Anaconda work force which represents a decided improvement over the earlier NCI groupings for purposes of quantitative risk assessment. Higgins established four distinct exposure intensity categories associated with non-overlapping quantitative exposure intervals. Analysis using Higgins' classifications demonstrated an excess lung cancer risk only among workers in the high and very high exposure groups (exposure levels 500 $\mu g/m^3$ or greater), but none in the medium and low exposure groups. In summary, workers who were not exposed to airborne inorganic arsenic levels of 500 $\mu g/m^3$ or greater for at least 30 days evidenced no excess respiratory cancer risk.

Both the qualitative and the quantitative characterization of risk are dependent upon accurate definition of distinct and separate exposure classes. Previous analysis showing an excess risk in the NCI "Light" exposure group probably reflected the inappropriate inclusion of workers with relatively high exposures. However, this analysis, using Higgins' exposure classes, suggests that a threshold may exist for the association between arsenic exposure and respiratory cancer.

KEY WORDS: Arsenic, Respiratory Cancer, Exposure Data, Threshold, Risk Assessment

Several epidemiological studies demonstrating an increased incidence of respiratory cancer among workers with occupational exposure to inorganic arsenic have served as data sources for the carcinogenic risk assessments for arsenic. The Anaconda smelter studies presented by Lee and Fraumeni and subsequently updated by Lee-Feldstein represent the largest known data base. These analyses used a qualitative exposure classification developed in the mid 1960s by the National Cancer Institute (NCI) that was based on unpublished exposure data by Hendricks and Archer. This exposure data has subsequently been lost and thus cannot be validated.

Higgins et al. have recently refined the exposure classification based on all known exposure data and have reassessed the Anaconda mortality risk. Using this well-defined quantitative exposure classification, analysis leads to a new interpretation of the respiratory cancer risk associated with inorganic arsenic exposure at the Anaconda smelter.

The initial epidemiological study of the Anaconda work force using the NCI qualitative exposure classification scheme was published by Lee and Fraumeni in 1969.[1] The cohort contained 8,045 male employees of the Montana copper smelter with at least one year employment between 1938 and 1956, whose mortality was observed from 1938 through 1963. The exposure classification is described briefly in one paragraph which states that based upon unpublished data of Hendricks and Archer, workers in the arsenic refinery, the arsenic roaster, the stack cottrells, and the main flue were classified as having "Heavy" exposures; workers in the converters, reverberatory furnace, copper roaster, and acid plant were considered to have "Medium" exposures; and "persons in all other areas" were classified as having "Light" arsenic exposures.

Subsequent updates of the Anaconda cohort have continued to use the same qualitative classification scheme. Lubin et al. (1981) reported the 1964-1977 mortality experience of those workers known to be alive in 1964, but continued to classify workers according to the NCI classification using the pre-1957 work history.[2] Lee (now Lee-Feldstein) also followed up the study group and in 1983 reported the mortality experience of the total group from 1938 through September, 1978, but used the entire work history through 1977 to assign workers to the NCI exposure classification groups.[3,4]

In each of these studies, workers were assigned to the highest exposure category in which they had spent at least one year. Table I demonstrates that significant excess respiratory cancer mortality risk was observed by Lee-Feldstein for all three NCI exposure groups in the total Anaconda cohort.[4]

In 1975, the NCI exposure classifications were reviewed by H. F. Morris, an industrial hygienist at the Anaconda smelter, and an average quantitative value was associated with each category. He concluded that "Lee and Fraumeni divided their exposure categories in _about_ (emphasis added) the correct manner."[5] Morris derived "average" exposure values of 290 $\mu g/m^3$, 580 $\mu g/m^3$, and 11,270 $\mu g/m^3$ for the "Light", "Medium", and "Heavy" categories, respectively (Table I).

These average values were subsequently assigned to the three NCI exposure groups by several researchers and adapted by them in the development of quantitative risk assessments.[6-9] These risk assessments were primarily performed to determine the risk associated with

TABLE I

NCI EXPOSURE CATEGORIES AND RESULTS FOR ANACONDA SMELTERMEN (1938-1977)

Category	Exposure Estimate - 8 hr TWA+		Respiratory Cancer Mortality Risk++	
	Average	Range of Measurements	O/E	SMR
Heavy	11,270 ug/m^3	(2,600-26,070)	33/6.4	514*
Medium	580 ug/m^3	(12-1,458)	93/20.9	446*
Light	290 ug/m^3	(30-2,590)	136/58.9	231*

+ Morris, 1975 (5)
++Lee-Feldstein, 1983 (4)
* p <.01

relatively low to medium exposure levels (0-500 $\mu g/m^3$), but were greatly dependent on the risks found at rather heavy exposures (up to 26,000 $\mu g/m^3$). The assessments applied the linear, no threshold model to the risk observed in the NCI exposure categories paired with the Morris "average" exposure values. This resulted in predictions of significant excess risk at even the lowest arsenic exposure levels. These risk assessments, however, assumed that Morris' data were applicable to the NCI exposure classifications and that these classifications adequately described the worker's exposure. Furthermore, they assumed that the data were used consistently, and grouped the workers into separate and distinct exposure categories. Review of Morris' data revealed that these assumptions did not hold.

There appear to have been some changes in the exposure classifications between the original Lee and Fraumeni 1969 report[1] and the updated Lee-Feldstein 1983 report.[3,4] Both studies reported analyses for the same 8,045 men. The earlier paper reported 3,257 men as having no greater than "Light" exposure levels in their work experience. The later paper reported 4,448 workers as having no greater than "Light" exposure levels in their work experience (Table II). This discrepancy has yet to be explained. The ascertainment that all workers in the "Light" exposure group only had "Light" exposure is, however, most important to the assessment of lung cancer risk at low

TABLE II

NUMBER OF WORKERS BY EXPOSURE CATEGORY FOR LEE AND FRAUMENI AND LEE-FELDSTEIN STUDIES

	Lee and Fraumeni+ (1969)	Lee-Feldstein++ (1983)	Percent Change
Heavy	402	451	+12%
Medium	1,526	1,585	+4%
Light	3,257	4,448	+37%
Excluded	2,862	1,561	-54%
TOTAL	8,047**	8,045	

+ Lee and Fraumeni, 1969 (1)
++ Lee-Feldstein, 1983 (4)
* These workers are described as having worked less than 12 months in their category of maximum arsenic exposure in both studies. However, recent personal communications reveal that other unidentified workers were excluded from the study as well.
** Two female workers included in the Lee and Fraumeni study were excluded in the Lee-Feldstein study.

levels of arsenic exposure. It may be that some work areas have been reclassified as "Light," having possibly been previously classified as having unknown exposure levels.

The "Heavy" and "Medium" NCI exposure groups represent more homogeneous groups than the NCI Light group. This is because these groups were restricted to those with at least one year of "Heavy" or "Medium" exposure. Workers in the "Light" group had widely varied exposures. For example, the exposure measurements compiled by Morris, as shown in Figure 1, reveal that the exposures of the masons were far greater than any others in the "Light" category, with an average of 2,900 $\mu g/m^3$. The 2,900 $\mu g/m^3$ average for the masons was derived from work area measurements ranging from over 500 $\mu g/m^3$ to almost 13,000 $\mu g/m^3$. Clearly, the mortality experience of this group is ill-suited to assess the risk associated with exposures of 0-500 $\mu g/m^3$.

Similar classification problems are apparent for work areas where quantitative measurements were not available. For example, the NCI scheme assigned maintenance, surface, and shop employees to the Light

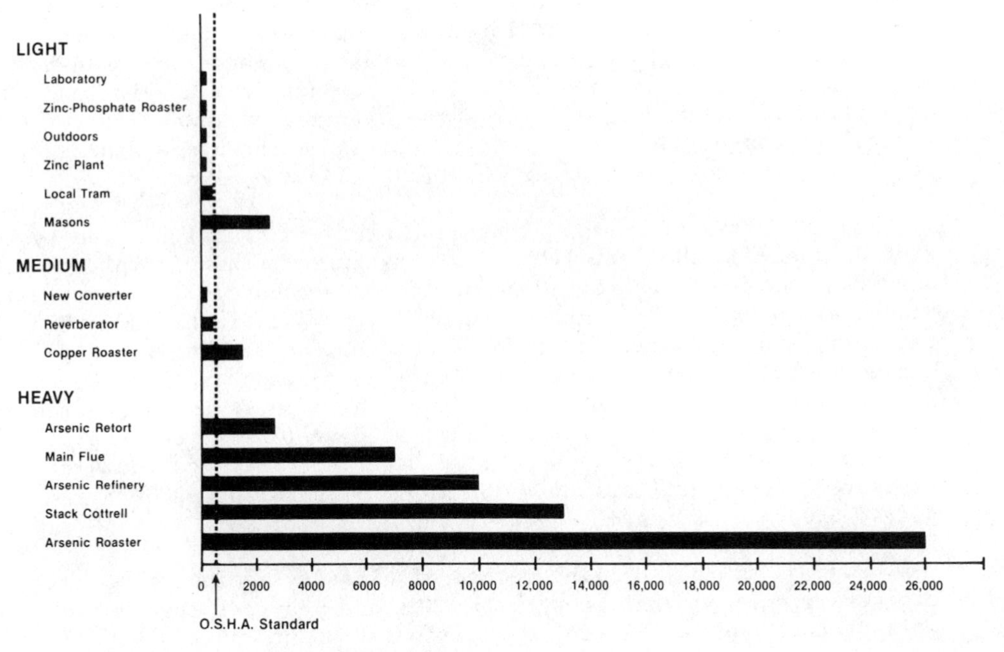

Figure 1

Area Average Arsenic Concentrations For Montana Smelter Departments

group, although they, similar to the masons, worked in many different areas of the plant, and would have been exposed to relatively high levels during the course of their work. The NCI classification schem also included slag workers in the "Light" exposure group, although th reverberatory furnace area where these employees worked was character ized by higher exposures.

Due to the imprecision in the NCI "Light" exposure category, the conclusion that the risk observed in that group is due to workers wit relatively low (0-500 $\mu g/m^3$) exposure is not well supported. In contrast, the quantitative classification system developed by Higgins is better suited to examine the risk associated with these low levels of exposure to airborne inorganic arsenic. Higgins divided jobs at the Anaconda smelter into four exposure groups with associated quantitative exposure values: low (0-99 $\mu g/m^3$); medium (100-499 $\mu g/m^3$); high (500-4,999 $\mu g/m^3$); and very high (above 5,000 $\mu g/m^3$).[10,11] In severa pages of text and tables, Higgins explained the extensive exposure data he obtained from the Montana State Health Department, the Anaconda Company, and other sources. He also described the system used for deriving average exposures for job areas and the assignments made for each area in the study. The average exposure in each of Higgins' fou groups and the range of averages within each is shown in Table III an Figure 2. Significantly, the ranges do not overlap and are much narrower than the NCI groups, indicating greater homogeneity of exposure experience within the groups. Table IV compares the Higgins and NCI classification schemes and reveals the differences in assignment of several groups of workers, including masons, maintenance, surface, shop, slag, unknown, converter, casting, and acid plant.

On the basis of this more precise exposure classification system Higgins analyzed the mortality of a stratified, random sample of 1,80 workers out of the entire 8,045 man Anaconda cohort. This sample included all workers who worked more than two years in the NCI "Heavy" category, and a 20 percent random sample of the remainder. Higgins linked individual workers to their entire in-plant exposure histories Workers were classified on the basis of the highest exposure area to which they were assigned for at least 30 days (30-day ceiling classification method). Workers were also separated into four cumulative exposure categories (less than 500 $\mu g/m^3$ - years; 500-2,000 $\mu g/m^3$ - years; 2,000-12,000 $\mu g/m^3$ - years; and 12,000 or more $\mu g/m^3$ - years).

Higgins' 30-day ceiling method of classifying individuals into exposure groups is analogous to the NCI one-year ceiling method, but greater care has been given to the separation of the exposure groups. Analysis of the Higgins data leads to a very different conclusion tha that of the earlier NCI classification scheme. While the NCI classification scheme indicates a rough dose-response relationship among th three intensity exposure groups (Figure 3), the Higgins exposure category analysis (Figure 4) makes clear that workers whose 30-day ceilin exposures did not exceed 500 $\mu g/m^3$ have <u>no excess lung cancer risk</u>

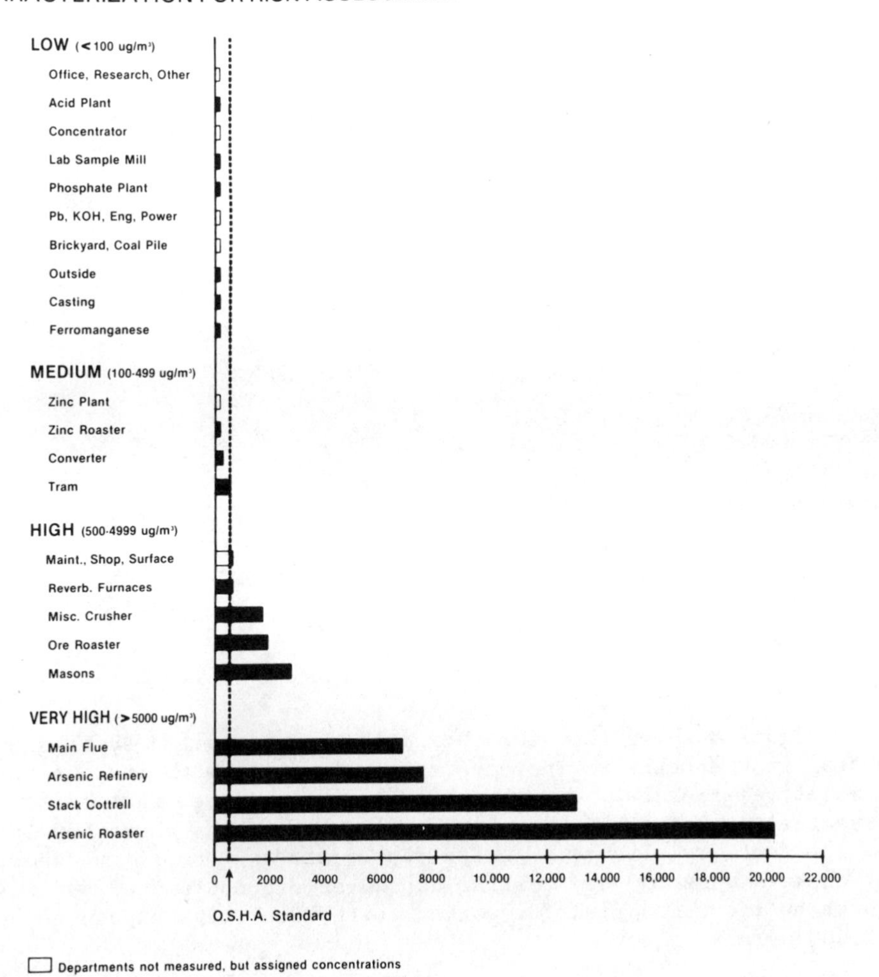

Figure 2

Average Arsenic Concentrations
For Montana Smelter Departments
By 30-Day Ceiling Arsenic Categories*

(i.e., SMR is not greater than 100). These data suggest that the excess risk observed in the NCI "Light" group is due to the inclusion of individuals who had high intensity exposures.

This result is confirmed when the Higgins data are analyzed in terms of the risk associated with various cumulative exposure levels.

TABLE III

HIGGINS LIFETIME CEILING
AND RESULTS FOR SAMPLE STUDY OF ANACONDA SMELTERMEN+

Category	Exposure Estimates - 8 hr TWA		Respiratory Cancer Mortality Risk	
	Average	Range of Measurements	O/E	SMR
Very High ($>5,000$ ug/m^3)	11,892 ug/m^3	(6,870-20,060)	34/5.5	617*
High (500-4999 ug/m^3)	1,387 ug/m^3	(598-2,589)	41/13.7	300*
Medium (100-499 ug/m^3)	255 ug/m^3	(111-415)	2/2.4	82
Light (<100 ug/m^3)	43 ug/m^3	(0.45-82)	3/3.9	77

+ Higgins et. al., 1982 (11)
* $p < .01$

Figure 5 shows that for the whole group reported by Higgins, the excess risk appears to increase rather linearly with increasing cumulative exposure. However, Figure 6 illustrates that this risk is essentially that of workers with ceiling exposures above 500 $\mu g/m^3$. On the contrary, Figure 7, which represents the mortality experience of workers whose 30 day ceiling exposures did not exceed 500 $\mu g/m^3$, shows no increased risk for workers with cumulative exposures up to 12,000 $\mu g/m^3$ - years.

The critical fact for quantitative risk assessment purposes is that workers with exposures below 500 $\mu g/m^3$ evidenced lower than expected mortality overall, and in all but the highest cumulative exposure group. The elevated SMR in the highest cumulative exposure category is determined by one case (with 0.1 expected): a man who was hired in 1903, worked for 16 years in low exposure jobs and then worked for 31.5 years in the tram (a department with an assigned exposure level of 415 $\mu g/m^3$, based on measurements from the 1950s).[12] During the course of this employment, it is very likely that this worker had exposures to levels well above 500 $\mu g/m^3$ for sustained periods. This analysis supports the observation that workers without exposures near or above 500 $\mu g/m^3$ do not experience excess lung cancer risk.

The foregoing analyses of the Higgins data suggest that exposure intensity appears to be the critical determinant of respiratory cancer risk and that exposures below 500 $\mu g/m^3$ are not associated with excess risk. This result (see Figure 4) is consistent with a non-linear "hockey-stick" or threshold model.[13]

The differences in the exposure assignments and classification system used by Higgins and by NCI may account for the very marked difference in the interpretive results for these studies. Risk assessments based upon the NCI classifications predicted excess lung

TABLE IV

COMPARISON OF HIGGINS ET AL. AND LEE-FELDSTEIN DEPARTMENTS BY EXPOSURE CATEGORY

NCI GROUPINGS		HIGGINS GROUPINGS			
Group	Exposure (ug/m3)	Very High (5000+)	High (500-4999)	Medium (100-499)	Low (<100)
Heavy	11270	Arsenic Roaster* Arsenic Refinery* Cottrells* Main Flue*			
Medium	580		Ore Roaster* (excluding acid plant) Reverberatory Furnace* (excluding slag workers)	Converter*	Casting* Acid Plant*
Light	290		Maintenance Surface Shops Unknown Masons* Slag	Zinc Roaster* Zinc Plant	Concentrator Ferromanganese Plant* Foundry Lead Plant Lime Processing General Office Phosphate Plant* Engineering Power Research+ Plant Office Brick Yard Industrial Hygiene+ Company Farm Coal+ Lab Sample Mill*+
Reclassified[2]			Materials Crushing* Bag House Electric Furnace	Tram *	Area 21

*Denotes Departments with actual exposure measurements from Higgins. There is no information about the measurements used in the NCI classifications scheme.

+Made into a separate department by Higgins

[1]Morris, H.F., 1975 (6)

[2]Departments newly identified by Higgins in areas considered by Lee-Feldstein to have neither medium nor heavy exposures.

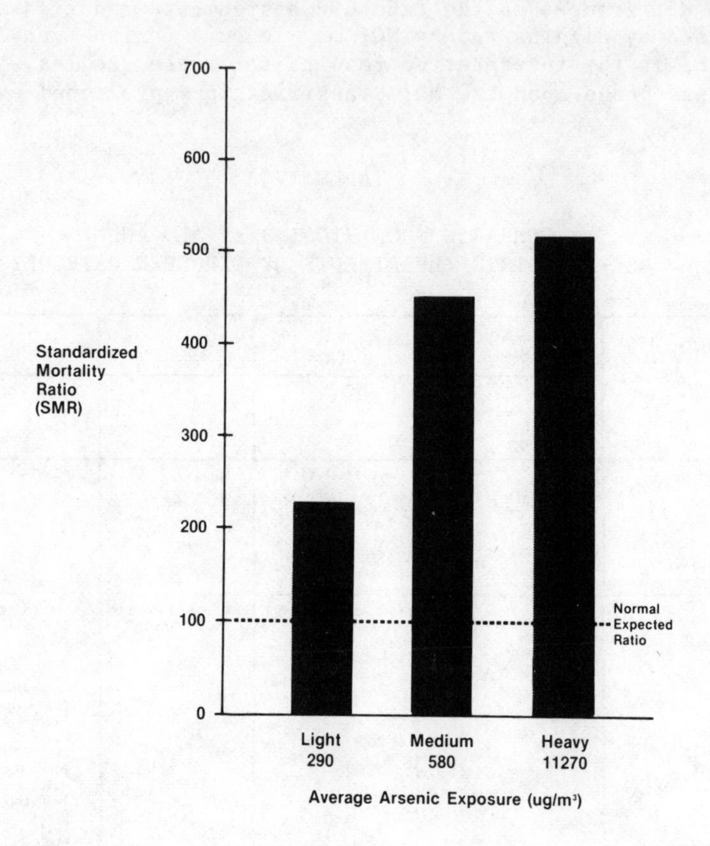

*Lee-Feldstein (1983)

Figure 3

Lung Cancer Mortality by NCI Exposure Category*

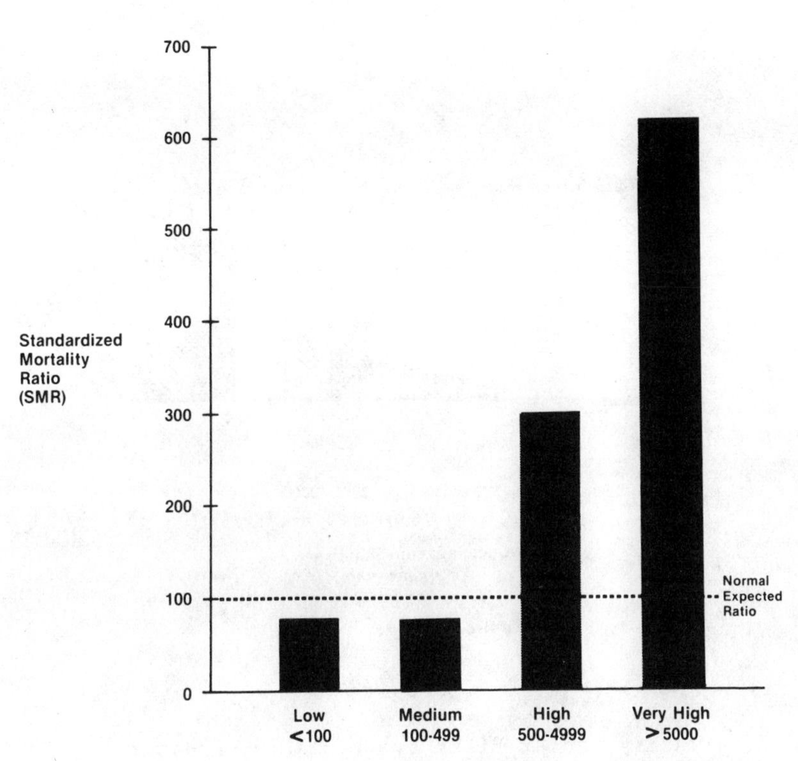

*Higgins (1982)

Figure 4

Lung Cancer Mortality by Higgins et al. Exposure Category*

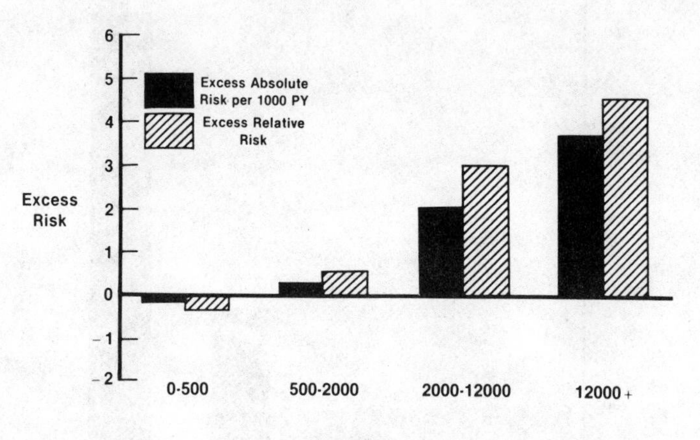

Figure 5

Respiratory Cancer Risks by Cumulative Exposures for Workers with Lifetime Ceiling Exposure Levels of 500 $\mu g/m^3$ or Greater or of Less than 500 $\mu g/m^3$*

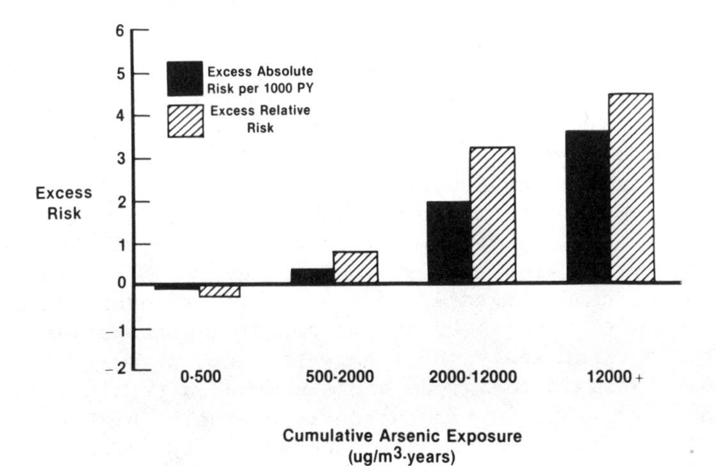

Figure 6

Respiratory Cancer Risks by Cumulative Exposures for Workers with Lifetime Ceiling Exposure Levels of 500 $\mu g/m^3$ or Greater*

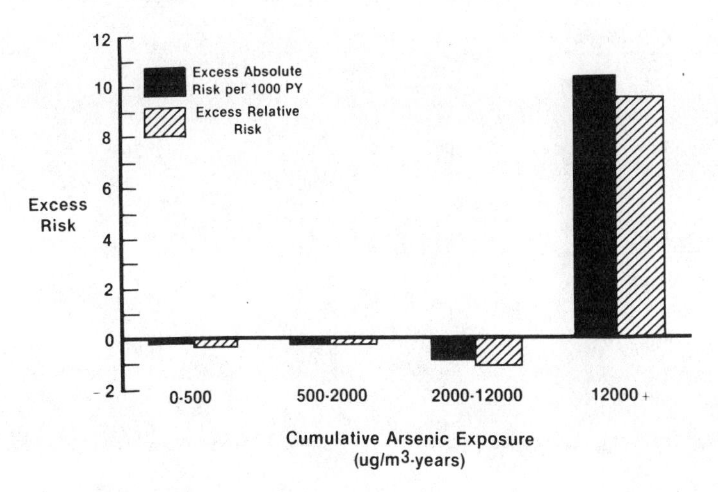

Figure 7

Respiratory Cancer risks by Cumulative Exposures for Workers with Lifetime Ceiling Expousre Levels of less than 500 $\mu g/m^{3}$*

cancer risk at low arsenic exposure levels (0-500 $\mu g/m^3$). However, analysis using the Higgins classification, on the contrary, strongly suggests that exposures to airborne inorganic arsenic at levels below 500 $\mu g/m^3$ are not associated with excess respiratory cancer risk.

These data suggest that a threshold exposure level may exist below which respiratory cancer risk is not increased. Because of their more precise quantitative exposure descriptions, the Higgins exposure classifications may be better suited than the NCI groupings for quantitative risk assessment at low exposure levels. The differences illustrate the need for careful attention to exposure description and quantification that is necessary to perform quantitative risk assessment. This current analysis of ours is based on only a few observed lung cancer cases in the group whose exposures remained below 500 $\mu g/m^3$ (Table V). Higgins is currently expanding his study to the entire Anaconda cohort and updating the mortality followup to 1980. The results of the larger extended study should serve well to test the hypothesis that we have developed based on observations from his initial study, namely that there appears to be a carcinogenic threshold for arsenic exposure approximating 500 $\mu g/m^3$.

TABLE V

CUMULATIVE LIFETIME ARSENIC EXPOSURE (g/m^3-yrs)*

	<500	500-2,000	2,000-12,000	$>12,000$
Exposures <500 ug/m^3				
Lung Cancer Cases	3	1	0	1
Expected Cases	4.5	1.3	0.5	0.1
Person-Years Obs.	9001	1589	563	87
Exposures ≥ 500 ug/m^3				
Lung Cancer Cases	1	8	27	39
Expected Cases	1.3	4.5	6.4	7.2
Person-Years Obs.	4954	9304	10656	8931

* Data from Higgins et al. presented at OSHA Arsenic Hearing (1982) by Lamm

REFERENCES

1. Lee, A.M. and Fraumeni, F.F., Jr., Arsenic and Respiratory Cancer in Man: An Occupational Study," J. Natl. Cancer Inst., 42:1045-1052 (1969).
2. Lubin, J.H., Pottern, L.M., Blot, W.J., et al., "Respiratory Cancer Among Copper Smelter Workers: Recent Mortality Statistics," JOM 23:779-784 (1981).
3. Lee-Feldstein, A., "Arsenic and Respiratory Cancer in Man: Follow-up of an Occupational Study" in Arsenic: Industrial, Biomedical, Environmental Perspectives, Lederer and Fensterheim (eds.), Van Nostrand Reinhold Company, 1983, pp. 245-254.
4. Lee-Feldstein, A., "Arsenic and Respiratory Cancer in Humans: Follow-up of Copper Smelter Employees in Montana," J. Natl. Cancer Inst., 70:601-609 (1983).
5. "Morris Exposure Data in the OSHA Hearing Record," Federal Register, 43:19595 (1975).
6. Environmental Protection Agency, Carcinogen Assessment Group, Final Risk Assessment on Arsenic (May 2, 1980).
7. Chu, K., Risk Assessment for Inorganic Arsenic (January 14, 1981).
8. Clement Associates, Assessment of Human Respiratory Cancer Risk Due to Arsenic Exposure in the Workplace. Prepared for OSHA (September 5, 1980).
9. Brown, C., Chu, K., "A New Method for the Analysis of Cohort Studies: Implications of the Multistage Theory of Carcinogenesis Applied to Occupational Arsenic Exposure." (In press).
10. Higgins, I., Welch, K., Oh, M., Burchfiel, C., "Mortality of Anaconda Smelter Workers in Relation to Arsenic and Other Exposures." University of Michigan, School of Public Health (April 12, 1982).
11. Welch, K., Higgins, I., Oh, M., Burchfiel, C., "Arsenic Exposure, Smoking, and Respiratory Cancer in Copper Smelter Workers," Arch. Envir. Health, 37:325-335, 1982.
12. Higgins, I.T.T., Personal Communication, June, 1982.
13. Hoel, D., "Implications of Nonlinear Kinetics on Risk Estimation in Carcinogenesis," Science, 219:1032-37, 1983.

LOWER-BOUND BENEFITS OF AIR POLLUTION REDUCTIONS IN THE MUNICH METROPOLITAN AREA (MMA)

Hans W. Gottinger

Institut fur Medizinische Informatik und Systemforschung
MEDIS der Gesellschaft fur Straheln - und Umweltforschung
mbH, GSF
Ingolstadter Landstr. 1
8042 Munchen-Neuherberg
Federal Republic of Germany

ABSTRACT

The mortality component of the marginal social benefit function is estimated based on a linear multivariable regression study of air pollution health effects in the Munich Metropolitan Area, Gottinger (1983). The resulting pollution-related disease specific mortality functions are monetized through the use of risk estimates of occupational risk to approximate an individual's willingness to pay for mortality decreases.

KEY WORDS: Mortality; Benefit Estimation; Risk Assessment; Air Pollution

INTRODUCTION

In a previous essay, Gottinger (1983), we investigated air pollution's aggravation effect on intra-urban mortality by a cross-sectional statistical analysis. A natural further step is to quantify these costs on the basis of modern approaches of cost-benefit analysis, Freeman (1979). The results of the present study could be important for the public decisionmaker, by providing him with a conceptually accurate quantification of the costs of pollution-aggravated mortality (excess mortality). The policy-relevant aspect of this contribution is to assist in establishing socially efficient standards for ambient air quality. We have actually provided estimates of physical social benefits, measured as potential decreases in mortality that can be expected from improvements in air quality, Gottinger (1983). The present paper considers the application of a method for quantifying these benefits in monetary terms.

THE MORTALITY COMPONENT IN COST-BENEFIT ANALYSIS

Thus far, there have been quite a few attempts at monetizing changes in life expectancy (mortality), advanced along various lines by Schelling (1968), Mishan (1972), Hirshleifer et al. (1974), Zeckhauser (1975), Jones-Lee (1976), and Arthur (1981). To be consistent with the foundations of cost-benefit analysis it is required that changes in welfare, induced by mortality changes, be expressed as either the compensation required to return an individual to his original utility level (or bundle of goods) after a decrease in welfare or, equivalently, as willingness to pay in order to maintain an increase in welfare. In particular, if an improvement in air quality reduces mortality, the correct social valuation of this benefit is the summation of the resulting positive changes in consumer surplus or the summation of what individuals would be willing to pay rather than forego this increase in welfare. Alternatively, if a reduction in air quality leads to increases in an individual's probability of death, then the correct social valuation of these costs is the summation of the resulting negative changes in consumer surplus or the summation of what individuals would require in compensation to regain their original level of welfare. In addition to benefits in the areas of public health, material preservation, and vegetation growth, there are also "spillover" benefits to friends, family, and society from decreases in the individual's probability of death. The relevant social benefit from a reduction in any individual's probability of death is the sum of what he, his friends and relatives, and society are willing to pay rather than forego the reduction.[1] Summing up these valuations for all individuals in society, the total societal benefit for any change in mortality can be determined.

MONETARY VALUATION OF RISK CHANGES

Although, in principle, the concept of consumer's surplus (willingness to pay) can be applied to the valuation of changes in mortality, the relevant question becomes whether reliable empirical estimates of this concept can be obtained. The possibility of simply asking individuals what they are willing to pay for changes in their probability of death or that of friends and relatives is the logical starting point. Serious problems, however, are inherent in this approach. First, individuals may not express their true willingness to pay because of the free rider problem of providing a public good. Second, even if individuals would express their valuation, most of the probabilities usually discussed are relatively small (e.g., one in a million), and it has been argued that individuals cannot adequately assess such small probabilities in an explicit way, Zeckhauser (1975, p. 444).

In an attempt to circumvent these problems, some economists have focused on the market system to explore the possibility that individuals reveal monetary valuations of changes in their own probability of death by actually trading such probabilities for changes in money income. Now a market system in valuing risky activities does exist in the form of explicit or implicit premiums paid to various occupations for assuming extra risk. These risk premiums may be in the form of direct income or indirect compensation (e.g., pension benefits). Zeckhauser (1975, p. 436) states some difficulties involved with the use of these premiums, arguing that they are not distribution-free and biased against the poor. Two major critical arguments against using occupational risk estimates for benefit estimations of ambient air pollution abatement strategies can be made:

- Compensation measures may not be transferable from one kind of risk (in the case of occupational exposure) to another kind of risk (in the case of ambient air pollution), because they may require more or less for compensation.

- The general population tends to be much more risk averse than an occupational selection of the population.

Both factors tend to be biased in the direction that the estimate of compensation definitely will be a lower bound, and is likely to be higher for the general population.

Besides these considerations the compensation scale is likely to be "highly non-linear" and even from a certain point (probability of death), people may outrightly refuse any monetary compensation in exchange for the likelihood of death. Monetary compensations may appear acceptable for low risk outcome gambles but not for high risk outcomes. In addition, it also depends crucially on the severity of the outcome. (Along this line, being forced to play Russian roulette, would you pay the same amount by reducing one bullet out of two in a six-shot revolver as compared to a Russian roulette where one bullet out of five is removed?)

In the light of these criticisms, market outcomes of compensation required to induce an individual to accept additional risk must always be qualified. Specifically, since such estimates only approximate one of the components of society's total valuation, any extrapolation of these results to monetization of the benefits of decreases in mortality, resulting from improvements in ambient air quality, must be interpreted as an extreme lower-bound estimate.

Figure 1 graphically depicts how the labor market reflects individual's monetary valuation of risk changes. Specifically, this figure shows how an individual can increase his welfare by rationally choosing between changes in risk and corresponding changes in compensation or payment. For example, assume an individual at point A has the opportunity to gain income by increasing his risk. If his trade possibilities are summarized by MM', the market equilibrating tradeoff curve between risk and income, he will maximize his welfare by trading income for security and move from A to B. The monetary value of this change in his welfare is BC. What is observed in the labor market, however, is the income differential Y_0Y_1 which is larger than BC since the marginal utility of money is not constant. Using the trade possibility curve MM', the marginal value of this market reflection of the compensation required to induce the individual to assume the small increase in his probability of death may be expressed as Y_oY_1/P_oP_1. Extrapolating this marginal valuation to the individual's valuation of life (e.g., OM) may not be valid if the individual has a high probability of death (low probability of survival) beyond which he will not make income-risk tradeoffs. This possibility is demonstrated in Figure 1 by the vertical line extending from the low probability of life P.

ESTIMATES OF SOCIAL BENEFIT

The gap between an acceptable theoretical determination of the willingness to pay for decreases in mortality and quantification of

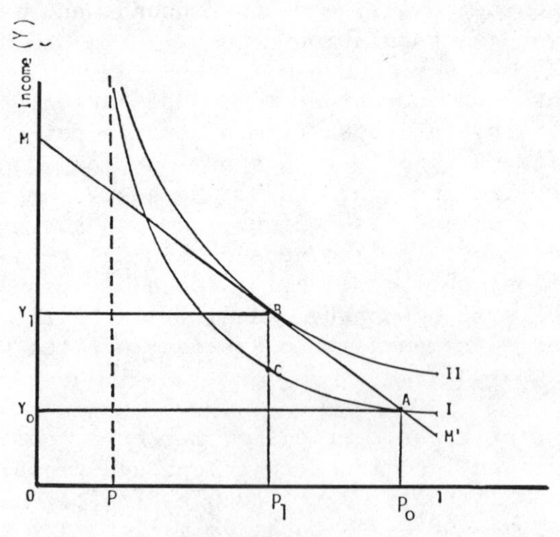

Figure 1
Market Reflection of Risk-Income Tradeoff

the abstract model is quite large. It appears that work by Thaler and Rosen (1976) has managed to obtain valid empirical estimates of the marginal valuation curve (MM' in Figure 1). By matching risk data which detailed the number of excess deaths by industry and occupation from the "1967 Occupation Study of the Society of Actuaries" with the "1967 Survey of Economic Opportunity" data on personal and industrial characteristics of individuals, Thaler and Rosen were able to construct a database consisting of 907 adult male heads of household. Using these data, the authors statistically controlled for "other" factors in addition to risks, which influence wage rates. Specifically, regional, demographic, social, and job characteristic differentials were controlled by fitting various hedonic demand curves to these data through the use of multiple regression analysis. Utilizing this technique, Thaler and Rosen were able to estimate a set of implicit marginal prices for risk acceptance. Extrapolating these marginal values to the point of a "statistical death" their estimates "lie in a reasonably narrow range of about $200,000 + 60,000," in 1967 U.S. dollars, Thaler and Rosen (1976).

Before proceeding to use these results to quantify in monetary terms the benefits of air quality-related mortality reductions, it should be noted that these risk premiums contain an inherent bias. Thaler and Rosen's sample was composed of relatively risky occupations.[2] Therefore, the individuals included in the sample were relatively less risk averse than the general population. This downward bias--coupled with the fact that these risk premiums only approximate the individual's own willingness to pay for risk reductions and not those of family, friends, and society--insures that any benefit estimates based on the Thaler and Rosen results will unequivocally represent a lower bound estimate. Let us start out with the results obtained from the MMA study on pollution-related disease specific mortality functions, as shown in Table 1.

For lack of empirical data on the risk-income tradeoffs of the reference population we use the detailed results of the Rosen-Thaler study. Even if it might be argued that the preference profile of the Thaler-Rosen population may hardly be in line with our population in numerical terms, it is highly likely that the orderings expressed by the choices are identical for both populations. The main reason for using this type of approach is that risk-income tradeoffs of this sort are "revealed preferences" and as such more justifiable than other measures of "willingness to pay" (for instance, as derived from questionnaires).

By selectively combining the average (marginal) compensation for risk estimates (0.001 = $200) derived from the Thaler and Rosen analysis with the various pollution-related disease specific mortality functions, estimated in Table 1, it is now possible on the basis of the risk preferences of the Thaler-Rosen population to monetize the pollution-aggravated mortality component of the marginal social benefit function.[3] Specifically, the marginal compensation premiums

TABLE I

MORTALITY FUNCTIONS ESTIMATES IN DEATHS PER 100,000
1974-1978 THE MUNICH METROPOLITAN AREA (MMA)

POLLUTION-RELATED DISEASE-SPECIFIC MORTALITY FUNCTIONS.

	<45 years of age		45 - 65		>65	
VARIABLE	MALE	FEMALE	MALE	FEMALE	MALE	FEMALE
constant	22.002	68.982	1128.087	354.199	1752.255	2077.786
percent of adult population with high school education (E)	-0.431 (3.945)[1)]	-0.067 (0.745)	-9.984 (6.619)[1)]	-4.492 (6.891)[1)]	-28.134 (5.651)[1)]	-9.165 (2.294)[1)]
TP ($\mu g/m^3$)	0.183 (2.145)[1)]	0.057 (0.880)	2.698 (2.167)[1)]	1.364 (2.616)[1)]	9.746 (2.117)[1)]	8.431 (2.241)[1)]
SO_2 (ppt/24 hrs)	0.026 (0.152)	0.76 (0.592)	1.538 (0.618)	0.321 (0.316)[1)]	2.457 (0.286)	5.111 (0.766)
CO (ppm/24 hrs)	0.244 (0.212)	0.182 (0.151)	1.192 (0.924)	0.746 (0.478)	1.423 (1.075)[1)]	1.560 (1.325)
# days precipitation >5mm	0.155 (0.704)	0.116 (0.672)	11.542 (3.387)[1)]	3.184 (2.150)[1)]	71.915 (5.555)[1)]	37.346 (3.604)[1)]
# days max. temp. $\leq 0^o$	0.500 (1.181)	-0.335 (1.032)	-4.283 (0.639)	1.828 (0.638)	-38.055 (1.515)	-25.890 (1.302)
population per hectar (P/ha)	0.006 (0.536)	0.015 (1.663)[2)]	0.361 (2.123)[1)]	0.152 (2.303)[1)]	0.080 (0.170)	-0.265 (0.746)
$\bar{R}^2$	0.366	0.210	0.554	0.566	0.703	0.669

[1)] significant at 0.05 probability level (two-tailed test)

[2)] significant at 0.10 probability level (two-tailed test)

Figures in the parentheses represent student t-values associated with each coefficient.

estimated by Thaler and Rosen may be interpreted as estimates of the compensation required to offset a negative change in an individual's consumer surplus (e.g., acceptance of increased risk) which is equal to the amount he would be willing to pay rather than forego an equivalent positive change in his consumer's surplus (e.g., a reduction in risk). The procedure used to combine risk and mortality estimates involves the following steps:

- Estimate changes in the various pollution-related mortality rates for a one percent improvement in each of the air pollution variables--this task was undertaken by using percentage changes around the mean values of SO_2, CO, and TP while holding all other explanatory variables constant at their mean values. The results of these estimates are presented in Table 2.

- Relate these changes in mortality rates to one statistical individual out of 100,000 and multiply by the Thaler and Rosen compensation estimate of $200,000--this results in an average (marginal) social benefit estimate (ASBE) for the various age-sex specific, pollution-related mortality rate changes. The results of this procedure are presented in Table 3.

TABLE II

CHANGES IN POLLUTION-RELATED DISEASE-SPECIFIC MORTALITY RATES RESULTING FROM ONE PERCENT DECREASE IN SO_2, CO, OR TP*

SEX	AGE	SO_2	CO	TP
Male	less than 45	0.009	0.089	0.204
	45 - 64	0.496	0.352	3.000
	65 and over	0.797	0.425	10.836
Female	less than 45	0.025	0.062	0.064
	45 - 64	0.104	0.275	1.517
	65 and over	1.648	0.485	9.374

* Remark: All mortality rate changes are per 100,000, all other variables held constant on their mean value in mortality rates.

TABLE III

AVERAGE (MARGINAL) ANNUAL MONEY BENEFITS FOR POLLUTION-RELATED MORTALITY CHANGES RESULTING FROM ONE PERCENT DECREASE IN SO_2, CO, OR TP

SEX	AGE	SO_2	CO	TP
Male	less than 45	0.018	0.178	0.408
	45 - 65	0.992	0.704	6.000
	65 and over	1.584	0.850	21.672
Female	less than 45	0.050	0.124	0.128
	45 - 64	0.208	0.550	3.034
	65 and over	3.296	0.970	18.648

o Multiply these average (marginal) social benefit estimates by the population under exposure to obtain the total social benefit estimate (TSBE) of pollution abatement, structured along age-sex specific estimates--this step involves multiplying the various age-sex values in Table 3 by the corresponding 1975 population of the MMA (Table 4). The results of these multiplications are presented in Table 5.

TABLE IV

COMPOSITION OF THE POPULATION OF THE MMA BY 1975

SEX	AGE	POPULATION	
Male	less than 45	160,000	
	45 - 64	190,000	
	65 and over	160,000	
			Total:530,000
Female	less than 45	150,000	
	45 - 64	210,000	
	65 and over	200,000	
			Total: 560,000
Total Male and Female 1,090,000			

TABLE V

TOTAL MARGINAL ANNUAL MONEY BENEFITS FOR POLLUTION-RELATED MORTALITY CHANGES RESULTING FROM A ONE PERCENT DECREASE IN SO_2, CO, OR TP*

SEX	AGE	SO_2	CO	TP
Male	less than 45	3	28	65
	45 - 65	188	134	1,140
	65 and over	285	153	3,888
TOTAL MALE		476	315	5,093
Female	less than 45	7.5	19	19
	45 - 64	44	110.5	637
	65 and over	659	119	3,720
TOTAL FEMALE		710.5	328.5	4,376
Male and Female		1,186.5	643,5	9,469

*All figures are in thousands of U.S. dollars ($1,000) in terms of 1967 prices.

From the results presented in Table 5 it can be concluded that the benefits of reduced mortality derived from an abatement of particulate matter are substantially higher than for sulfur dioxide abatement. That is, individuals in the MMA are willing to pay at least approximately $9.5 million annually (in terms of 1967 prices) in order to maintain TP at a level one percent below those experienced during the 1973-1977 period.

Relative dollar differentials were 1.2 million annually for a similar percentage reduction in SO_2, and 650 thousand annually for a reduction in CO. These relative dollar differentials were to be expected on the basis of the results from the mortality model for the MMA, especially, from the elasticity estimates, Gottinger (1983).

Thus, on the basis of risk evaluation in the limited context of occupational risk, some generalizations can be made on risks generated by air-pollution exposure for large urban populations. By monetizing the mortality component of the (marginal) social benefit function, more meaningful decisions about the socially optimum level of air quality can be made. Specifically, these money estimates will permit decisionmakers to more closely approximate the optimum level of air quality by permitting the direct comparison of mortality reductions, as well as other benefits of air pollution abatement, with the cost associated with such abatement policies. It should be emphasized that the ranking of abatement strategies, on the basis of TSBE, is robust against the values of life to start with. This would imply that other, more accurate and reliable values of willingness to pay measures, no matter how obtained, either revealed or by questionnaires, would not alter the ranking of abatement strategies though they might alter the monetary dimension of the problem.

CONCLUSIONS

The results of this study, therefore, lend support to the contention that an improvement in ambient air quality will produce social benefits in the form of decreased probabilities of death. Specifically, the results obtained for the MMA suggest that ceteris paribus efficient resource allocation would be enhanced by devoting relatively more financial resources to the control of particulate matter rather than sulfur dioxide, or carbon monoxide. The validity of this assertion is contingent upon existing technology. Specifically, the present marginal costs of controlling sulfur dioxide exceed the marginal costs of controlling particulate matter; and the estimated mortality benefits of decreasing particulate matter exceed the estimated mortality benefits of decreasing sulfur dioxide. Therefore, on both counts it would appear that relatively more resources should be devoted to the control of particulate matter rather than sulfur dioxide.

It must be emphasized, however, that this latter conclusion is based solely on the mortality component of the (marginal) social benefit function. The risk preference profiles and the resulting compensation schemes of the Thaler-Rosen study have been used as a first approximation to a willingness to pay measure of a large urban population. The results have to be qualified in this limited context, and may only be seen as an extreme lower bound for benefit estimates of pollution reduction for a population at large.

However, the general method of obtaining such estimates still holds true for other quantitative estimation procedures of willingness to pay measures. Even if quantitative results may differ and shifts in the relative weights of air pollution can be obtained by careful sensitivity studies, the qualitative ranking of the willingness to pay measures is not likely to change. Further work is required to test the reliability of both the mortality functions and willingness to pay estimates. It is not clear, for instance, whether the relatively large impact of particulate matter on mortality patterns is due to the specific way in which it is sampled in the MMA or due to the inherent physical-chemical properties of particulate matter.

REFERENCES

Arthur, W.B., "The Economics of Risks to Life," American Economic Review 71, 1981, 54-64.

Conley, B.C., "The Value of Human Life in the Demand of Safety," American Economic Review 66, 1976, 45-55.

Dardis, R., "The Value of a Life: New Evidence from the Market Place," American Economic Review 70, 1980, 107-108.

Freeman III, A.M., The Benefit of Environmental Improvement, Resources for the Future. Johns Hopkins Univ. Press: Baltimore, 1979.

Gottinger, H.W. (1983), "Air Pollution Health Effects in the Munich Metropolitan Area - Preliminary Results Based on a Statistical Model," to appear in Environmental International.

Hirshleifer et al., "Applying Cost-Benefit Concepts to Projects Which Alter Human Mortality" U.C.L.A. Eng. 7478 (University of California: Los Angeles), 1974.

Jones-Lee, M.W., The Value of Life. University of Chicago Press: Chicago, 1976.

McDougall, G.S. and C. Wright, "A Proposal for Improving the Measurement of Benefits from Pollution Abatement," Journal of Environmental Economics and Management 7, 20-29.

Mishan, E., "The Value of Life" in R. Layard (ed.) Cost-Benefit Analysis, Penguin: Middlesex, England 1971.

Rhoads, S.E., "How Much Should We Spend to Save a Life," The Public Interest 51, 74-92, 1978.

Schelling, T.C., "The Life You Save May Be Your Own," in S.B. Chase (ed.) Problems in Public Expenditure Analysis, Brookings Institution: Washington, DC, 1968.

Thaler, R. and S. Rosen, "The Value of Saving a Life: Evidence from the Labor Market" in N.E. Terleckyj (ed.) Household Production and Consumption. Columbia University Press: New York, 1976.

Zeckhauser, R., "Procedures for Valuing Lives," Public Policy 23, 419-464, 1975.

Kneese, A.V. and W.D. Schulze, "Environment, Health and Economics - The Case of Cancer," American Economic Review 67, 326-332, 1967.

FOOTNOTES

1. The value of a statistical life is linked intrinsically to the marginal survival probability (decrease of probability of death). The distinction between an individual's "value of life" and the "statistical value of a life" (or the "value of life-saving" for society at large) has been well presented for a particular situation by R. Dardis (1980, p. 1078).

"Suppose 1,000 persons require a compensation of $200 due to a decrease in survival probability of 0.001 then the estimated value of life is $100 thousand for individuals in the community. Thus, there will be one more fatality once the hazard is introduced and the total required community compensation is $100 thousand. However, this does not mean that any one individual will be willing to sacrifice his life for $100 thousand."

2. As S.E. Rhoads (1978, p. 82) correctly points out: "Thaler and Rosen do have data showing death rates by occupations, but only for very risky occupations, and there is reason to believe that those attracted to the job are more tolerant of risks... For this reason, the Thaler and Rosen figures will underestimate most people's willingness to pay for risk reduction."

3. The idea of using the Rosen-Thaler risk estimates as a reference to calculate damages associated with carcinogens in the environment originated in a review paper by A.V. Kneese and W.D. Schulze (1967), but to my knowledge, it has not been applied to any specific air pollution abatement situation, though various other methods have been tried, see G.G. McDougall and C. Wright (1980), B.C. Conley (1976). Using the Rosen-Thaler risk estimates as a basic reference has been initiated by the fact that data on the willingness to pay measures for the exposed population in the MMA are not available, and thus we are using the Thaler-Rosen estimates as proxy for preferences of the MMA population.

ENVIRONMENTAL RISK ASSESSMENT IN THE ENERGY FIELD: ISSUES AND APPLICATION TO POLICY*

Ian M. Torrens

Head of the Resources and Energy Division
Environment Directorate
OECD

INTRODUCTION

Risk assessment in the environmental field is a fascinating topic, mainly because it is at the frontier between the domain of the scientist and that of the politician. The scientist's inbuilt desire for exactitude and factual precision, and the politician's frequent need to blur the edges of the issues being dealt with, do not clash to the extent that might be anticipated a priori. In fact, the complexity of environmental analysis obligingly provides the blurred edges in the form of uncertainty, despite the sterling efforts of many scientists.

Except in some very specific and usually highly localized circumstances, policymakers in the environmental field do not usually look to risk analysts to provide unambiguous scientific reasons for their decisions. Experience over the past decade with attempts to define and quantify the effects of energy production and use on human health and other facets of the ecosystem, has made it abundantly clear that the findings of risk assessment in this field remain subject to interpretation in a wide variety of ways. Both proponents and opponents of a given policy initiative can often point to the same specific piece of work in risk assessment as a justification for their policy stances, depending on how they interpret the results.

Does this imply that risk assessment is not worth applying to the environmental issues related to society's exploitation of energy? I believe that provided both the limitations and the potential utility of the technique in its application to policy are clearly understood, any effort to narrow the ranges of uncertainty in this field will clearly be of value to decisionmakers.

The objective of this paper is to examine the relationship between environmental risk assessment and policymaking in OECD countries, mainly by reference to two specific pieces of work carried out by the OECD. These are the "COMPASS" project--a comparative assessment of the environmental implications of various energy systems[1]--and a project on the costs of coal pollution abatement[2-4]. Neither is a risk assessment per se, but both can teach us something about the interaction of risk assessment and policy.

Given the specific interest of this meeting in the interactions between the public and private sectors, the paper will also examine this aspect by reference to another initiative by the Coal Industry Advisory Board of the OECD's International Energy Agency, which examined the environmental issues related to coal use.[5] Again, while not a risk assessment, the CIAB's work provides some useful indications on how the public and private sector interact on environmental policy.

METHODOLOGICAL TRANSPARENCY IN ENVIRONMENTAL RISK ASSESSMENT

The environmental field is one of great complexity. Knowledge of the mechanisms involved in the many impacts of pollutants on man and his environment, particularly in quantitative terms, is still lacking. Uncertainty increases as one proceeds along the analytical chain from emissions of pollutants to describing their environmental impacts and to placing a value on these damages as an essential element of, for instance, a cost-benefit analysis.

The stages of comparative environmental assessment are:

- Definition of the energy technology
- Determination of the environmental pollutants (emissions to air, effluents to water, solid wastes, land use, water use)
- Determination and evaluation of the environmental pathways and impacts (air and water quality, ecological and health impacts)
- Evaluation of methods to minimize environmental impacts

Comparisons can take place at any stage of the environmental assessment. For example, air emissions of different pollutants can be compared, as can similar categories of impacts. The most comprehensive type of comparison involves valuation and comparison of unlike impacts. At successive stages, because the uncertainty involved in the evaluation increases, so does the judgmental element.

It has already been noted that formal assessments of a complex problem, such as that of environmental impacts of energy, can rarely provide the policymaker with an uncontestable answer as to what is the "best" decision. They can, however, provide a focus for debate among participants on policymaking, contribute to the formation of consensus on a given issue, and identify questions which require further study.

In order that comparative assessments may play a valuable and influential supportive role in the opinion-forming and decisionmaking processes, they must be transparent in their methods. Participants to the decision process must be able to discern the procedures by which data were transformed into the information from which conclusions were drawn. Assumptions must be made explicit, and the nature and degree of uncertainty identified.

The transparency and potential utility of a particular comparative assessment may be judged by the way it handles five key methodological criteria, described briefly below.

Bounding the Problem

No analysis could have the resources (financial, manpower, or time) to incorporate all the relevant aspects of a problem. Bounds must be set on the phenomena to be analyzed, and the choices made in this regard are important aspects of any analysis. In energy-environment studies, this includes the kinds of energy systems considered and the coverage of fuel-cycle stages for each energy system. It is also important to identify the kinds of environmental effects which are considered, and their boundaries in both space and time.

Disaggregation

Geographic Disaggregation. Comparative assessments differ in the way they disaggregate the analysis. Geographic disaggregation is severely constrained in most national assessments in spite of the fact that environmental impacts are often local in nature, and reveal significant locational variation.

Comparisons based on national average statistics can be useful in revealing the average consequences of technical choices and practices that have prevailed in the country for which the statistics were gathered. Such comparisons may direct attention toward differences in damage levels extreme enough to influence national decisions about environmental regulation and energy strategy. But regional and local choices concerning energy options or the focus of hazard-abatement efforts must be based on information applicable to the particular locality. The methods applied in geographically more aggregated assessments may be useful for deriving the information needed for location-specific purposes, but the quantitative data themselves will generally have to be derived again.

Temporal Variation. If the aim of an assessment is to determine the impact on society of past energy choices and associated activities up to the present, it is appropriate to work with the average effects of existing facilities and practices. If, on the other hand, the aim of the assessment is to help determine what energy options should be developed and/or deployed in the future, then one should assess the effects of new or projected facilities and practices.

The use of data corresponding to outdated technologies and practices has been responsible for many of the higher hazard estimates found in the coal literature--notably, estimates of coal workers' pneumoconiosis caused by coal-dust in underground mines (a problem that has been reduced significantly in the past 15 years by means of better dust-suppression practices), estimates of accidents in underground mining (being diminished by improved miner training and better enforcement of safety regulations), and estimates of public disease from coal-produced air pollution (shrinking with time as tighter standards and pollution-control technologies and practices to meet them are implemented).

For some types of effect, there is temporal variation in emissions or damages. Seasonal variation is one example, where changes in weather conditions throughout the year determine the extent of damages from a constant rate of emissions.

Classes of Damages. Deciding on the number of categories of damages to be considered in an assessment is not easy. Too few categories mean either serious incompleteness or expensive aggregation of qualitatively different kinds of harm into a small number of indices, amounting in either case to a loss of information significant in the energy choices. Too many categories make the findings difficult to absorb and magnify the problems of choosing between alternatives.

Uncertainty

Given the complex nature of environmental processes, this is one of the most important issues and needs to be recognized as such in any study. Uncertainties may reflect the normal confidence limits of statistical fluctuations (e.g., the impact of air pollutant emissions on local air quality); they may be a measure of imperfect knowledge of cause and effect (e.g., health impacts of a pollutant); or they may stem from an attempt to quantify the risk of an event not yet experienced (e.g., a serious nuclear reactor accident involving large release of radioactivity or an LNG tanker explosion). Uncertainties and the method of handling them need to be made explicit.

Valuation

In almost any assessment, considerable attention is paid to the complex process of estimating impacts. However, impact determination is seldom totally satisfactory. Impacts themselves need to be assigned weights which reflect their relative values to man.

Valuation, in addition to assigning relative importance to impacts, is also a further stage in data aggregation. For example, impacts like fish kills, altered recreational opportunities, radiation exposure risks, and risks of death by explosion could be placed in the same value unit at this stage of analysis. Considerable controversy exists about the feasibility of this. The aggregation of unlike impacts (the "apples and oranges" problem) is the source of most controversy in assessments in which valuation is included. For this reason, assigning values to environmental impacts is often considered to be more a political than a scientific task.

If it is felt that the valuation of environmental impacts should reflect the values and preferences of the individuals affected by the activity, then benefit-cost analysis may be appropriate. Another approach is to use values chosen by the decisionmaker who assigns, directly or indirectly, the weights to be used.

Another aspect of this question is the cost of performing the analysis. It is much more expensive to value impacts in terms of preferences of affected individuals than to use the preferences of decisionmakers.

An important issue is the value over time as some impacts may only be felt in the future, and thus require some form of discounting. It is generally believed that the present is more highly valued than the future. In addition to the psychological predisposition toward having something now rather than in the future, there is also the possibility of investing that reward in activities which generate larger returns in future. The valuation process must then include preferences about the timing of the benefits and costs of activities.

One aspect of time valuation which is especially difficult to handle satisfactorily is the potential impact on more than one generation, in terms of benefits and costs. Unborn generations cannot express their preferences about current decisions. Environmental issues that have intergenerational implications include the long-term storage of nuclear waste, depletable natural resources and biological diversity.

Documentation and Reproducibility

A key to understanding and evaluating a comparative assessment is documentation of data and methodology. Results cannot be reproduced if documentation is inadequate. Data sources, collection and aggregation procedures, and assumptions must be clearly identified. Documentation contributes to making a study transparent--a desirable if not essential property for its use in the policy process.

EVALUATION OF SOME RECENT COMPARATIVE ASSESSMENTS

Using the methodological criteria described in the previous section, a review was carried out for the OECD by Professor Holdren of the University of California in Berkeley, critically evaluating eight of the most important and representative of recent studies in the field of comparative environmental assessment.[6-13] The eight studies all reported, or themselves reviewed, quantitative findings, and some also surveyed methods. The review also traced the genealogy of data bases for comparing health and safety effects of energy technologies used in electricity generation. In fact, many studies are based to a substantial extent on a common and quite limited set of primary data.

With regard to the five methodological criteria described above, analysis of the chosen studies revealed some general points worth noting:

- Bounding the Problem: Though there was considerable variation in the choice of system boundaries, the studies examined did not fall into the "inconsistent boundary" trap--defining the boundary differently for different energy systems (e.g., including the environmental impacts of materials acquisition for one energy system but not for another).

- Disaggregation: Most of the studies used national average information, which can limit their usefulness, particularly as the the damages expected from specific emissions of air or water pollutants depend on factors such as dispersion processes and populations at risk, which vary from locality to locality.

  Temporal variation was a problem in several of the studies, e.g., where outdated data overestimate the future occupational health risk run by coal miners, or pollutant emissions factors do not accommodate a realistic degree of environmental control.

  Disaggregation of types of impact frequently posed problems. This was true particularly with regard to health risks, where public and occupational risks were combined to obtain "total" damage to health. Aggregating deaths while ignoring years of expected life lost was another, as was the combination of mortality, morbidity, and injuries into a single index of "working days lost."

- Uncertainty: Treatment of this issue was found to be generally unsatisfactory. Numbers were often cited to an unjustified degree of precision with no indication of their confidence limits. Where uncertainty was identified, it was sometimes unclear whether it was due to basic statistical variation, lack of knowledge, or to facility type, location, etc.

- Valuation: Few of the studies went as far as characterization of damages, let alone valuation, mainly due to the present state of knowledge about environmental processes. Attention focused on a small subset of impacts (basically human health) which has been judged relatively amenable to quantification. The emphasis on numbers has drawn disproportionate attention to those quantifiable impacts, at the expense of non-quantifiable categories of environmental damage which may be of greater importance.

- Documentation: In all the studies examined, some data were given, the origins of which were not clearly specified, and other information, although identified as to source, lost the assumptions that accompanied it originally. A part of the problem is hereditary, in that earlier work from which the data were taken was inadequately documented. However, generally the level of effort devoted by analysts to this area was lower than its importance would warrant.

The fact that these individual studies do not always meet the criteria is not surprising as they had different purposes and limited objectives. However, examination of how they measure up to the chosen methodological criteria can be of value both to the decisionmakers who need to judge the usefulness of their findings, and to future analytical efforts in the field of comparative environmental assessment.

THE TREATMENT OF HEALTH IMPACTS

When one speaks of comparing environmental impacts of different fuel cycles, the "bottom line" is frequently the impact on human health which is the most significant common denominator for mankind. This is the reason for the multiplicity of such studies.

The results of comparative health studies ought to be treated cautiously not only because of the uncertainty and biological variability of the data underlying them, but also because of the political and social sensitivity of public health issues. It is therefore fair to ask the following questions of studies of this nature:

- o Do they provide any clear conclusions, within acceptable limits of uncertainty, as to the comparative health risks of different sources of energy?

- o Does this type of study address the real political issues behind interfuel controversy or energy/environment conflict?

This paper discussed the cumulation of uncertainties at each stage of environmental impact analysis from emissions of residuals to valuation of impacts. Health risks are well down this chain of uncertainty, to the extent that not only the large range cited in studies for mortality or morbidity might be, and usually is, questioned, but in some instances the very existence of the specific risk (the nature of the dose/response relationship) may be queried by some experts in the field.

This is not to say that health impact analysis cannot be useful or should not be done. Similar criticisms may be applied to assessments of other major environmental issues. Health impact and risk analysis can be very useful in demonstrating major gaps in our knowledge where research could be fruitful and might be encouraged. It can also help to identify activities within a fuel cycle where better technologies or controls could diminish health hazards, and it can put the health impacts related to energy production and use into the perspective of the risks involved in other human activities. Indeed, the technique might be more usefully employed to encourage a balanced approach to energy development and rational use on the part of policymakers and the public, than to try to judge the relative merits of different fuel cycles from the point of view of their impact on human health.

ENERGY/ENVIRONMENT RISK ASSESSMENT AND POLICY: THE INTERFACE

The main conclusion to be drawn from the previous sections is that environmental risk assessment--while an inexact science and likely to remain so--can be a useful auxiliary tool for policymaking. It is one factor in a complex equation which contains many other interlocking factors: public opinion on the subject, the power and influence of the government department responsible for environmental policymaking, the strength of industrial lobbies affected by these policies, the general economic situation and energy prospects, etc. What risk assessment can do is to influence the so-called conventional wisdom on the topic in question, and modify the opinion of the public and of other parties to the problem.

It is worth noting that the role of risk assessment can also be tactical. For example, the inability to define the environmental risks clearly and unambiguously, as a consequence of scientific uncertainty, can be cited as grounds for taking no immediate steps to reduce these risks through strengthened environmental policies--even though there may be little hope that the range of uncertainty can be significantly reduced in the near future. In some instances this may be a sound policy stance; in others, it may not. The balancing of risks, costs, and benefits of various courses of action is a sound objective in principle, but its application can in some circumstances be counter-productive in the achievement of a better environment.

THE CASE OF CONTROL OF AIR POLLUTION FROM COAL COMBUSTION

This provides an interesting illustration of some of the observations made about risk assessment and policymaking. Several years ago the OECD undertook a major assessment of the costs and benefits of sulphur oxide control in the European region.[14] This of course involved not only combustion of coal, but of other fossil fuels as well. The results were not quantitatively conclusive, mainly because of the wide ranges of uncertainty. But the study was valuable in defining the problem more clearly and in pointing out which aspects needed further research and analysis, if policymakers were to be helped in deciding what measures to take to combat the effects of sulphur oxide pollution.

The OECD study showed that not only were the <u>effects</u> of sulphur oxides on the ecosystem imperfectly understood, but indeed so were the <u>costs</u> of various methods of control. Debate and controversy focused on the desulphurisation of fuel gases from fossil fuel combustion, the most effective means of sulphur oxide control (up to 95 percent of sulphur removal) but also the most costly of pollution control techniques, and one for which the technical reliability had been questioned as a result of some problems in its early applications.

Subsequent to the cost-benefit study, the OECD initiated an assessment of the costs of coal pollution abatement, it being evident that control costs were a significant factor influencing control strategies, and that the whole topic of costs had been, to some extent, shrouded in mystery, with estimates covering a very wide range. This work is now at an advanced stage, and has resulted in several OECD publications,[2-4] including the proceedings of a Symposium held in the Netherlands in 1982.[2]

Perhaps its most useful contribution has been to provide governments, and other interested parties, with a much clearer understanding of what technologies are available and commercially proven for coal pollution abatement, and what factors influence their costs. While not necessarily making pollution control cheaper, this has influenced the policy debate by removing one layer of opacity from a very comple: problem.

At the same time, the intense political interest in the acid deposition issue has stimulated a great deal of work in OECD countrie to try to better understand the effects of acid deposition. This worl is now beginning to bear fruit.** The situation has not yet reached the point where the risks, costs, and benefits can be brought togethe in a way in which unambiguous conclusions can be drawn--perhaps we ar still a long way from that. But the assessments made over the past few years, coupled with an intensification of public interest in acid deposition, are causing a definite shift in the political attitudes t this issue. Although there are still important links in the cause-effect chain missing, the circumstantial evidence that air pollution from fossil fuel combustion is transported over long distances in the atmosphere and can cause damage to forests, crops, lakes, and materia is mounting up. At the same time knowledge and experience in the field of control of emissions of the main air pollutants is evolving rapidly. The juxtaposition of these is putting the policymaker into better position to decide on the most appropriate action, even though such decisions are still made in conditions of relative uncertainty.

THE INTERFACE BETWEEN GOVERNMENT AND INDUSTRY

It is sometimes thought that the government/industry interaction is bound to be one of conflict. In fact, both governments and privat industries are heterogeneous entities, and their interactions cover the spectrum. As far as environment/energy risk assessment is concerned, the most important consideration is a shared perspective on what is known and what is not known about the problem, and a shared perspective on what can be done about it and what this costs. One very important precursor of this shared perspective is information dissemination and education of both governments and industry, preferably transcending national frontiers.

A very interesting example of this process of information dissemination was the work by the IEA's Coal Industry Advisory Board on "Coal Use and the Environment." This exercise included 31 case studies in 10 OECD countries covering a wide field of environmental protection during coal use. As well as making recommendations to IEA Governments and putting before the public a vast body of useful information and data in the IEA publication which resulted from the project,[5] industrial companies participating in the work gained much useful insight into pollution control techniques, regulatory practices, and government/industry interactions in other countries--all o which will influence their perspective on the government/industry interface in their own country, and perhaps help the process of consensus.

ENVIRONMENTAL RISK ASSESSMENT: A SCIENCE OR AN ART?

It will now be clear to the reader that a particularly brilliant piece of work in the field of risk assessment is not likely to come along and make policymakers see the light. They will probably continue to struggle to discern more of the broad features of the environment/energy problems emerging dimly from the fog of imperfect knowledge. The analysts should be encouraged to continue their efforts to dispel this fog, and in doing so will provide an important scientific contribution to the policy debate.

Perhaps some of the criticism which has been directed toward a number of environmental risk assessments in the energy field owes its origin to the failure to distinguish adequately between the task of the scientific or economic analyst and that of the decisionmaker. Any scientific analysis which incorporates a value judgment will inevitably be seen as trespassing in the policy field, perhaps rightly so, since the independent analyst cannot easily judge the different pressures on the policymaker faced with the need to make a difficult decision on environmental policy.

Where science ends, therefore, and art begins is in the interpretation of the state of knowledge for policymakers and the public. Here some degree of judgment is essential, not only to decide how the gaps in the knowledge should be handled, but also to insert the results of environmental assessment into the bigger picture which includes all the other influences on policymakers, so that the latter may decide whether or not the time is now ripe for action.

FOOTNOTES

1. "Environmental Effects of Energy Systems: The OECD COMPASS Project," OECD, Paris, 1983.
2. "Costs of Coal Pollution Abatement," edited by E. Rubin and I. M. Torrens, OECD, Paris, 1983.
3. "Coal: Environmental Issues and Remedies," OECD, Paris, 1983.
4. "Coal and Environmental Protection: Costs and Costing Methods," OECD, Paris, 1983.
5. "Coal Use and the Environment: A Report and Case Studies," by the Coal Industry Advisory Board of the International Energy Agency (IEA, Paris, 1983).
6. Comar, C. L. and L. A. Sagan. 1976. Health Effects of Energy Production and Conversion. Annual Review of Energy 1: 581-600.
7. Foell, W. K., ed. 1979. Management of Energy/Environment Systems: Methods and Case Studies. New York, NY: Wiley-Interscience.
8. National Academy of Sciences, Committee on Nuclear and Alternative Energy Systems. 1979. Energy in Transition 1985-2010. San Francisco: W. H. Freeman.

9. United Nations Environment Programme. 1979, 1980. The Environmental Impacts of Production and Use of Energy: Part I, Fossil Fuels; Part II, Nuclear Energy; Part III, Renewable Sources of Energy. Nairobi, Kenya: UNEP.
10. Cohen, A. V. and D. K. Pritchard. 1980. Comparative Risks of Electricity Production Systems: A Critical Survey of the Literature. Report of the United Kingdom Health and Safety Executive. London: Her Majesty's Stationery Office.
11. Hamilton, L. D., ed. 1974. The Health and Environmental Effects of Electricity Generation--A Preliminary Report. Report BEAG-HE/EE 12/74. Upton, NY: Brookhaven National Laboratory, Biomedical and Environmental Assessment Group.
12. Hamilton, L. D. 1980. Comparative Risks from Different Energy Systems: Evolution of the Methods of Studies. International Atomic Energy Agency Bulletin 22(5/6): 37-71 (October).
13. U.S. Department of Energy, Office of Environmental Assessments, 1981. Environmental Information Handbook: Volume 1, Energy Technologies and the Environment; Vol. 2, Technology Characterizations. Reports DOE/EP-0026 and DOE/EP-0028. Springfield, VA: National Technical Information Service.
14. "Costs and Benefits of Sulphur Oxide Control," OECD, Paris, 1981.
15. Proceedings of the 1982 Stockholm Conference on Acidification of the Environment, published by the Swedish Ministry of Agriculture, Stockholm, Sweden.

*The opinions expressed in this paper are those of the author and do not necessarily represent the views of the OECD or the governments of its Member countries.

**A recent major conference was held in Stockholm, Sweden, in June 1982, on Acidification of the Environment. The conclusions have now been published[15] and the work done by experts in that meeting provides a timely and professional review of the state-of-the-art in this field.

RISK MANAGEMENT IN HOUSEHOLD DETERGENT CONTROL

Tohru Morioka

Department of Environmental Engineering
Suita, Osaka, 565
Japan

INTRODUCTION

The use and discharge of synthetic household detergents have had two major impacts on our environment--an accumulating stimulus by excessive nutrients in the builder of detergent, and toxicological effects of surface active agents to organisms in the aquatic ecosystem.

Lake Biwa is the largest lake in Japan with a water volume of 27.5 km^3. Thirteen million people depend on the lake for their drinking water. In order to improve the trophic level and eliminate the odor problem in Lake Biwa, the locally authorized eutrophication control program, since 1980, has prohibited inhabitants in the lake watershed from the sale and consumption of synthetic detergents which contain phosphorus. The prohibition may be the most drastic regulatory measure for environmental preservation, but the administrative decision was scientifically supported by use of a deterministic ecological model of phosphorus cycle and a semi-quantitative macroscopic dose-response relationship between the discharged nutrient amount and the water quality level.

The grass-roots movement of citizens against environmental hazard had retained and circulated the principle to use soap for washing, which is recommended from the viewpoint of environmental policy. The synthetic detergent with zeolite instead of phosphorus compounds has begun to be widely used because of its easy handling and high solubility in low water temperatures.

Some experimental data in laboratory study on the effects of detergent use support the affirmative judgment of the ecotoxicological question, and others support the negative. Therefore, the decision-maker cannot draw any clear conclusion by using the similar deterministic approach as in the phosphorus regulation program. The potential risk of toxicological effects of surface active agents such as Linear Alkyl Benzen Sulfate (LAS) has increased by nearly 20 percent per year since 1980 and cannot be ignored from the viewpoint of environmental managment. It is important to find the intangible risks in the cycle of surface active agents and their aggregate effects rather than scientific reductionistic experiments.

Environmental Hazard and Risk of Surface Active Agents

The representative risk originated by discharging surface active agents in synthetic detergents is separated into the following two groups:

- Dermatitis and health effects through drinking water
- Environmental hazard and ecological negative effects

In this paper the latter is discussed, but the risk management model developed here will be applied to the elimination of the first risk by the appropriate collection and processing of experimental and epidemiological data.

The toxicological effects on biotic systems can be measured using the standard testing methods, for example in the OECD guidelines for testing of chemicals, in terms of degree of algal growth inhibition IC_{50}, the degree of Daphnia species reproduction, and the acute toxicity on fish LC_{50}. These standard procedures to evaluate toxicological effects suggest that we adopt an intensive systematic procedure in which organisms are exposed to chemicals in experiments or conceptual models.

The laboratory studies are not yet able to completely identify the long-term effects. We can hardly obtain sufficient information about the simple dose-response relationship in terms of LC_{50}, IC_{50} of species which are dominant or significant for the fishery in Lake Biwa.

Systems analysts can often estimate the succession or degradation of the specific biotic group in a given environmental condition using the model simulation approach and reported values of parameters without the large scale laboratory experiment or field survey as a particular case study. But the environmental manager has to decide on a future course of action in some way, even if perfect information could not be provided by ecotoxicologists. The risk of no action until the best control program is established should also be taken into account.

MEANING OF RISK AND IDENTIFICATION OF PROBABILISTIC PROCESS

Specification and Identification of Probabilistic Process

The control of use and fate of household detergents requires ecotoxicological impact assessment, fate analysis and evaluation, and consideration of social aspects. Table 1 shows research strategies and models as effective instruments for risk identification in three subsystems. "Detergent Discharging Process with Social Dimensions," implies the phases of production, distribution, consumption, and discharge which are described by the regional parameters and inhabitants' life style patterns. "Fate of Detergent including Transport, Transfer, and Transformation," is affected by the properties of a surface active agent and varies in different environmental conditions. "Ecotoxicological Process in Ambient Environment," is evaluated by judging the response of a specific biotic organism in a specific condition characterized by a set of DO, temperature, hardness, and other factors in water.

TABLE I. RESEARCH STRATEGY AND MODEL IDENTIFICATION

Sub-system	Research Strategy	Conceptual and Instrumental Models	Significant Uncertainty and Risk	less risk ←→ more risk
Detergent Discharging Process with Social Dimensions	questionnaire and interview survey on household detergent consumption and feasibility test of reduction or alternative use of detergent	original unit method with consideration of disaggregative factors such as life style pattern	preference of synthetic detergent to soap in each social group and consumed quantity, discharge intensity due to demographic condition	low population density ←→ densely inhabited preference of soap ←→ preference of synthetic detergent
Fate of Detergent including Transport, Transfer and Transformation	field survey on spacial distribution and fluctuation of chemicals and computor simulation	non-point source runoff model and other hydrological models with transport, transfer and transformation terms	dilution and peak moderation effect in specific water basin, degradation and sink after dischage in specific season	dilution or moderation ←→ concentrated easily decayed ←→ less biodegradable or more active
Ecotoxicological Process in Ambient Environment	summarizing laboratory data followed by modeling and forecasting of dose-response relationship	operational dose-response relationship model with probabilistic parameters	response of sensitive species or bio-indicators in environmental condition with additive stimuli	resistive biota ←→ sensitive bio-indicator counteraction or masking ←→ environmental condition with additive stimuli
Utility Measurement Stage	questionnaire survey on resources utilization and socio-economic study on resources allocation	multi-attribute utility function model	public awareness of real estate and amenity service of environmental resources	less concern ←→ intensive concern low-priced ←→ high-priced

Mortality of Biota and Change of Biomass for Evaluating Physical Environmental Impacts

The fundamental concept of the physical model for estimating the magnitude of undesirable effects to biota and damage or degradation of biotic resources are as follows:

- A mortality profile function in the standard condition in Figure 1-a represents a dose-response relationship without the threshold.
- The short-term effects and the long-term effects can be evaluated by using the lethal concentration LC_{50} and the ratio of decreased biomass without any dose to unit biomass, respectively.

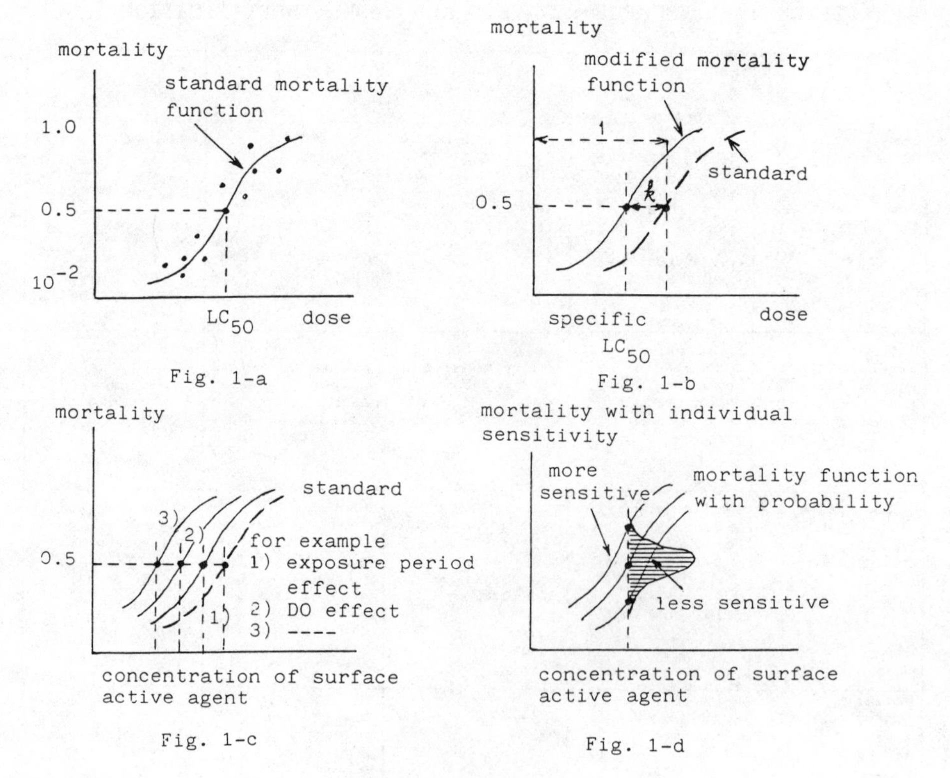

Figure 1 Dose Response Relationship

The following steps are designed and can be applied in detergent control in the Lake Biwa Basin:

- The biotic species, life stage, exposure period, temperature, and other water quality indices are identified as physical systems parameters, assuming that those parameters are independent of each other.

- The value of LC_{50} in each specific case is compared with that in the standard testing condition.

- The mortality function in each specific case is shifted parallel to the standard mortality function corresponding to the ratio of the specific LC_{50} value to the standard LC_{50} value as shown in Figure 1-b. The shifted length corresponds to the discount rate k which is discussed later.

- The LC_{50} value is modified in each biotic and environmental condition using linear additive formula for exposure period, biotic life stage, chemical structure in surface active agents, biotic species sensitivity, dissolved oxygen, temperature, and hardness as shown in Figure 1-c.

- The individual difference of organisms is reflected in the normal distribution around the mean value determined by the mortality function as shown in Figure 1-d.

- The mortality of young carp, which are chosen as the standard bio-indicator in the range of 0.2-0.5 ppm-LAS, is estimated to be 10^{-2}, which has been neglected in the traditional approach using LC_{50}. The value of LC_{50}, which means the dose intensity for the mortal probability of 0.5, is in the range of 2-3 ppm-LAS. These data are used for designing the standard mortality function.

- The biomass balance equations in differentiated life stages are then formulated, in which the survival rate in the same stage and the growth rate toward the next stage are involved.

A systems analyst still cannot deterministically identify features of physical environment and social regional characteristics because of the existence of probability and uncertainty in various indicators as systems parameters. Therefore, the decreased benefits or utilities related to detergent control as a final output caused by toxicological dose are expressed in a conceptual manner of integration with parameters α_i in the supply stage, β_j in the fate stage and γ_κ in the biotic reaction stage and δ_ℓ in the utility measurement stage as follows.

The relative decreased utilities =

$$\int I(\tau,\alpha_i,\beta_j,\gamma_\kappa,\delta_\ell)\cdot P(\tau,\alpha_i,\beta_j,\gamma_\kappa,\delta_\ell)\cdot U(\tau,\alpha_i,\beta_j,\gamma_\kappa,\delta_\ell)\, d\tau\cdot d\alpha_i\cdot d\beta_j\cdot d\gamma_\kappa\cdot d\delta_\ell$$

Here, I, P, U and τ denote environmental impact, probability of parametric event and utility of unit impact and sequential time, respectively.

This formula does not relate a real mathematical integral equation, but implies a conceptual convolution with probabilistic parameters in the serial impact process. An alternative integration can be formulated by the method of how to specify environmental impact in the receiving water. The key concept of impact stream is interpreted variously as follows:

- When the decisionmaking is possible by means of identification of the magnitude and range of deterioration of some bio-indicator in the ecological system, social parameters in the "Discharging Process" are taken into consideration and then the risk management is limited in the fate and ecotoxicological problem.

- When people in the basin have wide and uneven concerns and interests in biotic resources, the risk of marginal spacial distribution and variation of resources must be evaluated through the weight modulation process. In an extreme case, parameter δ_ℓ implies the economic alteration and the preference for utilization of biotic resources.

- The expression of risk and probability of parameters in the fate stage has a special significance, when the region under consideration has any of the non-uniform hydrological conditions, uneven load intensity of the chemical, and different water-related amenity service in subregions.

- The degradation of quality of environmental resources[1] in the utility stage must be expressed, including not only the foaming trouble or dermatitis caused by water contact, but also the decreased recreation opportunity, when assessing the environmental impact in the field of amenity services at water-front.

This report aims at: (1) establishing a model to estimate the change of biotic resources which are influenced by severe damage with low probability, (2) making an inventory of risk in the qualitative use of synthetic detergents, and (3) designing an orientation of social control of the chemical using the risk management technique.

SYSTEMS SIMULATION FOR RISK EVALUATION

Modeling Response of Nekton for Estimation of Fish Resources

The framework conceptually shown in Figure 2 illustrates the role of sub-systems in attaining the total objective. The function of each sub-system is formulated in the concise, consistent, and well-balanced manner as follows:

Standard Mortality Function. The standard mortality function is designed as a straight line on the normal probabilistic sheet in the standard condition, in which carp as an indicator organism are exposed for 24 hours to LAS(C=12) at $20^{o}C$ with 0 ppm-hardness and 8 ppm-DO. Figure 3 shows the standard mortality function which is revisable in each biotic and environmental condition.

Time Step for Model Simulation. An appropriate unit time step for calculation is a week in which fluctuated concentration of LAS in receiving streams is considered and response of biota may be aggregated quantitatively with respect to time. Judging from experimental data in different exposure periods, the discount rate k of LC_{50} in the exposed period beyond 24 hours is 0.03 - 0.13 for LAS.

$$LC_{50}(x \text{ days}) = LC_{50}(24 \text{ hours}) \cdot (1.0 - kx) \qquad \text{eq.(1)}$$

The additional effects of an exposure period longer than 24 hours are formulated by using the time-concentration index. The time-concentration index means that the additional effect in higher concentration of LAS is equivalent to one in the longer exposure by the time calculated as follows:

Mortality (x days less than a week)

$$= \sum_{i=1}^{x} \text{Mortality } (i\text{-th day}) \cdot (1-f(i)), \qquad \text{eq.(2)}$$

where f(i) represents the degree of acclimatization of fish to stimuli in i-th day.

The effect in a week is assumed to be independent of those in the subsequent weeks. Figure 4 shows the equivalent mortality in the time scale in less than a week. The idea of equation (2) is based on the hypothetical filling process in damage boxes of a particular fish as in Figure 5.

Life Cycle of Fish. At least four stages of fish life cycle should be distinguished in this simplified model. The first stage is reproduction, which includes spawning, navigation of spermatozoa, fertilization, and hatching; and the second, third, and fourth are fry, young, and adult fish, respectively.

Figure 6 shows the summarization of the relative response of each life stage of a fish.

The material balance equations are expressed as follows:

$$\begin{pmatrix} x(1) \\ x(2) \\ x(3) \\ x(4) \end{pmatrix}_{t=t+1} = \begin{pmatrix} \mu(1) & 0 & 0 & \lambda(4) \\ \lambda(1) & \mu(2) & 0 & 0 \\ 0 & \lambda(2) & \mu(3) & 0 \\ 0 & 0 & \lambda(3) & \mu(4) \end{pmatrix} \cdot \begin{pmatrix} x(1) \\ x(2) \\ x(3) \\ x(4) \end{pmatrix}_{t=t} \cdot \begin{pmatrix} 1 - m(1) \\ 1 - m(2) \\ 1 - m(3) \\ 1 - m(4) \end{pmatrix} \quad \text{eq.(3)}$$

Here, $\lambda(i)$ denotes the growth rate of unit mass of biota per week in the i th stage to the (i+1)th stage, and $\mu(i)$ denotes the survival rate of biota in the th stage which remain in the same stage per unit mass of biota in the i stage in a week in normal environment. Therefore $\lambda(4)$ means reproduction rate. The values of $\lambda(i)$ and $\mu(i)$ exceed 1.0 in some cases, because those rates are defined in the condition with no artificial chemical dose. The modified mortality based on the standard mortality function and its discounting is denoted as m(i).

Although the values of the growth rate and the survival rate vary over time in each water basin, a preliminary simulation with constant parameters or those with probabilistic distribution will give us important insights to help establish the model.

Environmental Effects on Mortality. The additional effects of environmental condition on the mortality are formulated as follows:

Dissolved oxygen effect--the discount rate k is 0.12/1 ppm-DO

$$LC_{50}(DO=x\ ppm) = LC_{50}(DO=8\ ppm) \times (1.0-k(8-x)) \quad \text{eq.(4)}$$

Hardness effect--the discount rate k is 0.01/1 ppm-hardness

$$LC_{50}(Hr=x\ ppm) = LC_{50}(Hr=0\ ppm \times (1.0-k\cdot x) \quad \text{eq.(5)}$$

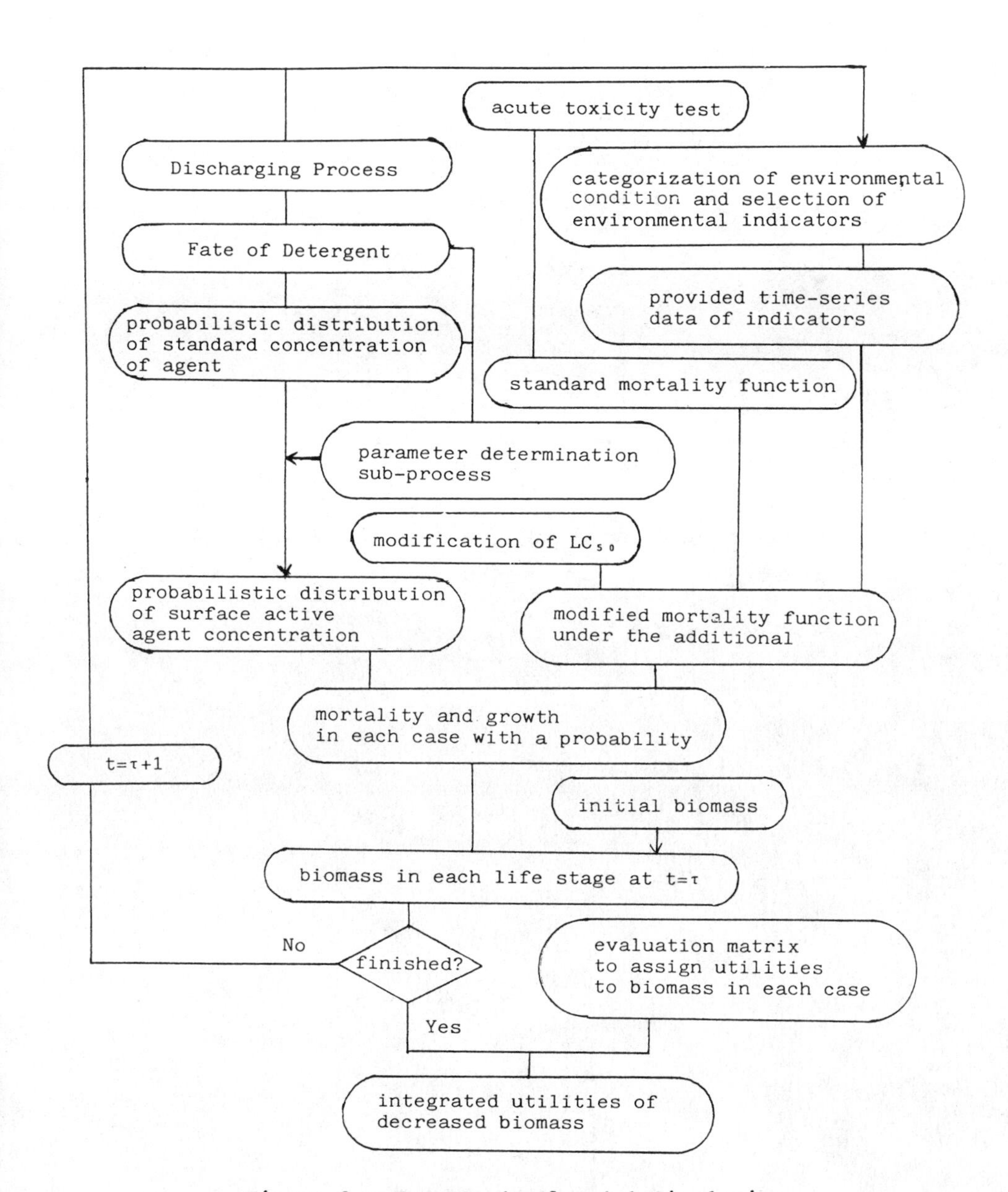

Figure 2. Framework of Model Simulation

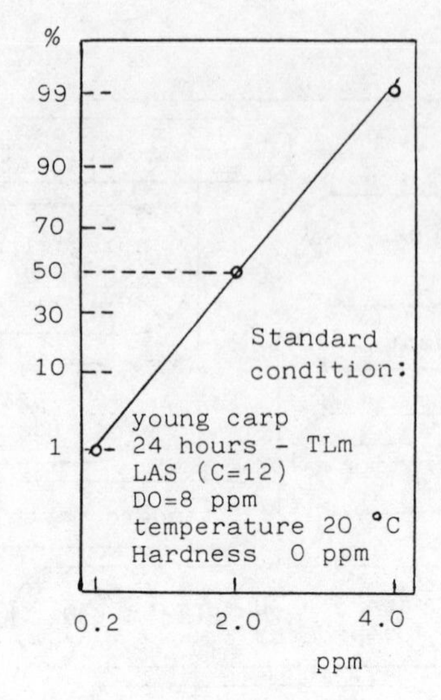

Figure 3. The Standard Mortality Function

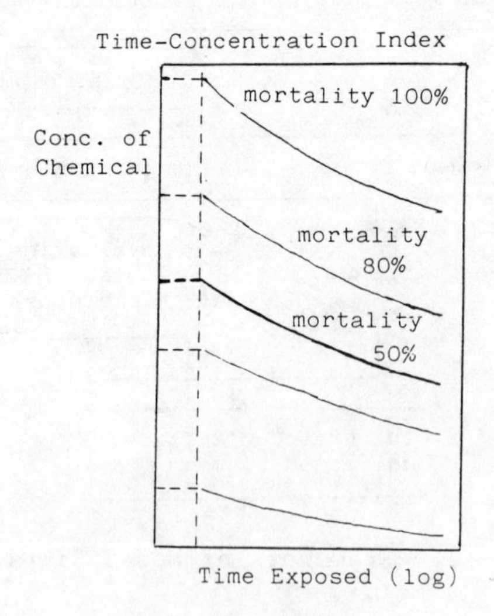

Figure 4 Equivalent Mortality in a Week

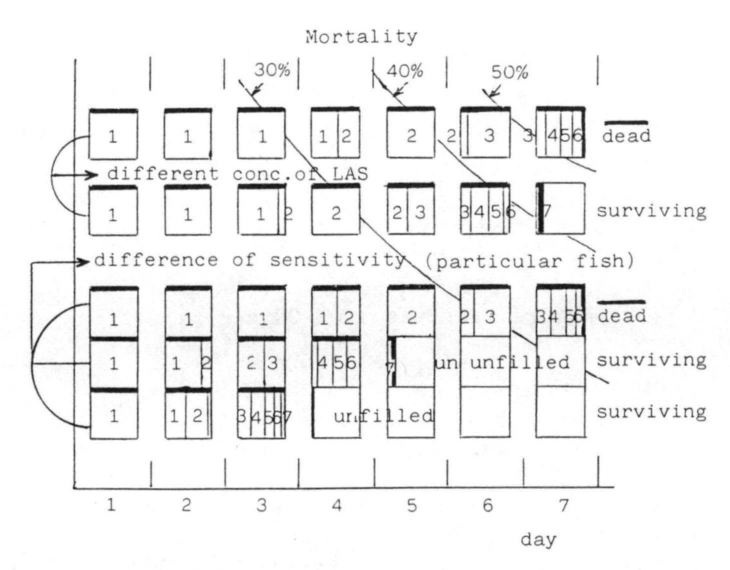

Figure 5. Hypothetical Filling Process in Damage Boxes

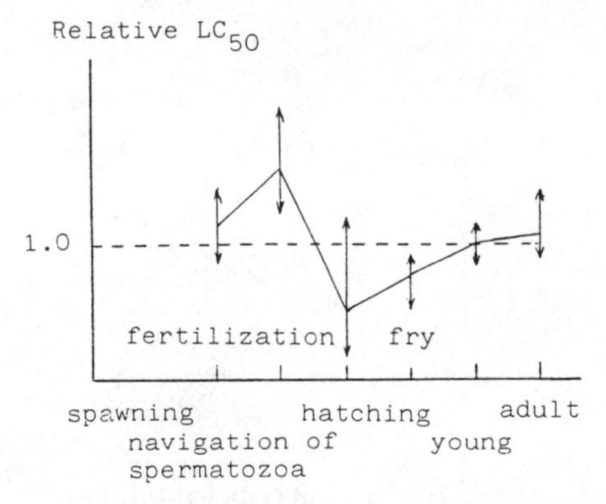

Figure 6 Relative Sensitivity of Each Life Cycle of Fish

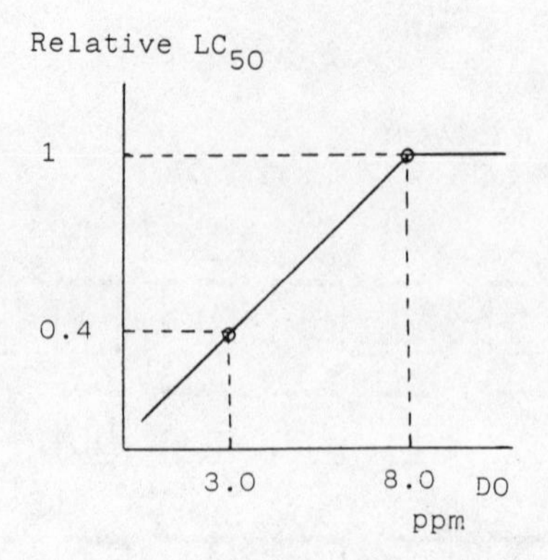

Figure 7. Dissolved Oxygen Effect

Temperature effects--the discount rate k is 0.02/1 ^{o}C

$$LC_{50}(Tm=x\ ^{o}C) = LC_{50}(Tm=20\ ^{o}C) \times (1.0-k(x-20)) \quad \text{eq.(6)}$$

These effects are illustrated in Figures 7, 8, and 9.

Species Sensitivity in Tolerance Level. The sweet-fish ayu (Plecoglossus altivelis) holds the first place in the fishery production in Lake Biwa. It is one of the fish most sensitive to the chemical dose in Lake Biwa, and has the large discount rate of LC_{50}. The tolerance levels of other fish are shown in Figure 10.

$$LC_{50}(i \text{ type}) = LC_{50}(\text{standard type}) \times (1-k) \quad \text{eq.(7)}$$

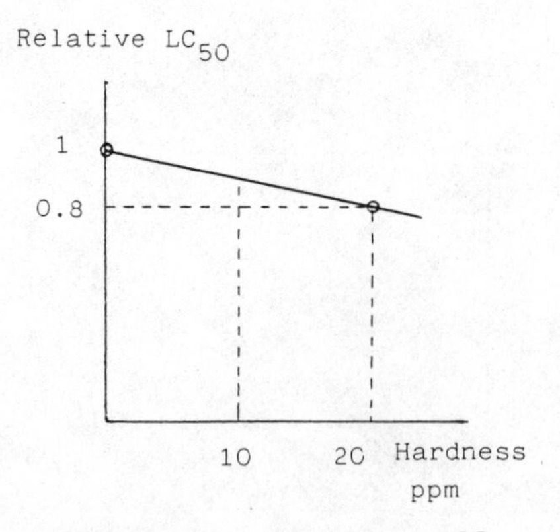

Figure 8. Hardness Effect

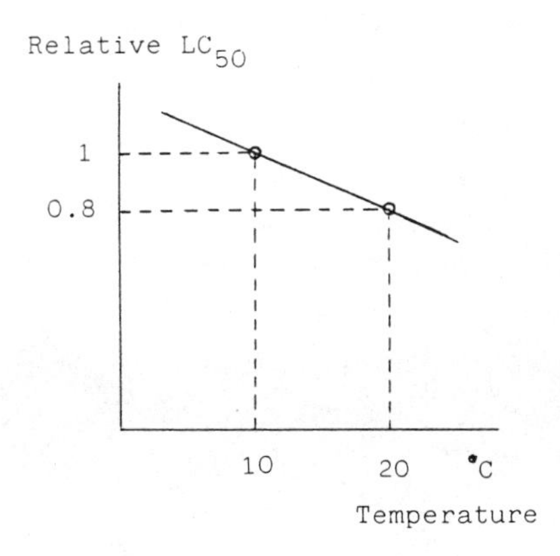

Figure 9. Temperature Effect

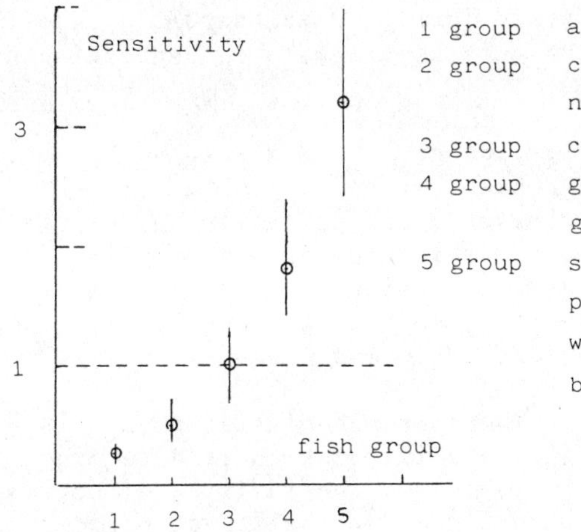

1 group ayu(plecoglossus altivelis)
2 group codfish, flatfish, common nibbler, tide pool goby
3 group carp, bluegill
4 group gold olfe, emerald, schner, goldfish
5 group small mouth bass, northern pike, fathead minnow, white sucker, common shiner, black bullhead

Figure 10. Sensitivity of Fishes to Surface Active Substances

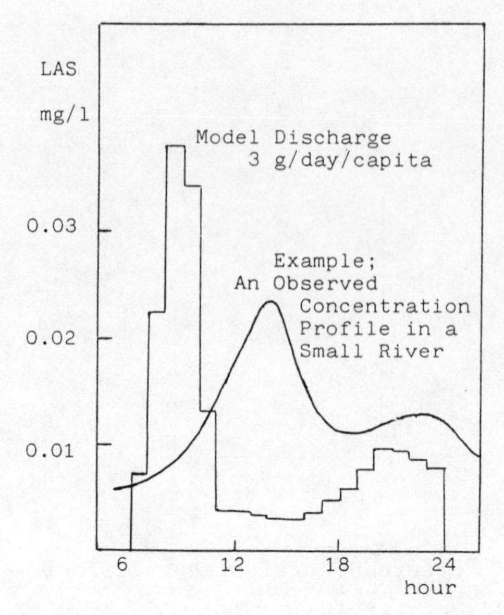

Figure 11. Typical Pattern of Fluctuation of Surface Active Substances

Expression of Fluctuated Concentration of Surface Active Agents. In general, daily concentration fluctuation of surface active substances in streams near discharged points shows two peaks as shown in Figure 11 in accordance with the use pattern of household detergent. A fate model which is valid to illustrate the distribution or spacial profile of surface active substances would involve a standardized runoff model modified by a set of parameters such as the mean concentration and the maximum/mean ratio.

Assuming that the degree of influence of LAS over time is determined only by the magnitude of mean concentration and the max./mean ratio, the mortality in each time step can be easily calculated quantitatively in each life stage. In this process, the max./min. ratio is used for the multiplication of the effects which are accelerated in high concentrations of the chemical; such as an acute action to branchiae causes the ulcerous degradation of mucosae.

Modified Mortality Function. The response of fish to LAS is quantified by use of the mortal probability which is determined by the additive effect matrix shown in Table 2. The additive effect matrix represents the range and the occurrence probability of parameters, which is expressed by means of the error function as in Figure 2-d, corresponding to the same value of the discount rate k.

Subdividing Receiving Streams. The basin under consideration is subdivided into zones which can be labeled by several regional parameters. The overall retention time reflects the scale of subdivided basins and determines the degradation or transformation of LAS in the stream. The ratio of population in each zone to the volume of receiving stream indicates the intensity of chemical dose over the zone. The other variables and parameters which have to be provided for systems simulation in each zone are: the initial quantity of fish in each life stage as resources for fishery and recreation, the aerial distribution of the discharges LAS, and so on.

The subdivision of the basin must also be based on the hydrological and geological zoning technique. The region labeled in the same zone in Table 3 have the same specific concentration level of LAS and the same hydrological type. The former ranges in $<$ 0.005 ppm, 0.005 - 0.05 ppm, 0.05 - 0.1 ppm, and 0.1 ppm $<$ and the latter, includes Lake Biwa water in the bulk sense, coastal water and environment in river mouth easily affected by detergent use activities, river water in down streams and rivulets, and brooks for rural irrigation and urban drainage.

Preliminary Model Simulation

The result of a preliminary model simulation is summarized as follows. The reasons why the final output of simulations are not introduced here are (1) the lack of even rough estimation of fish resources in each subregion, and (2) less reliability of a runoff model of the chemical. The relative comparison of simulated results in the simplified condition, however, provides a perspective for more adaptive model development and more relevant risk identification.

TABLE 2. ADDITIVE EFFECT MATRIX

discount rate k		20%	--	10%	-	0
DO	range	6.4ppm		7.2ppm		8ppm
	probability	8%/1ppm		10%/1ppm		20%/1ppm
hardness	range	20ppm		10ppm		0ppm
	probability	12%/1ppm		4%/1ppm		1%/1ppm
temperature	range	10°C		15°C		20°C
	probability	3%/1°C		5%/1°C		4%/1°C
----	----	---		---		---

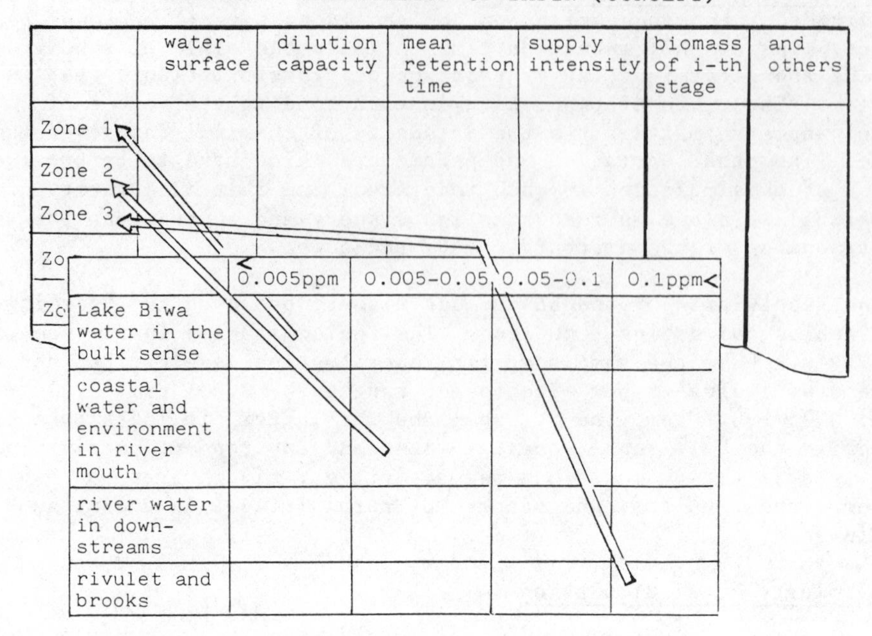

TABLE 3. SUBDIVISION OF BASIN (CONCEPT)

The change of fish resources simulated under the condition of the streamlined parameter setting is in a smaller range than expected. This is attributable to the fact that there are plenty of fish in the water which has sound environment throughout the year and where original contaminated influents have been diluted by a great deal of clean water in Lake Biwa. When the navigation of ayu upward in rivers is considered in the material balance equation (3), the severe damage of high concentrations of LAS in the river is faithfully reflected as the visible decrease of ayu in Lake Biwa in the behavior of the revised model.

There is a distinct difference among simulations with the mortal effects only beyond the threshold of stimulus, with the modified mortality function of annual mean environmental and/or biotic indices, and with the advanced mortality function of probability distribution with respect to parameters in the "discharging," fate," and "ecotoxicological" process. In the latter model the effects of LAS in an extreme case with low probability such as a phenomenon in the continuous dry weather are involved more exactly.

In the proposed model, all events in a gaming technique which determines environmental condition occur hypothetically in accordance with the normal distribution function. A more notable repercussion of risk would be detected, if the logarithmic normal distribution would be introduced in an advanced modeling, which had been often used in the hydrological modeling process.

In the "fate" process, the deviation around the mean of degradation rate is not yet separated from other systems parameters behind a macroscopic probability function, which are time-distance from discharged points, the dilution rate and so on. The transition of fish resources in each local sub-basin must be expressed in detail through the approach of differentiating those systems parameters in the "fate" process.

FURTHER DISCUSSION ON RISK MANAGEMENT

The comprehensive view to identify the environmental risk of synthetic surface active agents, especially one of LAS in the Lake Biwa basin, is introduced. A practical probabilistic model in the dose-response relationship, i.e., the ecotoxicological stage, is established for the risk evaluation in a future detergent control program. The basic structure and its dynamic behavior of the three-stage sequential model are revealed by the preliminary simulation.

Besides the physical modeling above described, the basic evaluation methods for risk analysis and risk management are discussed as follows:

- The elevated risk method has a little significance in the case that the index for monitoring and goal attainment in detergent control is designed in the manner of MBAS (methylene blue active substances) which contain those from natural lands.

- The comparative risk method gives a wide perspective toward a comprehensive modeling and assessment of potential environmental hazard originated by various household chemicals such as bleach and spray.

- The risk benefit method seems to be indispensable for detergent control because of the necessity of surface active detergents. However, the additional benefit and risk over ones for the traditional soap must be estimated and assessed.

- The balanced risk method is more important, when we prefer a surface active agent with the reasonable efficacy for washing, and relatively less risk.

- The decisionmaker directed his attention, in the first phase of detergent control, to biodegradability, a deterministic decay model and a simple choice among alternatives. However, in the next phase, we have to turn our attention to relative benefits, relative risk, and the public acceptance.

ACKNOWLEDGEMENTS

The author would like to express his appreciation to Professor T. Sueishi for the advice and staff of a problem-solving research group. This research group, which the author has taken part in, is executing a total research program that includes an intensive field survey on urban runoff and the total fate, a biodegradability test oriented toward estimating assimilation capacity.

FOOTNOTES

1. Environmental resources serve public recreation and amenity-receiving activities such as fishing, boating, swimming, collecting insects and plants, and so on.

SECTION 3: ETHICS AND VALUES IN RISK ANALYSIS

ETHICS AND VALUES IN RISK ANALYSIS: INTRODUCTION

Jeryl Mumpower

National Science Foundation
Washington, DC 20550

INTRODUCTION

Attempts to cope with risks from modern technological hazards run inescapably into thorny problems of ethics and values. These problems are broad in scope, ranging from fundamental questions of morality and justice, to questions of practical ethics and political theory, to questions concerning the proper role of the scientific enterprise in risk analysis and policymaking. Some of these problems are peculiarly endemic to risk analysis in the private sector, while others are generic to risk analysis in any setting.

Although the six papers in this section do not attempt to define or delimit the entire scope of the ethical and value issues in risk analysis, taken together they indicate the breadth and range of such issues and go a long way toward dispelling any notion that risk analysis can be conducted in a philosophical or ethical void. Any effort to cope with risks must inevitably also attempt to cope with questions concerning what is good or bad and what is desirable or undesirable. Such questions are inevitable because risk analysis is at heart concerned with human health, welfare, and safety. Because risk analysis cannot be divorced from these fundamental human concerns, it cannot be divorced from questions of ethics and values.

Each of the papers in this section makes a unique contribution to the study of ethics and values in risk analysis.

In his previous work, May has examined the special obligations and responsibilities associated with the use of cost-benefit analysis by for-profit firms in the private sector. In the present paper, May extends his analysis to those situations in which an individual user or consumer of a product may prevent or reduce injury or harm by taking appropriate action. The paper focuses on fundamental questions concerning the assignment of responsibility and liability in the area of product liability.

In the second paper, Ravetz and Funtowicz describe three types of activities that are ordinarily referred to by the common label "risk assessment." These range from "applied science" when the degree of uncertainty and the stakes are both low, to "technical consultancy" when both are moderate, to "total-environment assessment" when both factors are high. As Ravetz and Funtowicz emphasize, each of these different types of risk assessment gives rise to somewhat different types of ethical and value problems.

Jasanoff examines the allocation of risk assessment and management functions between the public and private sectors. She finds significant differences in the pattern of allocation in Europe versus the United States, differences which she attributes primarily to different social, political, and ethical values regarding (a) the role of scientific expertise in dealing with technical uncertainty; (b) public participation; and (c) accountability.

In the fourth paper, Sterling and Arundel examine how risk analyses may be inappropriately influenced or biased by interests and values associated with the practical consequences flowing from such analyses. They discuss three examples in which either relevant information is omitted or the validity of unwanted observations is denied, and consider the question of whether some method is required to hold scientists and other experts accountable for inadequate, or incompetent, performance.

The paper by Derr, Goble, Kaspersen, and Kates examines the differentials that frequently exist between the level of protection afforded the general public for a technological hazard and the protection afforded workers for the same hazard. The scope, extent, and health toll of such differentials are reviewed, and the authors critically discuss four potential arguments in justification of such a double standard. For the most part these justifications are found wanting, and the authors suggest guidelines for rectifying what they argue is a morally unacceptable situation.

Finally, Cox, O'Leary, and Strickland describe the views on risk assessment of a major representative of the private sector, the Chemical Manufacturers Association (CMA). CMA's policy on regulatory impact analysis of health, safety, and environmental regulations is reported, and the authors discuss some of the key questions and value issues that influenced the adoption of this policy position. This paper makes a particularly fitting closing piece to this section, explicitly tying together ethical and value issues in risk analysis with the overall theme of the volume, risk analysis in the private sector.

PRODUCT LIABILITY: WHEN DOES THE INDIVIDUAL ASSUME THE RISK?

William W. May

Program in Business Ethics
University of Southern California

THE PROBLEM

When, on what grounds, and to what extent do we or should we hold individuals responsible for their own actions in product liability cases? How do we separate cases in which the victim could not have exercised any control over the harm or injury suffered from those in which the victim could have avoided or minimized the harm? How do we assign degrees of responsibility and consequent liability?

In June, 1980, Richard Hogard, a pipe insulation worker at the Long Beach, California, Naval Ship Yard won a judgment of $1,200,000, subsequently reduced to $250,000, against Johns-Manville and Raybestos-Manhattan, two manufacturers of asbestos materials.[1] About 18 months ago, Manville Corporation (formerly Johns-Manville) declared bankruptcy under Chapter 11. Manville argued that although it is financially healthy now, it could not continue to be so due to continued legal exposure to asbestos cases, which could potentially number in the thousands.

Manville's action and Hogard's recovery point to two social goods that come into conflict in product liability cases: continued economic viability of manufacturing firms (with the attendant jobs, return to investors, contribution to taxes, etc.) on one hand, versus protection of the health and welfare of consumers and users of products on the other hand.

The focus in this paper is on those situations in which the individual consumer/user may prevent or reduce injury or harm by appropriate actions. In an earlier paper, the writer used the Ford Pinto case to argue that for-profit firms have a special obligation and responsibility in the use of Cost-Benefit Analysis and that the entire burden rightfully fell on the manufacturer. Put simply, the argument there was (1) that all of the significant knowledge about, and control of, fuel systems or other complex technological design was in the hands of the manufacturer and (2) that all of the benefit (profits) went to the manufacturer while all of the costs (death and injuries) went to the consumer. All choices were made by the manufacturer. Thus, according to ethical analysis, the use of CBA by for-profit firms puts special burdens and responsibility on the firms.

In this paper, I want to address the situation in which the victim, or potential victim, can control, at least to some extent, whether injury or harm occurs. The Hogard case is an interesting example to raise the question about individual responsibility for several reasons:

- Hogard began work at the shipyard in 1965, which was after the presentation and publication of the major study on shipyard workers and asbestos-related disease. The importance of this is that the manufacturer, Navy, shipyard, and union personnel were aware of the significant hazards.

- At the Long Beach shipyard, the industrial hygienist offered periodic classes, supposedly to include all workers, on asbestos hazards and safety precautions to take. In addition, the union encouraged spread of information to workers and began its own educational efforts in the late 1950s.

- Hogard, by his own admission, failed to use a respirator much or all of the time despite regulations to the contrary. (Rejection of the respirators was common among the workers.)

- Hogard, a long-time cigarette smoker, continued to smoke between a pack or two a day despite the information that smokers had a much higher incidence of asbestosis.

- Hogard received hazard pay for working with asbestos, which underscores his knowledge of the dangers.

Despite these factors, Hogard won the quarter of a million dollar judgment, plus additional amounts from other defendant companies which settled out of court.

The case raises basic policy questions about the law and the assessment of liability; it also raises important questions for manufacturers of dangerous or potentially dangerous products, with important social implications as well. What kind of costs should be factored into a CBA calculation to determine whether to proceed with a product? As a society, how much protection can we provide consumer/users without seriously jeopardizing our economic health? How much responsibility should we demand from individual consumer/users?

I think that we need to distinguish among several different kinds of situations, or, more precisely, extend and refine the kinds of distinctions we now make. We can envision a continuum from no individual responsibility to complete individual responsibility. At the no individual responsibility end are products like automobiles. We cannot, and should not, expect individual consumers (purchasers and passengers) to assess the quality of complex technology like fuel, brake, and steering systems. This end of the continuum would include prescription drugs, surgery and much medical treatment, and products and services which require professional training and certification for manufacture and/or administration of the product. The other end of the continuum where total responsibility falls on the individual is harder to illustrate at the extreme, but is approached in such things as mountain climbing, skydiving, and scuba diving where risk is great and substantial responsibility for use and maintenance of equipment falls on the user. Perhaps the best example is that of cigarette smoking in the post-Surgeon General warning days.

The problem cases fall in between the polar points of the continuum and raise the most interesting questions for us. The case of the employee who could, but doesn't, use available safety or preventative measures is one such case. Another type of case is that of the misuse of a potentially dangerous product, such as a power saw with a safety guard. One could also point to negligent use of a product, such as an automobile or prescription drug, where the burden is primarily on the manufacturer. One dramatic example of this type of situation, and one that points to the need for policy change, was the payment by General Motors of several hundred thousand dollars to survivors of the victim of a rear end collision. In this case the driver, while inebriated, stopped and fell asleep in the middle of the freeway. The vehicle was struck at high speed by a van, causing a ruptured fuel tank and fire. On what rational grounds should the manufacturer be culpable?

To be sure, the law tries to deal with the basic issue of allocation of responsibility, in either case or statutory law, in a variety of ways including duty, breach of duty, assumption of risk, contributory negligence, indemnity, comparative fault, comparative negligence, proximate cause, foreseeable harm, last clear chance, and causation. Nonetheless, in civil law the burden tends to fall most heavily on manufacturers, especially in the last two decades since strict liability has become the norm.

The prevailing pattern today is one in which public policy as established through case law seems to support holding manufacturers liable in almost all cases on two grounds: (1) they are more likely to have the resources to insure that victims are compensated (the deep pocket), and (2) they are best equipped, financially and with knowledge, to prevent or correct a problem. Clearly, this results in spreading the costs to consumers generally through higher prices, so "society" in the form of consumers plays the role of insurer for victims.

The economic implications of present policy, as illustrated by the asbestos cases, are substantial either in terms of increased product cost, removal from the market, or a decision not to introduce a potentially valuable product. If public policy were such that manufacturers of potentially dangerous substances would know that claims against them would be stopped or reduced if preventive measures were not used, product decisions and costs would be affected, perhaps drastically. RBA or CBA should be a useful tool in refining such decisions.

The suggestion that individuals should be held responsible for product use in employment situations similar to the shipyards and asbestos is subject to criticism on several grounds: (1) if manufacturers are absolved of liability or if exposure to liability is substantially reduced, then preventive measures and research on safety related procedures will be reduced or eliminated; (2) that employees are under substantial pressure to take risks in the workplace at the threat of losing their jobs; (3) that individuals who have suffered serious and real injury will not be compensated; and (4) that it is far too difficult and too expensive to try to establish allocation formulae for various substances and products.

These criticisms are weighty and deserve careful consideration. Any effort to move a large portion of responsibility and liability to the individual, may quickly be attacked as an attempt to blame the victim. In the extreme, it would be likened to the judge who blames the rape victim for wearing suggestive or revealing clothes.

The substantive counter to the criticisms is to focus on the concept of individual responsibility for decisions and actions in those situations where the individual can freely exercise control over measures that prevent or reduce injury or harm. Autonomy is very highly valued in our society. Individuals should be able, we say, to make or participate in decisions that affect their lives. Indeed, autonomy and participation were important components of the argument the writer made in assessing the use of CBA by for-profit firms. The very lack of autonomy and participation strengthened the argument that a higher degree of obligation and responsibility accrued to firms than is the case in either decision by government (proxy participation) or by individuals in undertaking risky activities.

If autonomy is to be used to argue for individuals in affording protection against firms when autonomy cannot be exercised, then the obverse is true and firms (manufacturers) should be able to expect individuals to exercise responsible choice within the parameters of action open to them.

Hogard and the recipients of Pinto and Firestone 500 recall notices, were able to choose actions that would mitigate or remove hazardous conditions admittedly caused by others. The possibility of individual choice and action certainly should not be used to remove all responsibility and liability from Manville, Ford, Firestone, or any other manufacturer. It is my contention that the possibility of individual choice and action should play a much larger role in determining the extent of liability by manufacturers. Moreover, such limitation must be known prior to decisions being taken by manufacturers where that is possible.

THE SHORTCOMINGS OF RISK ANALYSIS

Risk analysis literature tends to ignore or avoid the question of individual responsibility, especially in product liability cases. Indeed, risk assessors tend to focus on protecting the public. In one sense, this is not surprising, since risk and cost benefit analysis are based on aggregates with unknown victims. The literature on determining the value of a human life, likewise, deals with statistical deaths and necessarily with average values for all people or certain classes of people if the potential victims can be easily categorized. The closest thing that I have found to factoring in individual intention or motive (which could be a basis for considering willingness and choice when consumer/use volition is possible) are studies such as Thaler and Rosen or Smith in dealing with safety and job market choices.

The lack of attention to individual responsibility in certain types of situations where RBA or CBA is to be used strikes me as a significant missing component in that holding or not holding individuals partially or wholly responsible has a tremendous impact on the value of various factors in a CBA equation. Asbestos has been replaced by other materials today, but the problems with asbestos and many substances are the same. In the case of asbestos, a variety of measures were available to clean the work site up to a point, but mandatory use of respirators, non-smoking, and other measures were essential to reducing, if not eliminating, asbestosis and mesothelioma.

How can risk analysis help resolve product safety cases, and what role is there for the risk analyst? An example of the kind of thing that risk analysts can do is exemplified by Posner in his academic and judicial writings. In his _Journal of Legal Studies_ article on strict liability, he argues that the supports given for the strict liability approach failed to analyze the economic consequences of that principle correctly. The Manville Chapter 11 bankruptcy action may bear him out. In any event, Posner has applied his ideas in his work as a United States Appeals Court Justice.

In *United States Fidelity and Guaranty vs. Plovidba*, a 1982 United States Circuit Court of Appeals Case, Posner applied risk benefit analysis in arriving at a decision. He uses Judge Learned Hand's formula from an earlier case. The decision says, "Under formula for determining negligence in maritime cases, a shipowner or other alleged tortfeasor is negligent if the burden of precautions was less than the harm if the accident occurred multiplied by the probability that it would occur; the higher the probability that if the precautions were not taken the accidents would occur and magnitude of the loss if the accident occurred, and the lower the burden of the precautions necessary to avert the accident is, the likelier is a finding of negligence

Hand's formula, simply put, is negligent if $B < PL$, where B = burden of precaution, P = probability of accident without precaution, and L = Loss. Posner comments that although mathematical, the formula is imprecise because B, P, and L need to be quantified. Nonetheless, he still holds this to be "a valuable aid to clear thinking."

The point that I want to raise is that I think that this kind of thinking can and should be extended to RBA and CBA assessments in order to build in a factor for the precaution of the user when the injury/loss is (to some large degree) in his/her hands.

If some such approach had been available to Manville, either in its distribution of asbestos to the United States Navy in its shipyards or as a defense in court suits, especially by relatively recent workers such as Hogard, the number of cases would be vastly reduced and Chapter 11 might have been avoided. My point here is that risk analysis should be an available tool to use in cases involving large numbers of individual users, as employees or consumers, who, through the use of available preventive measures, can substantially reduce the incidents of accident or disease and the resultant legal exposure of the manufacturer. In this approach, risk analysis would be used as a way to counterbalance the present tendency of civil law to move entirely to holding manufacturers liable, seemingly regardless of possibilities of the intervention by the affected individual in some cases.

WHAT IS NEEDED

Again, the area of products liability presents us with two sometimes conflicting social goals: protecting individuals against harm and protecting the economy against undue burden.

What is needed is the development of risk assessment that will account for individual responsibility as well as manufacturer liability. Such assessment needs to be incorporated into risk benefit or cost benefit analysis used in production decisions. Present legal theory and case law focuses too heavily on after the fact recovery and individual claims for compensation. If it could be established that failure to use available safety measures, such as respirators in the case of asbestos, would be a bar to recovery for avoidable injury, individuals might take more responsibility for their own welfare and there would be economic benefits.

Some states, indeed some nations, have mandated seat belt usage for drivers. California is one state that has recently passed child safety seat and restraint legislation. Violators may be subject to fine and, in some cases, imprisonment. Some insurance companies offer reduced premiums for non-drinking, non-smoking policy holders. These examples demonstrate that both penalties and rewards can be attached to individual actions in large-scale calculations.

There are obvious problems in establishing the kind of risk assessment being urged here. Chief among the problems would be establishing which safety measures are reliable, what proportion of responsibility to attach to individuals, and how to protect workers and consumers against exploitation by manufacturers.

The development of strict liability in tort law as public policy has strong historical roots. My contention is that it needs some counterbalancing by putting more weight on individual responsibility, and that such accounting needs to occur before the fact to be of economic benefit. We need to provide a mechanism where Pinto-like victims will be protected while Hogard-like plaintiffs will recover less or nothing at all. Posner presents one example of the application of risk assessment to the problem. It is an example that needs more attention and development by risk assessors.

FOOTNOTES

1. Richard Hogard vs. Johns-Manville and Raybestos Manhattan, Superior Court, Los Angeles, June, 1980.

   The Manville case is complicated by a number of factors extraneous to the issue being discussed. One major item is the whole history of the asbestos industry in the United States and the extent to which information about asbestos-related disease was suppressed or distorted. That is a serious issue for the World War II shipyard workers, who make up the bulk of the victims and plaintiffs, as well as for workers in the 1950s. Hogard is interesting and distinctive because he began work after the 1964 Selikoff report which established the hazard to shipyard workers for the first time. The unwillingness of the U.S. Navy and U.S. Government to assume any liability for the asbestos victims, despite the control that they exercised over the Navy yards is a major factor in the Chaper 11 decision by Manville. Manville has sued the federal government over this issue.

THREE TYPES OF RISK ASSESSMENT: A METHODOLOGICAL ANALYSIS

Silvio O. Funtowicz
Jerome R. Ravetz

Department of Philosophy
The University
Leeds LS2 9JT
United Kingdom

INTRODUCTION

The study of risks and the impact of new technological systems in our society and environment is now accepted as a legitimate subject of research.

As a consequence of the dual character of technology--a benefit, but also a threat--policy debates on innovations in technology are increasingly focused on "risk." This is seen most clearly in the case of the nuclear industry, where long and expensive enquiries are now necessary before new installations are permitted (as is currently the case with Sizewell B in the U.K.).

The issues are frequently divisive, involving heavy costs, both economic and political. In such debates, the established methodology of industrial risk assessment (or "Applied Science Methodology") is stretched beyond its limits of applicability and effectiveness. The techniques whereby a particular installation or process can be analyzed and its hazards identified and logically displayed, do not extend to systems which are novel, large, complex, and few in number. The hazards and environmental impact of these systems are incapable of description, measurement, or forecast which the precision experts have, until a few years ago, assumed to be possible and necessary for decisionmaking.

In these new circumstances, quantitative statements about risks and safety, which by their form seem the product of a rigorous scientific methodology and mathematical techniques, are frequently challenged as dubious or biased, and can sometimes be proved totally erroneous.

Now it is not only militant environmentalist groups that challenge the scientific character of risk assessment. Leading experts and theoreticians echo their doubts; the result is a severe loss of confidence in the whole field. The optimism of earlier years has given way to confusion and pessimism among many experts. This new mood may make it more difficult to achieve those genuine results of which the field is capable. This work is intended to remedy this situation by identifying the different types of practice in risk assessment, with their characteristic limitations and strengths.

STATE OF THE ART

Because of the lack of an accepted methodology that is both successful and reliable, almost all research work published in the theory of Risk Analysis is about "Foundations." Basic terms such as "probability" and "risk" are argued everywhere (and it is difficult t find the same definitions twice); consequently, the estimation techniques differ considerably. Mandl and Lathrop provide an example of this problem in the context of Liquefied Energy Gas Terminals.[1]

> "The best way to develop a definition of risk is to start by quoting some of the definitions from the risk assessment literature.
>
> SAI-USA: "Risk is the expected number of fatalities per year resulting from the consequences of an accidental event."
>
> CREM-UK: "Risk is the probability of an injurious or destructive event, generated by a hazard, over a specified period of time."
>
> BATTE2-OTH: "Group risk is defined as 'the frequency at which certain numbers of acute fatalities are expected from a single accident'. The risk to society as a whole is defined as 'the expected total numbers of acute fatalities per year resulting from accidental events in the system'."

Although they emphasize the differences among the measures, we may note that three different terms for "probability" are used, each implying its own underlying theory of the measurement.

The optimistic expectations, that sophisticated probabilistic techniques would solve the problem,[2] have not been fulfilled; on the contrary, the experience of the last years has been of increasing uncertainty and debate. The concept of risk in terms of probability has proved to be so elusive, and statistical inference so problematic, that many experts in the field have recently either lost hope of finding a scientific solution or lost faith in Risk Analysis as a tool for decisionmaking. A recent authoritative report by an American committee comments:[3]

> While, as emphasized in the first part of this report, analysis has a limited role in decisionmaking, it still can be a powerful tool for categorizing risks, assessing them, evaluating a given set of alternative coping strategies, and devising new alternatives. The actual role of risk analysis in particular situations is variable, as are the methods used. There is also the fact that in some decisionmaking on risk, analysis has no role; in other situations analysis is useful only if it is designed so that it provides information that the decisionmaker cannot obtain.

These confusions and disappointments are familiar to many experts in the field; it is enough for us to indicate them briefly. One recent theoretical response to them was an advocacy of "Bayesian statistics," where the "subjective probabilities" based on the professional experience of experts are assigned to those hypothetical events which are incapable of effective empirical study. These are then (in theory) modified by a formal calculus, when experience changes.[4] But the variability of such expert judgments[5] and the inadequacy of the formalism in the case of very-low-probability judgments (which are the ones of concern) render this approach less fruitful than was originally hoped. Further, there is the discouraging experience that the serious accidents that do occur have tended to be of the sort where the mathematical theory of risks is irrelevant (Seveso), inadequate (Three Mile Island), or misleading (Salem, New Jersey).

Another response to this crisis in risk assessment has been to re-emphasize the "human" and managerial aspects of the creation of hazards. However, there has also been a tendency to a multiplication of traditional efforts in formal studies of risks. Both are represented in a comment by the _New York Times_:[6]

> The (Nuclear Regulatory) commission has two ways of analyzing safety--field reports of utility performance and probability studies of risk. Recently, it has been emphasizing risk assessment, which predicted the automatic scram system would fail once in 33,000 reactor years--hardly a red flag for the problem. But the Salem plant had been cited three times for below-average management. Risk

assessment may help compare risks; for real-world accidents, empirical evidence would seem a clearer guide.

Nuclear power is essential to an energy policy that seeks diverse sources. But the public will not respond in measured fashion to another core damage event; it may simply reach for the ax. The safety margin that prevented disaster at Salem is a thread too thin for a whole industry to depend on.

The same point is put even more forcefully in a leader in the New Scientist of London, a journal generally committed to the cause of civil nuclear power. Under the heading, "Nuclear Incompetence," it says:

Perhaps the best thing that could happen to the world's nuclear power industry would be the closure of every nuclear power plant in the United States. Then we might see an end to the series of stupid mistakes that cast a shadow over nuclear power in other countries.

Is it any wonder that some level-headed Americans aren't too keen on nuclear power, especially if it is in the hands of tin-pot utilities that cannot even keep track of their mail? ...We can only hope that the U.S. does not drag every one down with it in a complete meltdown of nuclear power's already shaky credibility.[7]

Such reflections give rise to renewed consideration of the cultural aspects of risk management. While it is only common sense to be cautious about installing nuclear reactors in such places as Zaire or Belfast, it is a sobering thought that the United States of America may not be an appropriate milieu for what was, until recently, accepted as a basic, universal energy technology. For an understanding of this phenomenon we need a theory of regulation, which is beyond the scope of this present study.[8]

Thus the "art" of the assessment of risks of complex technological systems is at an impasse. The early hopes that it could be reduced to a science are frustrated. And the crisis of confidence among the public in the assessment of such risks inevitably extends to the institutions that regulate the industries and also to the industries themselves. Although some individuals and agencies redouble their efforts in the old ways, others are tending to introduce the "human" and "cultural" factors. The question now becomes, to what extent should these predominate? Would it be to the reduction or exclusion of the "scientific" aspects? For, as we shall see, if the perceived phenomena of "risks" are interpreted as lacking all objective content or being merely a small part of some total cultural configuration, then there is no basis for dialogue between opposed positions on such problems. For this reason alone, such "relativist" arguments must be taken seriously. We analyze them in the next sections.

THE SUBJECTIVIST FALLACY

The very first issue of the first journal to be devoted to Risk Analysis contained as its main theoretical contribution a paper on the quantitative definition of risk. Appearing in such a position, the paper inevitably carries the authority of the community of experts who are represented by this journal. As a philosophical contribution, it therefore has an importance for policy debates which does not depend on the scholarly credentials of its authors. We therefore shall discuss it by the criteria of philosophical analysis, although the authors themselves may not have intended it as a contribution to a dialogue on the metaphysics of risks.

By "radical subjectivism" we mean the position which goes beyond the use of subjective probabilities and Bayesian statistics. Its proponents believe that there is a need to justify philosophically the use of subjective techniques, and they think that the justification would consist in showing that risk is a subjective phenomenon. An example is provided by Kaplan and Garrick.[9]

> Connected to this thought is the idea that risk is relative to the observer. We had a case in Los Angeles recently that illustrates this idea. Some people put a rattlesnake in a man's mailbox. Now if you had asked that man: "Is it a risk to put you hand in your mailbox?" He would have said, "Of course not". We, however, knowing about the snake, would say it is very risky indeed.
>
> Thus risk is relative to the observer. It is a subjective thing--it depends upon who is looking. Some writers refer to this fact by using the phrase "perceived risk". The problem with the phrase is that it suggests the existence of some other kind of risk--other than perceived. It suggests the existence of an "absolute risk". However, under attempts to pin it down, the notion of absolute risk always ends up being somebody else's perceived risk. This brings us in touch with some fairly deep philosophical matters, which incidentally are reminiscent of those raised in Einstein's theory of the relativity of space and time.

Their argument can be summarized as follows:

- o Example, then Risk is relative to the observer.
- o Risk is relative to the observer, then Risk is subjective.

Let us consider the first argument. It is clear that the letter box being a hazard depends on the presence of a snake (unless everything is a hazard, making trivial the concept of hazard). Putting a hand in a letterbox will be risky given the presence of a snake; information about the presence of a snake will change the judgment abou the situation and will, consequently, produce different behaviors. The letterbox with a snake is "risky" independently of the informatio possessed by a particular observer. What is dependent on the informa tion available to a particular observer, given that his attention has been drawn to a particular situation, is his judgment and subsequent behavior, not the situation itself. The snake is in the letterbox, and not in the mind of the victim. Thus, this particular example doe nothing to establish "relativity" in the notion of risk, at least wit the meaning that Kaplan and Garrick assign to "relative."

A simple change in the example provided will show that there is no difference, in principle, between "risk" and "weight" as attribute Instead of a letterbox, let us assume we have an empty dumbbell. If someone replaces the hollow dumbbell with a similar one, but filled with iron, we have the same situation in the two examples. If you ar asked: "Is the dumbbell heavy?" the answer would have been, "Of course not." (This is to paraphrase our authors.) We, however, knowing about the iron, would say it is very heavy indeed.

In the two cases there is a hidden factor, snake and iron. The aprioristic guesses about "riskiness" and "heaviness" will be based o experience and a state of ignorance of the observer about the real situation. An experiment will settle each question: for example, to look inside the letterbox, or to weigh the dumbbell. The aprioristic judgment does not necessarily have any predictive value, whereas the second kind of estimation could be used as a firm basis for future behavior.

We should notice that the experiment need not be a quantitative measurement. Just as we can speak of the "heaviness" of an object without weighing it precisely, so we can speak of the "riskiness" without going through the procedures necessary for a quantitative estimate.

We can now consider the second argument: "If x is relative to the observer" the "x is subjective." The following example will show its falsity:

> We have Planet 1 (red) with an observer 1 and planet 2 (blue) with an observer 2. There is an astronaut in a rocket, outside the visual range of the observers. Both observers are "Ptolomeic," whereas the astronaut is trained in Analytical Philosophy. For observer 1, planet 2 orbits around planet 1, and vice versa. For the astronaut, the problem of which planet is orbiting around which is a pseudo-problem, but as a matter of convenience we could decide that the blue one orbits the red one.

The movement of planet 2 is relative to the position of the observer 1 and vice versa; each observer will judge the situation differently and will behave accordingly, but the movement of the planets is not "subjective."

THE "SOCIAL-REDUCTIONIST" HAZARD

"Social-reductionism" is a methodological position which assumes that every debate over technological risks is really a conflict among contradictory "ways of life" and that the awareness of this would be enough to settle the question. An example is provided by B. Wolfe.[10]

> Someone who believes that the future welfare of society is dependent on new domestic energy supplies will see large advantages to the development of nuclear power, off-shore oil resources, and new sources of coal, even at some risk and inconvenience. Those who believe that society suffers because we already use too much energy will not accept even minimal risk or inconvenience in order to supply more energy. A public discussion of energy development between groups with these opposing views is like a discussion of pork processing among farmers, meat processors, and orthodox Jews and Muslims. One may talk about humane slaughtering techniques, but the underlying issue is whether or not pork should be eaten. The issue may be couched in technical terms of "spent-fuel disposition", but in fact it is an argument over the morality of eating pork.

The issue could be, of course, "an argument over the morality of eating pork" but the general public and authorities are also concerned with the possibility of "trichinosis" and this concern is shared by the pork industry because cases of disease from eating pork is (at the very least) bad publicity and financial loss for themselves. One may talk about "humane" slaughtering techniques and sometimes "the underlying issue is whether or not pork should be eaten," but it is also about public health, honest or dishonest trade, and controls.

This relativistic position is unlikely to provide a useful solution to the problems faced by risk assessors and decisionmakers. It has not solved the methodological problems of anthropologists, social scientists, or philosophers either, and it is in these fields where it is strongly opposed.[11] Moreover, it can be used as an excuse to deny the possibility and need of a dialogue and a rational debate on basic problems of technological risks and choices.

The most significant contribution to this approach is the recent book by Douglas and Wildavsky.[12] Combining the approaches of political science and anthropology, they produce a powerful case against the pretensions of a "scientific" approach to the management of the many risks to which high-technology civilization is subject. They show that a program of managing all possible risks is impossible; that many significant risks are ill-understood; that the selection of risks for popular concern in America is definitely not done on a scientific basis; and that those who campaign most intensively on particular hazards are motivated by a crusading zeal for a vision of a radically different civilization.

So far their argument expresses, in a particularly erudite and insightful way, the dilemmas of environmental risk management, as now agreed on by nearly all who are involved in it. We may say that risks are now revealed to lie outside the province of Science, and are more like those of politics, or of life in general. But Douglas and Wildavsky press their analysis further; they wish to explain why this strident environmentalist movement arose at a particular time in the United States. For this they construct a theory of three sorts of social institutions, "bureaucracy" and "market" (occupying the "centre") and "sects" (occupying the "border"). The "centre" institutions have failings of their own, familiar to social critics and well retailed here. Less familiar are those of "sects"; their peculiar constructed-world of moral absolutes is derived from the psychodynamics of closed voluntary societies. The history of the classic closed religious communities in North America is here used as the sole evidence of this argument, summed up in the assertion that "border theories emanate from its social predicament of social voluntariness." (p.189)

The conclusion from this analysis is that the "sects" enter into the political process only sporadically in response to a crisis or scandal where the centre cannot cope, and there is never any true dialogue between the mutually exclusive cultures. The policy implications of this conclusion are not made explicit by the authors, for their concluding discussions are of philosophical relativism and of "resilience" as a strategy for coping with present and future risks. However, their use of the term "sectarian" for the organized critics gives a hint, as does their comment, "The remedies most easily proposed in such organizations are to refuse to compromise with evil and to root it out, accompanied by a tendency towards intolerance and drastic solutions." (p.11) It is hard to see how they could disagree with the analysis and conclusions of B. Wolfe, mentioned above. Nor would it be easy to escape the practical conclusion that these new "sectarians," like all others in that historic category, are not entitled to respect as participants in a policy debate. For they join it for ulterior motives, and given their absolute and intolerant commitments, will only abuse its procedures for their own ends.

These conclusions (nowhere asserted but nowhere denied by our authors) could have serious consequences for environmental politics, and so deserve careful thought. They also run counter to the recent American experience of pressure-group politics, on the environment and other issues. On this latter ground alone, we are entitled to scrutinize the authors' models as a reflection of social reality. Of course the three sorts of organizations are "ideal types," and their authors analyze their real-life manifestations with great insight and subtlety in respect of their internal workings. But their theory of the psycho-dynamics of "sects" is based on closed communities that isolate themselves from society. What resemblance do these have to the politically active national organizations that rely on a mail-order membership, or even local activist groups which are only NIMBY ("Not in My Back Yard") coalitions? Not very much at all, except sometimes in a sort of style of rhetoric, which on occasion is even shared by the United States Congress! (p.163)

Although the authors discuss the activist groups, their concentration on their own model inhibits any real insight into politics as it is practiced in America. While they analyze the interactions of the two sorts of institutions of the "centre," they have very little to say on those between "centre" and "border". The reader is left with a picture of total non-communication between a "complacent" centre and an "alarmist" (or rather, fanatical) border.

In this way the authors have missed the opportunity to analyze real roots and stylistic tendencies in American political life in relation to their problem. Better than "sects," we can think of such terms as "populist," "evangelical," "crusading," "factional," and "litigious" to describe the traditional style of American reforming politics. Such great historic campaigns as anti-slavery, free-silver, and prohibitionism had their "sectarian" beginnings, and rose to positions of significant strength. More recently, some radical campaigns have failed (as on the hazards of recombinant DNA research), while others have gained surprising strength (such as opposition to nuclear weapons strategy.) This is the stuff of politics in America; all the skills of political science and cultural anthropology would be relevant to its study. But the authors are quite unconcerned with this level of reality, and their analysis is correspondingly abstract and impoverished.

Seeing the risks agitation nearly as a unique sectarian phenomenon, they tend to become influenced by the cross-cultural study of risks. Fascinated by the bizarre collections of beliefs about risks which are the stock-in-trade of anthropology, they convey the impression that all risks are equivalent, on the classic lines of "pollution is in the nose of the beholder." Thus, asbestos hazards are described as follows:

> Why is asbestos poisoning seen to be more fearsome than fire? Asbestos was developed to save people from burning; asbestos poisoning is a form of industrial pollution whose toll of deaths by cancer justified a particular anti-industrial criticism more strongly than does loss of life by fire.(p.7)

In view of the well-known history of willful neglect of the industrial and environmental hazards of asbestos, such an analysis is more characteristic of a "border" (though <u>not</u> the one criticized by the authors) than of a "centre."

Only an important book merits such detailed criticism; the authors have made a real contribution, through their theoretical insights and case studies. Their solution to the problem of civilized dialogue on risks is, in our view, excessively sceptical. But the problem exists, and they have laid down a challenge to all those who believe in the possibility of a solution. We hope to make a beginning on this in our three-fold division of risks problems in the next section.

A SOLUTION: THREE TYPES OF INQUIRY

Although no one asserts now that Risk Assessment applied to complex technological systems is a science, science continues to be the desideratum, as a model of practice, among a majority of experts. But is the methodology so successfully used in research science suitable to the sort of enquiry found in Risk Assessment? There are two reasons why it is not. First, it is well known that the ideal of a science consisting of exact quantitative "public knowledge"[13] is frequently unattainable in this sort of work. This failure is commonly ascribed to the degree of complexity, uncertainty, and ignorance in crucial problems. But the difference between this field and research science is not merely one of degree. Here, a second dimension, containing the so-called "value aspects," is an essential component of the problems. In this respect this field is qualitatively different from research science.

The "values" dimension does come into scientific practice, as in peer-review of research projects or in general setting of priorities. However, outputs of research as realized in papers are, formally at least, unidimensional; purely factual. Any methodology that pretends to be successful in Risk Assessment has to take into account both dimensions, facts and values.

We think we can solve the difficulty created by the contradiction between the ideal of "public knowledge" science and the characteristics of the problems encountered in Risk Assessment, without falling into sectarian relativism or social reductionism. The first step is to distinguish among the different types of problems in Risk Assessment and then to apply an appropriate methodology to each kind. The problems would be analyzed as functions of the two dimensions mentioned above: "systems uncertainty" on the technical side, and "decision stakes" on the values side; both varying independently.

"Systems uncertainty" contains the elements of inexactness, uncertainty, and ignorance encountered in the scientific and technical studies; whereas "decision stakes" involves the costs and benefits to all the interested parties, of all the various available policy options, including delays in decision of some definite or indefinite duration. It also includes possible costs to the consultants themselves of advice which is unwelcome to the client!

The diagram (which should be seen only as a convenient sketch) indicates the realms of application of three different methodologies and the names in it indicate them (Figure 1). When the uncertainties of the system and of the decision stakes are small, and the data bases are relatively large and reliable, we have the sort of problems where the "Applied Science" methodology has proved to be successful. Even some relatively rare events, involving quite high decision stakes, can be brought under control by "applied science"; aircraft accidents and disasters are a good example of this. One reason for this success is the availability of a good database, in the records of the extremely large number of movements and the large number of recorded incidents of varying degrees of severity. This is supplemented by empirical knowledge of the ratio of non-reported incidents to those reported. In such a case, it is possible to construct a "risks pyramid," giving ratios of empirical frequencies of related sets of incidents, accidents, and disasters. By their means, control-measures may be devised to reduce incidents and their more serious consequences in salient cases.[14]

When both the systems uncertainty and the decision stakes are considerable, but where professional expertise can still operate, we will define a different model of practice, the "Technical-Consultancy" methodology. It still involves the use of quantitative tools, but these are explicitly supplemented and interpreted on a qualitative basis by experienced judgment. The result is not intended to function as an element of "public knowledge" science; it is not designed to be fully testable and reproducible, nor to be applied mechanically to other similar problems. In this sense, this kind of inquiry fails to satisfy the epistemological criteria for research science. Rather, it is an input to a process of decisionmaking in which risks are one among several factors. The values at issue condition the inquiry at every stage, because of its function in a particular decision process.

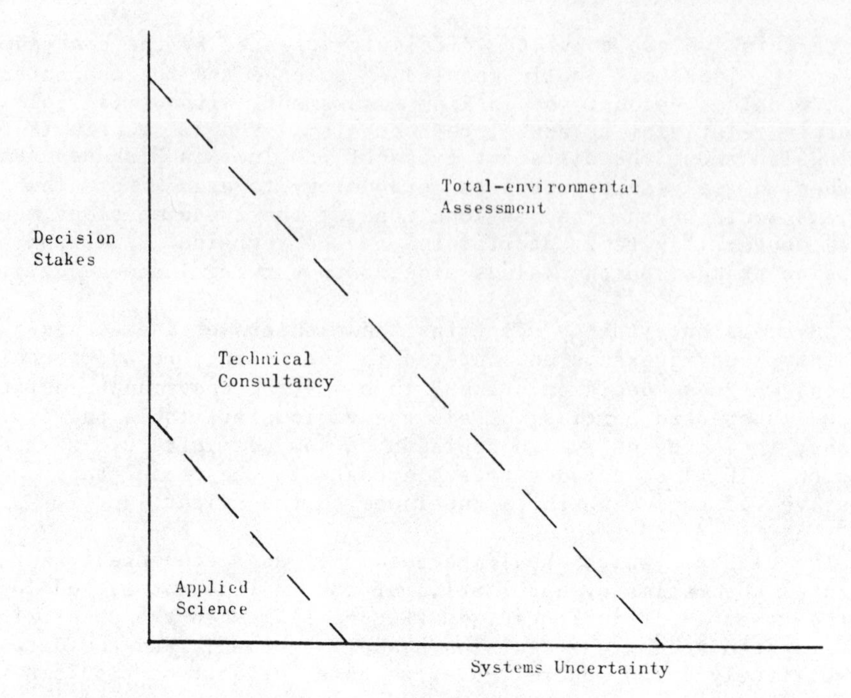

Figure 1

Classification of types of enquiry involved in Risk Assessment

Outstanding examples of the "technical-consultancy" methodology are the two studies on the petro-chemical installation on Canvey Island, by the Health and Safety Executive.[15] There they dealt with a particular site, in terms of its special problems. They claimed no general results, and gave cautions about the inexactness of their probabilistic conclusions, in terms of the varying degrees of reliability of their data. Although they concluded with a policy recommendation, they had sufficient detail on many local hazards for their prescription to be applied to a drastic improvement of safety on the island.

Nowhere would the reports claim that the island could be made "safe," or that their conclusions were "scientific." Indeed, the two reports were commissioned and received in a highly-charged political atmosphere, and criticized in proceedings of an adversarial nature. But in terms of the stated tasks of the inquiries, and also as examples of how such work is to be done, they stand as solid accomplishments.

Finally, when decision stakes and systems uncertainty are very high, we have a "total-environmental assessment," which is permeated by qualitative judgments and value commitments. Its result is a contribution to an essentially political debate on larger issues, though no less rational in its own way for that. The inquiry, even into technical questions, takes the form largely of a dialogue, which may be in an advocacy or even in an adversary mode.

Although only a small proportion of risk assessments fall into this category, they are frequently those of greatest practical and political significance. It is to them that the analysis of "social reductionism" is most plausibly applied.

It is very important to realize that a "total-environment" problem is, by definition, not simple or static. Although value-commitments may shape arguments and inquiries on many important aspects, the debates are not all in the realm of metaphysics, nor are the decision stakes purely in the area of sectional power struggles. Thus, to take the example cited above, the issue of asbestos hazards is not simply a case of disliking industry more than fire. Even on more complex and speculative issues, as energy supplies, there is a steadily growing area of scientific knowledge and technical expertise. This complements the value-considerations, reducing uncertainties on particular critical decisions, and perhaps securing agreement among experts on all sides concerning particular technical problems. The hearings on the PWR at Sizewell is an example of how, in spite of an overtly adversarial style of debate, common accumulated experience yields an evolution of the issues if not to policy agreement, then at least to consensus on salient areas of debate. This last is all we require for a problem in this domain to be excluded from the "social reductionist" model of environmental debates. This evolution, toward rationality and dialogue, may indeed take years to accomplish; in its early stages a "total-environmental assessment" may really seem to be a clash between incommensurable world-views. But such debates tend to stimulate the production of knowledge, of relevant facts and of value-commitments, which eventually enable such problems to be resolved by political debate rather than by civil war. It is important to note that the classification is not static but dynamic, in the sense that any changes in any (or both) of the dimensions could push a particular problem completely from one realm to another, consequently imposing a change in the way it is handled. The most notable recent example of this (welcome) tendency is the problem of possible physical and psychological harm to children resulting from atmospheric pollution by petrol-derived lead.

We should also observe that all these types of problem-solving are different from the practice of "safety management" or "loss control" in an installation that is in operation. There the "scientific" work of survey, analysis, and design is organized around the influencing of human behavior by various means. The problems of this field have been analyzed in terms of a "risk triangle," illustrating the varying valuations and perceptions among those who impose, endure, or regulate the risk.[16]

CONCLUSION

This study began with a review of the crisis in confidence in the assessment of technological risks. The earlier optimism in the power of scientific method to assess all risks has given way to confusion and loss of confidence in any risk assessment. This tendency to despair can have serious consequences for the urgent tasks of social management of the technological risks of the present and future. Attempts to identify new philosophical foundations for risk assessment have proved fruitless. We have therefore concluded that the construction of a general theory applicable to all sorts of risks is not the appropriate way forward.

Our path to a solution lies in a distinction among the different types of risk. It is not merely that some are inherently more difficult to assess. Rather, the tasks of assessment for the various cases are qualitatively different. By analyzing the appropriate methodology for each case, we are able to identify the areas that are particularly problematical and challenging, and we can also explain and reinforce successful practice where it exists. Also, by showing what sort of "knowledge" is actually needed from any particular type of risk assessment, we avoid the error of excessively severe demands and high expectations. By doing this, we prevent the disappointment and confusion that occurs when these are not fulfilled.

We have shown that there is a large and important area where scientific techniques can be successfully applied to complex technological risks even where important policy decisions are at stake; we have called this "technical consultancy." Its success, when properly done, can be understood when we appreciate the nature and function of its outcome. Its conclusions are not intended to be a weaker version of the "public knowledge" resulting from research science. Rather, they are an input to a decision process, constrained by the particular problem set by the client, and validated both by the scientific techniques and by the skill and judgment of those who apply them.

Even in the case of "total-environmental assessment," the problems are not an undifferentiated mass of scientific speculations and cultural values. Such issues always include components capable of study by "technical consultancy" or even "applied science." Also, public debate on such problems stimulates research whereby the "scientific" component is strengthened, perhaps eventually to the point where the value-commitments themselves are changed in the light of new evidence and new experience.

The assessment of technological risks is a very new field. It is only natural that it should go through a cycle of early over-optimism followed by disappointment and pessimism. Our work has been intended to build toward the next phase of this practice, involving a more mature appreciation of its real strengths and limitations. We have done this partly through identifying and reinforcing successful professional practice where it exists, and using this as the basis for further study.

FOOTNOTES

1. C. Mandl and J. Lathrop, "Comparing Risk Assessments for Liquefied Gas Terminals - Some Results". In The Risk Analysis Controversy. An Institutional Perspective, H. Kunreuther and E.V. Ley (eds.) (Springer-Verlag, 1982) pp. 42-43.
2. C. Starr, "Social Benefit Versus Technological Risk," Science, Vol. 165, 19 Sept. 1969, pp. 1232-1238. G. Apostolakis, "Probability and Risk Assessment: The Subjectivist Viewpoint and Some Suggestions," Nuclear Safety, Vol. 19, No. 3, May-June 1978, pp. 305-315.
3. Risk and Decision Making: Perspectives and Research. Committee on Risk and Decision Making Assembly of Behavioral and Social Sciences, National Research Council, National Academic Press, Washington, DC, 1982, p. 61.
4. G.N. Parry and D.W. Winter, "Characterization and Evaluation of Uncertainty in Probability Risk Analysis," Nuclear Safety, Vol. 22, No. 1, January-February 1981, pp. 28-42.
5. D. Okrent, "A Survey of Expert Opinion on Low Probability Earthquakes," Annals of Nuclear Energy, 2, 1975, pp. 601-614.
6. "The Axman and Nuclear Power," (Leader) New York Times, 20 March 1983.
7. "Nuclear Incompetence," New Scientist, 19 May 1983, p. 434.
8. J.R. Ravetz, "The Safety of Safeguards," Minerva, Vol. XII, No. 3, July 1974, pp. 323-325.
9. S. Kaplan and B.J. Garrick, "On the Quantitative Definition of Risk," Risk Analysis, Vol. 1, No. 1, 1981, p. 12.
10. B. Wolfe, "Is the Energy Debate Really About Energy?", IAEA Bulletin, Vol. 24, No. 4, 1983, pp. 28-32.
11. M. Hollis and S. Lukes (eds.), "Rationality and Relativism," B. Blackwell, Oxford, 1983.
12. M. Douglas and A. Wildavsky, Risk and Culture, University of California, 1982.
13. J.M. Ziman, Public Knowledge, Cambridge U.P., 1968.
14. J. Tye, "On the Safety Beat", Safety (London), February 1982.
15. Health and Safety Executive. "Canvey: An Investigation of Potential Hazards from Operations in the Canvey Island/Thurrock Area," 1978, and "Canvey: A Second Report," 1981 (London: HMSO).
16. J.R. Ravetz, "The Political Economy of Risk," New Scientist, 8 September 1977, pp. 26-27.

LEGITIMATING PRIVATE SECTOR RISK ANALYSIS:

A U.S.-EUROPEAN COMPARISON

Sheila Jasanoff

Cornell University
Ithaca, NY 14853

INTRODUCTION

Since the Second World War, the governments of the advanced industrial nations have been called upon to assume increasing responsibility for protecting the public against technological risks. The spate of environmental and health and safety laws enacted in many Western countries during the 1970s attests to the widespread popular demand for a strong governmental role in risk regulation. Although these laws reflect similar public concerns in Europe and the U.S., they differ substantially in the way they allocate risk assessment and management functions between the public and the private sector. I use the terms risk assessment and risk management here as defined in a recent National Academy of Sciences (NAS) report:

> Risk assessment is the use of the factual base to define the health effects of exposure of individuals or populations to hazardous materials and situations. Risk management is the process of weighing policy alternatives and selecting the most appropriate regulatory action, integrating the results of risk assessment with engineering data and with social, economic, and political concerns to reach a decision.[1]

On the whole, European regulatory statutes offer considerable scope for involving private sector organizations in both aspects of risk control. American law more rigorously defines the public role in risk regulation and explicitly assigns both assessment and management functions to government agencies. Reviewing specific examples of the two regulatory approaches, I would argue that the European balance of public and private responsibilities greatly reduces the economic and political costs of decisionmaking. However, this approach rests upon social, political, and ethical assumptions that appear fundamentally at odds with prevailing American views about uncertainty and expertise, participation and public accountability. Accordingly, the prospects for incorporating some of the most attractive features of European risk analysis into U.S. rule-making appear limited.

TWO FRAMEWORKS FOR RISK ASSESSMENT

American risk regulation statutes of the 1970s delegate to public agencies the twin tasks of risk assessment and risk management. Laws regulating toxic and hazardous substances, for example, require the designated agency to identify a threatening product or process, to see if the risk it presents meets standards for intervention laid down by law, and to devise an appropriate regulatory response. These substantive delegations of authority are hemmed around by procedural constraints designed to prevent bias and arbitrariness in administrative decisionmaking and to promote rational policy outcomes. Such provisions give to American regulatory law a complexity and formalism unmatched by European statutes.

Three features of the American regulatory model carry particular implications for the private sector's role in risk assessment. First, U.S. health, safety, and environmental laws generally contain detailed standards telling regulators when, how, and in what manner to intervene. For example, the Toxic Substances Control Act (TSCA) requires the EPA Administrator to determine whether chemicals present an "unreasonable risk" to health and the environment. It also creates a timetable for action and specifies what control measures the Administrator may employ. Second, American laws prescribe procedures for obtaining data pertaining to risk assessment from private parties with a stake in the regulatory process. These procedures are both formal and adversarial, permitting parties in the debate not only to present their own analyses of scientific, economic, and legal issues, but to challenge contrary evidence submitted by their opponents. Third, through citizen suit provisions and related measures, American law provides generous access to the courts for those seeking review of both substantive and procedural aspects of agency risk determinations.

This complex legal framework has the effect of relegating the private sector to, at best, a supporting role with respect to risk assessment. It is not the private sector's estimate of risk, but the independent assessment of a "neutral" public decisionmaker that forms the basis for management decisions. Though private sector groups are expected to contribute data and analyses, their submissions are presumed to be colored by self-interest. It is the task of the public agency to listen to different private interest groups, often holding widely divergent views about risk, and to fashion from these an assessment that steers clear of any particular set of biases. Since the responsibility for mediating conflict in this process belongs solely to government, private interest groups have little incentive to abandon partisan analyses of risk in favor of more "centrist" assessments. Indeed, with the final outcome of the regulatory process often turning on a lawsuit, private interests are encouraged to prepare the way for a maximally favorable outcome by advancing exaggeratedly high or exaggeratedly low estimates of risk.

The European approach to risk assessment could hardly be more different. Environmental and public health regulation is usually carried out under broad enabling statutes that leave important details of implementation to be worked out at the discretion of administrative agencies, including procedures for risk assessment. European laws rarely specify timetables or procedures for rulemaking. Statutory public participation requirements are usually limited to provisions listing which private interests should be consulted during policymaking. While legislation often prescribes the forms of possible intervention, it seldom establishes conditions under which regulators _must_ act. Finally, European statutes almost never include special provisions facilitating judicial review of administrative rulemaking.[2]

An illuminating, though admittedly extreme, example of the European legislative style is Britain's Health and Safety at Work Act of 1974 (HSW Act). This multi-purpose statute corresponds both to the U.S. Occupational Safety and Health Act and to TSCA, in that it authorizes premarket control of chemicals. The HSW Act was conceived as a conscious attempt to deregulate the field of worker protection. Parliament accepted the argument that there was "too much law" in this regulatory sector and enacted a statute that returned much of the responsibility for developing occupational safety and health standards to private parties.

From an American perspective, the most striking feature of the HSW Act is that it creates no affirmative obligation for public officials to assess workplace hazards or to develop protective standards. Instead, the law commits both risk assessment and risk management to employers, manufacturers, and other private parties who control conditions in the occupational environment. Thus, the general duty to discover potential hazards, assess risk, evaluate the costs and benefits of alternative control options, and choose an appropriate regulatory strategy _all_ rest with the employer. For toxic substances in the workplace, the law also creates a general duty for manufacturers to identify hazards and to inform users how these can be controlled.

Explicit delegation of risk assessment and management functions to private interests is virtually unknown in Continental European legislation. But French and German chemical control laws resemble the HSW Act in other important respects. They provide few, if any, mandatory rulemaking procedures, leaving the public authority unlimited discretion to structure the private sector's input to regulatory risk assessment. The most common approach is to rely on scientific advisory bodies for the necessary expert analysis. These can be of two types: either an "independent" committee of highly qualified research scientists, or a "representative" committee that includes members from the major private interest groups affected by regulation. A prestigious example of the former type is the Commission of the Deutsche Forschungsgemeinschaft, Germany's scientific honor society, that establishes tolerance levels for toxic substances in the workplace. Though this body includes experts from both industry and academia, selection of members is based on scientific competence alone. Professional preeminence effectively safeguards the committee's risk assessments against serious challenge by either government or organized labor. Examples of the more common representative type include the German Labor Ministry's advisory committee on toxic substances and its counterparts in France and Britain.

The use of advisory committees for risk assessment shapes the public debate on regulatory policy very differently from the formal, adversarial procedures of American regulation. Unlike most American advisory bodies, European expert committees wield effective control over both risk assessment and risk management, since administrators seldom deviate from their findings and recommendations on either aspect of decisionmaking. The knowledge that their decisions will be final gives committee members a strong incentive to resolve professional and political differences and to reach accommodations that all are willing to live with. Pluralistic advisory committees offer special advantages for consensus-building by providing divergent interest groups a private forum for negotiating compromises on risk assessment and management. Such committees also permit technical and political issues to be raised in a unified process, so that decisionmakers do not have to strive for the difficult separation of fact and values, subjective and objective considerations that U.S. regulatory agencies have repeatedly sought to achieve. Pluralistic committees facilitate informal negotiation on regulatory costs and benefits, with the result that European rulemaking has generally avoided the technical and ethical problems, as well as the financial costs, of formal economic analysis.[3]

It appears then that leaving room for autonomous private sector involvement in rulemaking, particularly with the aid of structured pluralistic committees, carries a number of advantages that are lost in the American approach to risk regulation. But does this mean that the American administrative process can or should be modified in the direction of the European model? My own view is that the private sector will never enjoy as authoritative a voice in American federal regulation as it does in Europe. The European pattern of risk management is founded on views of science, public participation, and political accountability that seem inconsistent with dominant traditions in American politics.

EXPERTISE AND SCIENTIFIC UNCERTAINTY

Public sophistication about scientific uncertainty has increased to the point where neither Europeans nor Americans any longer expect science to provide definite answers to questions about risk and regulatory policy. Since uncertainty cannot be eliminated, critical choices of regulatory policy are often formulated in institutional terms: What individuals or societal groups should be responsible for making decisions when the factual basis is controversial or uncertain? And having assessed the relevant risks, who should decide how to integrate the results with the social, political, and economic concerns addressed in risk management? These questions have received very different answers in Europe and the U.S.

In most European regulatory schemes, the resolution of scientific uncertainty is still regarded basically as a task for experts. Government's reliance on technical advisory committees reflects the prevailing view that a properly constituted group of experts is better qualified to deal with uncertainty than a civil servant or bureaucrat. The American approach to risk assessment proceeds from a much more skeptical attitude towards expertise. In particular, risk assessment has long been viewed in the U.S. as involving "trans-scientific" issues, which cannot be resolved by expertise alone, since their resolution entails value judgments in addition to technical determinations.[4] This view has been embraced not only by social scientists and administrators, but by the federal courts, which have held for purposes of administrative law that decisions at the frontiers of scientific knowledge should be treated as legislative or policy determinations rather than mere fact-finding.[5]

Recent demands for improving the quality of science in American regulation have refined, but not fundamentally altered the national perspective on scientific uncertainty. The NAS risk management report cited above continues to stress the fact that policy choices are integral to <u>assessment</u> as well as <u>management</u>, even though the policies in the assessment phase may be "related to and subservient to the scientific content of the process."[6] In the NAS panel's view, this intertwinement of science and policy demands that risk assessment should be performed by the same agencies that are responsible for the sociopolitical choices involved in risk management. In other words, the resolution of scientific and technical uncertainty in regulation should remain a prerogative of government. Thus, although the NAS report recommends a larger role for independent expertise in the U.S. regulatory process, it falls far short of endorsing the kind of autonomy granted to expert advisory committees in Europe.

RISK ASSESSMENT AND PUBLIC PARTICIPATION

Public participation rules offer another notable point of divergence between American and European risk assessment. In each system, the decision as to what members of the public should participate is linked to ingrained cultural views about uncertainty and the role of expertise. The European tendency to see uncertainty as the domain of experts has a narrowing effect on opportunities for participation. The subject matter of risk assessment is regarded as primarily technical, hence beyond the analytical competence of most members of the public. As a result, expert advisory committees are often selected exclusively from the ranks of the technically qualified, as, for example, in the case of Germany's independent commission on chemicals in the workplace. In cultures where the policy implications of scientific uncertainty are neither identified nor debated, such committees can maintain their credibility simply by demonstrating the highest possible technical expertise.

Even when European regulators seek to incorporate political balance into their advisory committees, they seldom open the doors to participation by the general public. Interest groups are selected more or less *ex officio*, as representatives of well-established consumer, environmental, labor, or industrial associations with a recognized right to be heard in rulemaking. In many cases, these representatives bring to their task a substantial formal training in relevant scientific disciplines. Those who have no special technical background often acquire expertise in risk assessment through membership in standing public advisory committees.

American rules of participation in risk assessment appear wide open by comparison with the European approach. Regulatory statutes provide opportunities for comment by any interested person on proposed administrative decisions. Information is freely made available pursuant to specific statutes and the Freedom of Information Act, so that any reasonably energetic member of the public can inform himself about the substantive issues involved in rulemaking. Subscribing to a policy of wide citizen participation, Congress has even authorized some federal agencies to fund public interest interveners who appear capable of making fruitful contributions to rulemaking proceedings.

Open information and liberal public participation are consistent with the vision of democracy that Jefferson articulated some two hundred years ago. The survival of these characteristics in today's complex universe of technological regulation suggests that Americans regard the twentieth century's scientist policy-makers with no less distrust than they have historically directed at their political officials. Mistrust of expertise is expressed in the American political environment as a tendency to emphasize the value conflicts underlying scientific controversies, sometimes to the point of denying that there is any legitimate technical basis for disagreements among experts. It follows that the risk assessment is perceived as a process of accommodating divergent values and preferences rather than a technical exercise to be completed through expert deliberations. The emphasis on values provides a rationale for granting broad access to the public. Without open participation, there is always the danger that decisionmakers will be insufficiently informed about the social and political conflicts inherent in different approaches to risk assessment.

Inevitably, the decision to grant the public liberal participatory rights in risk assessment drives the process towards greater procedural complexity. To promote full participation, the law has to ensure that all points of view receive equal access and that none are arbitrarily excluded. Further, participants must be given the opportunity to test each other's views and those of the government regulator in order to illuminate the values at stake in technical decisions. To make sure that participation does not turn into a mere public relations game, there must be additional constraints on the decisionmaker that force him to respond to substantive issues raised by private intervenors. Finally, procedural provisions must be backed up by judicial review, which serves in the American process as the ultimate check on administrative discretion. All these procedural requirements widen the gulf between European and American regulation, forcing U.S. regulators towards ever more formality and methodological rigor in the evaluation and assessment of risk.

THE GOVERNMENTAL ROLE

I have argued that the European view of risk assessment stresses the technical nature of the problem and the need for specialized expertise in dealing with uncertainty. In contrast, the American view emphasizes the political and value-laden aspects of decisionmaking at the frontiers of science, and uses liberal public participation to throw value conflicts into high relief. These differences go a long way towards explaining the different allocation of risk assessment responsibilities between the public and private sectors in Europe and the U.S.

In a system that treats risk assessment as the preserve of experts, political officials may legitimately take the back seat, influencing policy through indirect means. Administrators remain powerful, but they use power in less visible ways than their American counterparts. Primarily concerned with consensus-building, European regulators devote their energies to institutional matters. Their goal is to create the kinds of institutions that will best further communication among private parties with a high stake in risk assessment and management. The techniques they employ are informal, but often highly effective: controlling the flow of information, selecting members of advisory committees, excluding marginal groups from pre-regulatory consultation, behind-the-scenes bargaining, and, finally, threatening to intervene if the private parties fail to reach a satisfactory accommodation among themselves.

By dwelling on the values at stake in risk assessment, the American regulatory process, by contrast, underscores the need for a neutral public decisionmaker. Only public agencies, duly authorized by law, have the moral authority to strike compromises among the conflicting values underlying risk assessment. Operating under public and judicial scrutiny, government agencies are expected to engage in more responsible decisionmaking than unsupervised committees of scientific experts. Equally important, committing risk assessment and management to clearly identified public entities means that someone can always be held accountable for policy decisions of great political significance. Unambiguous accountability, a prized characteristic of American regulation, is lost in systems that entrust risk decisions to elite scientific bodies or negotiating groups consisting of large, organized private sector interests.

CONCLUSION

What I have said thus far in no way detracts from the American private sector's real and extensive contributions to risk assessment. Historically, private sector groups in the U.S. have played an indispensable, though not always highly visible, role in risk assessment, primarily through standard-setting. According to one count, "there are more than 400 private organizations--trade, technical, professional, consumer, and labor--that have written or sponsored approximately 20,000 current commercial standards."[7] Government agencies, operating with restricted financial, technical, and temporal resources, could never hope to take over the massive task of risk assessment and management reflected in these private standards.

At the same time, we can expect that when risk assessment is tied to policy issues of particular concern--nuclear power, genetic engineering, toxic chemicals--the U.S. public will not settle for a process that leaves technical decisionmaking to the private sector. American views about the relationship between technical controversy and value conflicts will necessitate a clearer separation between public and private sector risk assessment than is considered desirable in Europe. Unfortunately, the insistence on open, publicly accountable decisionmaking carries substantial costs. It places government and the private sector in antagonistic positions, induces procedural formalism and rigidity, encourages litigation, and offers private sector groups incentives to put self-interest ahead of the public interest. At the limit, the U.S. approach to risk assessment threatens the credibility of science itself, by postulating, in effect, that all disagreements among experts are political in origin.

The European approach, however, also carries costs that should not be overlooked. Confining risk assessment to technical committees, even when these are nominally "representative," promotes public ignorance and indifference on issues of wide social significance. Technical committees run the danger of falling into particular "pathologies" of risk assessment[8] that could be prevented through public debate, such as underestimating certain kinds of risk systematically, or overlooking certain benefits of regulation. Pluralistic advisory committees also create a potential for the cooptation of less informed and less organized interests by powerful or sophisticated ones, a danger that is more easily averted in the open and procedurally structured American decisionmaking process.

In the end, a drastic reformulation of the private sector's role in risk assessment and risk management appears improbable in the U.S. A recent episode at EPA illustrates the socio-political obstacles to reform. William Ruckelshaus, newly reappointed to head the agency, attempted to secure greater public involvement in a difficult risk management decision confronting EPA. According to agency experts, the ASARCO copper smelter in Tacoma, Washington, will be unable to reduce its arsenic emissions to levels generally considered safe even after installing advanced technological controls. This leaves EPA with two options: either to tolerate emissions producing an excessive increase in cancer risks or to order the plant to close down. Ruckelshaus issued a direct appeal to the citizens of Tacoma requesting their opinion on the policy choices available to EPA, as well as their suggestions about other less drastic control options.

Ruckelshaus' call to the public was widely misunderstood. A lead editorial in the New York Times denounced Ruckelshaus as a modern-day Caesar, imposing on Tacomans an "impossible choice" between a plant shutdown and a five percent increase in the risk of cancer.[9] Though based on serious misinterpretations of the Administrator's objectives, the editorial accurately reflected a widespread public reaction. For many Americans, EPA's effort to draw the public into a complex regulatory decision was not a valid exercise of discretion, but a wholly unjustified abdication of the agency's duty to enforce the law, however painful the consequences. This attitude leaves U.S. regulators in a sensitive position. Not only must they *make* the difficult policy choices integral to risk assessment and management, but they must be *seen* to do so. Unlike their European counterparts, they cannot avoid full public accountability by orchestrating a compromise among interested private parties. Such accountability, as we have seen, carries a sizeable price tag, for it can only be secured through elaborate procedural safeguards. The reward, however, is a governmental process that is less likely to be removed from popular control than one sometimes fears in Europe.

FOOTNOTES

1. National Academy of Sciences, *Risk Assessment in the Federal Government: Managing the Process*, National Academy Press, Washington (1983), p. 3.
2. For a detailed comparison of European and American legislation in the area of toxic substances control, see Ronald Brickman, Sheila

Jasanoff, and Thomas Ilgen, Chemical Regulation and Cancer: A Cross-National Study of Policy and Politics, NTIS No. PB-83-206771 (1982).

3. Sheila Jasanoff, "Negotiation or Cost-Benefit Analysis: A Middle Road for U.S. Policy?", The Environmental Forum, Vol. 2 (July 1983), pp. 37-43.
4. Alvin M. Weinberg, "Science and Trans-Science," Minerva, No. 10 (1972), pp. 202-222.
5. See, e.g., Amoco Oil Co. v. U.S. Environmental Protection Agency, 501 F. 2d 722 (D.C. Cir. 1974).
6. National Academy of Sciences, op. cit., p. 37.
7. Samuel Florman, Blaming Technology, St. Martin's Press, New York (1981), p. 110.
8. For an elaboration of this concept, see Guy Burgess, "Social and Political Pathologies of Risk Decision Making," in Dean Mann, ed., Environmental Policy Formulation, Lexington Books, Lexington, MA (1981), PP. 149-159.
9. The New York Times (July 14, 1983) p. 22.

ARE REGULATIONS NEEDED TO HOLD EXPERTS ACCOUNTABLE FOR CONTRIBUTING "BIASED" BRIEFS OF REPORTS THAT AFFECT PUBLIC POLICIES

Theodor D. Sterling
Anthony Arundel

Faculty of Interdisciplinary Studies
Simon Fraser University
Burnaby, B.C., Canada V5A 1S6

ABSTRACT

Direct testimony, briefs, and committee work of scientists and scholars influence public policy in numerous crucial issues. Unfortunately, scientific testimony is frequently biased, either as a result of the omission of relevant information, or by denying the reality of unwanted observations. Three examples of biased evidence are discussed: a series of submissions to the FDA and the EPA by industrial scientists; a report by the NAS on indoor air pollution; and the treatment of occupational carcinogenesis in the Surgeon General's Reports on Smoking and Health. Presently, there are no recognized standards or restraints to hold experts accountable for clearly inadequate, incorrect, or incompetent performance. Methods are needed to ensure accountability.

INTRODUCTION

Critical analysis of the problems in deriving risk estimates have tended to stress two major limitations:

- The difficulties often encountered in assigning values for risks and/or benefits

- The frequent inadequacy of data for statistical anslysis (such as an insufficient number of observations or the occurence of confounding factors)

However, when risk analysis takes place against a background of social and economic conflict, it may be, and often is, influenced by interests and values that supercede an unbiased assessment of risk. Thus a third problem in deriving risk estimates may be due to biases resulting from the consequences associated with whatever inferences are drawn. The common formulation of the statistical decision problem as

$$\text{consequence} = F(\text{inference})$$

may be reversed in a biased situation to take on the form of:

$$\text{inference} = G(\text{anticipated consequence})$$

In other words, an inference may be biased in order to avoid an unfavorable consequence or to obtain a favorable consequence. These biases may be understood to be systematic selections among available data points.

The problem of bias in the assessment of risk problems has been masked by the belief that experts/scientists disagree only because of honest differences of opinion on how to interpret "facts." Such a belief rests on the supposition that expert testimony is similar to scientific inference. After all, it is well established that scientists can draw different conclusions from the same data base in the process of formulating and testing hypotheses. However, the public regulatory process primarily requires information about a state of nature. In most cases the participation of the scientist/expert does not involve creative inference or hypothesis testing. For example, the regulatory process is often triggered by the outcome of laboratory tests required under federal regulations for the registration of chemical substances. The tests follow largely inflexible protocols which specify both the type and number of animals to use and the method of statistical analysis. They are designed to produce unequivocal answers to questions concerning the physiological response of animals to possible carcinogens, teratogens, or toxins, and do not require the scientist/expert to generate speculative hypotheses to explain the results. Essentially regulatory agencies and investigative committees simply require the expert to inform (to inform on the current state of available information from the existing work of scientists and from fact-finding trials) and not to speculate about possible states of nature.

Insofar as the public policy process in our society ought to rest on an objective evaluation of risks as well as benefits incurred by policy alternatives, a serious attempt is needed to deal with instances where scientists/experts introduce bias into risk evaluation procedures. This requires the development of criteria to aid in the identification of instances where bias occurs. There seem to be at least two methods that are used to introduce bias and to carve out adversary positions:

- The omission of the results of studies that are in conflict with some specific claim or assertion

- The denial of certain observations by some subterfuge that gives the appearance of offering a scientific explanation for unexpected or unwanted results

Three examples will be discussed here that illustrate the occurrence of omissions and misrepresentation of data in important expert submissions used for the public policy process.

EXAMPLE ONE: MISREPRESENTATIONS IN REPORTS TO A REGULATORY AGENCY BY SCIENTIFIC EXPERTS FROM INDUSTRY

The toxic properties of a commonly used herbicide, 2,4,5-Trichlorophenoxyacetic acid (2,4,5-T) became the subject of reevaluation in the late 1960s. Suspicion that 2,4,5-T or one of its contaminants were teratogenic was based on two independent sources: the FDA-sponsored series of tests of chemicals commonly used in the home; and reports of stillbirths and malformations among children of Vietnamese women exposed to 2,4,5-T in the defoliant "Agent Orange." In order to obtain additional data the AAAS sent a special study team to Vietnam which found considerable evidence of an increase in stillbirths in Vietnam hospitals serving areas exposed to heavy spraying of Agent Orange, but the data was not thought to be sufficiently clear cut to conclude that 2,4,5-T was teratogenic. As part of the evaluation process, scientists from a number of companies producing the herbicide 2,4,5-T performed animal experiments to test the teratogenicity and general toxicity of 2,4,5,-T and of its most notable Dioxin contaminant, 2,3,7,8-Tetrachlorodibenzo-P-dioxin (TCDD). Results were submitted to the FDA, NIEHS, and a special committee of the National Academy of Sciences. Although some of the studies duplicated the teratogenic findings of the NIEHS scientists, most of them tended to stress aspects of the data which indicated that 2,4,5-T was harmless, or else concluded that the environmental background levels of 2,4,5-T and TCDD were safe. One such report compared the offspring of female rats which had been exposed to various amounts of the herbicide to controls.(16) The report is one of a series of documents submitted to regulatory agencies which denied that 2,4,5-T posed a teratogenic hazard. The authors describe the results of this study as follows:

> "Skeletal and visceral examination...as well as histopathologic examination of certain fetuses failed to reveal teratogenic or embryotoxic effects."

However, inspection of the data in the report revealed that this statement is grossly incorrect (see Table 1). The offspring of the exposed group compared to those of the control group showed a sevenfold increase in skeletal abnormalities of the fifth sternebra. Skeletal abnormalities had also been reported from Vietnam and in experiments using other animals.

In their discussion, the authors advanced several curious explanations for their conclusion that the experiment "failed to reveal" any teratogenic effects including the statements that:

- Too few "poorly ossified" sternebrae had occurred among controls (with reference to unpublished data)

- Sternebrae were poorly ossified and not unossified

- Removal of the fetuses had been poorly timed

Certainly blaming control animals for having too few abnormalities is a unique explanation.[2] If the authors had taken seriously their own explanations, they would have been obliged to replicate their experiment. Otherwise their claim of no effect and their explanations for it were clearly unsupported speculations. Their results, as they ought to have been described to the regulating agencies, were that an excess of teratogenic instances had been observed.

TABLE I
REPRODUCTION OF PART OF THE DATA SUBMITTED BY DOW CO. TO THE FDA, NIEH, AND A NAS/NRC ADVISORY COMMITTEE. DATA OF INTEREST ARE IN TABLE 6, PAGE 13, OF SUBMISSION TO NAS/NRC COMMITTEE (2)

DETAILED SUMMARY OF SKELETAL AND VISCERAL ABNORMALITIES OBSERVED IN PUPS FROM FEMALE RATS ADMINISTERED ORAL DOSES OF 2,4,5-TRICHLOROPHENOXYACETIC ACID DURING ORGANOGENESIS (DAYS 6 THROUGH 15)

Dose Level	Control (Vehicle)	24 mg/kg
Number of Fetuses Examined/Group	103	103
Abnormality (Skeletal)		
Accessory Ribs		
Right	4	4
Left	6	7
Bilateral	20	14
5th Sternebra Poorly Ossified	4	29
2nd and 5th Sternebrae Poorly Ossified	0	1
Incomplete Ossification of Parietals (medial margin)		2
Malaligned Sternebrae		1
Centrum of 11th Thoracic Vertebra Poorly Ossified		1

It is not possible to determine to what extent the series of submissions by industry scientists delayed suspending the registration of 2,4,5-T in 1979. However, it is important to note that the suspension of 2,4,5-T was a result of documented reports of an increase in miscarriages and an increased prevalence of soft tissue carcinoma in human populations. Adequate data existed in 1970 to conclude that 2,4,5-T was carcinogenic and teratogenic without waiting for additional epidemiological evidence.

EXAMPLE TWO: OMISSIONS OF DATA IN A REPORT PREPARED BY AN NAS COMMITTEE FOR A REGULATING AGENCY

In response to an increase in health-related complaints by occupants of modern airtight buildings, the National Academy of Sciences was asked by the EPA to appoint a committee to determine the present status of knowledge about the amounts and effects of indoor air pollutants.

All products of combustion were of interest. Of special interest was the amount of carbon monoxide in houses due to gas range emissions, especially as modern materials and insulation make these houses airtight. Evidence had been accumulating that dwellings using gas ranges for cooking also reported a higher incidence of respiratory disease, especially among children, than dwellings cooking with electricity.(46) Thus the possibility existed that unusually or unsuspectedly high levels of carbon monoxide and other combustion byproducts were occurring in dwellings with gas ranges. The crucial upper limit for exposure was 35 ppm (parts per million) of carbon monoxide based on studies by the EPA that prolonged exposure (more than eight hours) to carbon monoxide was hazardous at levels exceeding 35 ppm. The NAS committee found that carbon monoxide levels in homes were typically below a peak of 35 ppm.

> "Typical average indoor carbon monoxide concentrations in residences vary between 0.5 and 5 ppm; observed peak values reach 25 ppm."(31)

This claim was possible only because the committee omitted references to at least two published studies (in the major journal dealing with air pollution findings) that showed, respectively, an average of 40 ppm carbon monoxide and 61 ppm carbon monoxide, measured under realistic conditions of cooking with gas ranges in 17 dwellings in two cities three thousand miles apart.(45,46) Also omitted was the finding that in approximately 50 percent of a sample of rent-subsidized dwellings in New York, the gas range was used as a supplemental heater during the cold months of the year and so the occupants (mostly Blacks and Latins) were exposed to unknown but high levels of carbon monoxide.(13,46) While the existence of these studies was brought to the attention of the committee, unpublished reports by two members of the

NAS committee seem to have been used as the major support for claims about the low carbon monoxide levels due to gas range use. A number of members of the NAS committee have their work supported by utility companies. So far, NAS has ignored the problem of possible conflicts of interest.

The production of toxic by-products by gas ranges is perhaps the most potentially adverse finding in the literature on indoor air pollution. Present U.S. policy is to increase the use of natural gas in order to reduce dependence on oil based generation of electricity. Secondly, the acceptance of the evidence linking an increased prevalence of respiratory disease among children in homes with gas ranges may lead to liability problems on the part of suppliers and distributors of gas.

EXAMPLE THREE: SELECTIVE REPORTING AND INADEQUATE INTERPRETATION OF DATA IN AN ONGOING FACT-FINDING GOVERNMENT REPORT

The U.S. Department of Health and Human Services (and specifically the Surgeon General) is required by Congressional mandate to produce annual follow-ups to the 1964 Report on Smoking and Health. These reports are of primary importance for the consideration of environmentally related health problems, and their recommendations and conclusions have served as the basis for a variety of legislative proposals. This is especially true for problems of occupational exposure because smoking is most prevalent among blue-collar workers.

The assessment of the relative roles of occupation and smoking is of special importance for a number of reasons, not the least of which is the need to define exposures to suspected carcinogens in the workplace. There is a growing debate concerning the extent to which smoking by workers has been used to divert attention away from hazardous occupational exposures.(3,17,44) The question often is: Does smoking kill workers or does working kill smokers?(43) Yet, it was not until 1979 that the Surgeon General's Report on Smoking and Health included more than a few brief references to studies that analyzed both the smoking habits and occupational exposures of various populations.(48) Subsequent reports did not discuss occupation as a separate topic, but the 1982 Surgeon General's Report(49) discussed the evidence of the association between smoking and cancers of various sites (e.g., lung, larynx, pancreas, bladder, etc.). These sites have also been consistently associated with occupational exposures to a wide range of substances including asbestos, radon, rubber and dyes, organic chemicals, and chromium and nickel compounds.(37)

It was apparent to investigators familiar with occupational factors related to cancer that the 1982 Surgeon General's Report did not include a considerable body of important data. The Report ought to have referred to all studies that provided adequate data on the smoking habits of populations with occupational exposures to suspected carcinogens in order to provide an unbiased review of available information that would increase the understanding of the health effects of smoking among working populations.

To investigate this problem further, a review of the epidemiological literature available to the authors of the 1982 Surgeon General's Report was undertaken. The search was limited to reports presenting new results from original research and published in peer reviewed medical and scientific journals before December 31, 1980, to ensure that they were available for review in the 1982 Report. In a few cases, several published reports referred to the same body of research data and were counted as one study. A routine search through publicly available information[3] yielded 34 studies on cancer etiology which analyzed the effects of both occupational exposures and smoking habits for the same study population.[4] Of the 34 studies identified, two did not report a statistically significant occupational effect. Each of the remaining 32 studies ought to have been included in a review of the smoking and health literature because they presented relevant data about the complex association of smoking and occupation to various cancers.

The 1982 Surgeon General's Report failed to cite 18 out of the 32 relevant studies. In addition, the Report failed in most cases to discuss or even refer to the occupational findings of the studies which were cited (see Table 2). For example, the two relevant bladder cancer studies were cited without reference to the occupational causes of bladder cancer that were examined in the two studies. The poorest coverage was for cancer of the lung and of the larynx. Six studies on larynx cancer were identified, but only one was cited and no reference was made to occupation. Twenty-five studies were available for lung cancer, but only six were cited with a reference to the occupational effect. All six occupational references occurred in Table 12 of the Surgeon General's Report. This table referred to 14 reports which "demonstrate an interaction with cigarette smoking," but in fact, only the six previously identified studies were cited. The remaining references in Table 12 consisted of five duplicate citations to the six studies; two references were made to review articles, and one reference had nothing to do with either smoking or occupation.

Table 2 summarizes the omissions in the 1982 Surgeon General's Report. Each study counted in Table 2 contained the data and analysis necessary to evaluate the effect of both smoking and occupation on one or more cancers. Each study also found a significant association between an occupational exposure and cancer. The omissions indicate that the discussion of smoking as an antecedent of cancer was based on a review that neglected most relevant studies which found an association between cancer and occupation.

TABLE II
THE NUMBER OF PUBLISHED STUDIES ON CANCER ETIOLOGY WITH ADEQUATE DATA ON OCCUPATIONAL EXPOSURE AND SMOKING, RELEVANT TO THE DISCUSSION OF CANCER IN THE 1982 SURGEON GENERAL'S REPORT, SMOKING AND HEALTH. EACH OF THE RELEVANT STUDIES WAS PUBLISHED PRIOR TO DECEMBER 31, 1980.

Organ Site	Number of Studies Showing an Association Between Cancer and Occupation and Containing the Necessary Data to Evaluate the Association to Cigarette Smoking	Number Cited in the 1982 SGR Report	Number of Citations Identifying the Occupational Effect
Bladder[a]	2	2	0
Esophagus[b]	1	1	0
Kidney[c]	1	0	0
Larynx[d]	6	1	0
Lung[e]	24	9	6
Pancreas[f]	1	0	0
Stomach[g]	2	1	0

References are (numbers in boxes are citations for the same study):

a 29, 54

b 18

c 21

d 9, [30, 40], 36, 39, [41, 42], 53

e [1, 25], 2, [5, 6, 47], [7, 32], 8, 9, 10, 11, 14, 15, 18, 22, 23, 24, 26, 27, 28, 33, 34, 35, 36, 38, 39, [50, 51]

f 39

g 39, 52

SUMMARY

Science and scientists now play a large role in public policy, especially in the regulatory process. Sometimes that role is decisive. Where different segments of society stand to gain or lose economically or politically, that role may be even adversary. With acknowledgement of this central role has come increasing concern in cases where there is conflicting scientific evidence.

Policy has consequences, often considerable risks to population segments and sometimes to society at large. Risks cannot be estimated without the careful examination of a problem by the rigorous application of proven scientific methods in order to produce the "best" facts. When human ingenuity is at a loss, when the right questions have not been put to experimentation, or if lack of time or finances (or usually both) has prevented conclusive investigation, then the assessment of risk is not possible. However, the failure of science to produce good data may be relatively rare, especially when well researched problems impact on areas of cultural, economic, political, or social conflicts. For instance, a recent discussion of the availability of an adequate science base for current regulatory needs, sponsored by the editors of Environment,(4) found the five invited policy specialists in agreement that the necessary data was available for most of the current public health issues, but somehow not in the form where it could be used to support regulatory policies (including, of course, a policy not to regulate). There was agreement that the lack of "neutrality" among scientists acting as experts was at the core of the risk assessment problem. Such a conclusion is hardly world shaking anymore and has been the basis of various discussions such as Kantrowitz's(20) call for special science courts or Jasanoff and Nelkin's(19) plea for the regular justice process deciding among conflicting testimonials.

If the lack of neutrality among scientists is one major obstacle against adequate risk assessment, are there no ways to assure a quality of expertise that makes neutrality irrelevant or at least neutralizes scientific bias? Before answering that question it is necessary to decide if it is possible to reach a consensus that bias has been introduced in some cases of expert assessments intended for the public policy process.

The three examples show that the quality of expert performance by scientists can be analyzed and concretely evaluated. Example 1 is an instance of a report that does not accurately report the outcome of a required experiment/investigation to a regulating agency. Example 2 is an illustration of a literature review that omits the results of studies that conflict with the review's conclusions.

Example 3 is perhaps the most difficult instance for deciding if bias has been introduced into public policy. The Surgeon General's Report on Smoking and Health is not directed to a specific regulatory body, but to the U.S. Congress. Therefore, while the Surgeon General's Report has no immediate effect on policy, it does exert a constant influence on a wide variety of government actions through congressional action and thus should be an objective and accurate document.

The final question that clearly needs to be faced is, what to do next? Can expert bias be tolerated in evaluating a risk benefit situation? Perhaps it is necessary to seek some sort of mechanism to hold the scientists/experts to account for the quality of expert input to the public policy process. Such a mechanism should specify consequences for those who do not meet acceptable standards. Where the quality of performance directly affects the process by which public policy is formed, consequences might range from open discussion of the problems to officially excluding from future participation those scientists/experts who have been found to introduce bias. A beginning in this direction is the Federal "debarment" regulations, first invoked recently in a case in which an investigator submitted fabricated data to the FDA.(12) As in other fields (such as law or medicine), the possibility of holding scientists to account might itself have prophylactic values.

REFERENCES

1. Archer, V.E., Wagoner, J.K., Lundin, F.E. "Uranium Mining and Cigarette Smoking Effects on Man" Journal of Occupational Medicine 15(3): 204-211. 1973.
2. Armstrong, B.K., McNulty, J.C., Levitt, L.J., Williams, K.A., Hobbs, M.S.T. "Mortality in Gold and Coal Miners in Western Australia with Special Reference to Lung Cancer" British Journal of Industrial Medicine 36: 199-205. 1979.
3. Ashford, N.A., Analyzing the Benefits of Health, Safety, and Environmental Regulations Cambridge Center for Policy Alternatives, M.I.T., CPA-82-16, 1982.
4. Ashford, N.A., Davies, J.C., Dowd, R.M., Russell, C.S., Josie, T.F., "Examining the Role of Science in the Regulatory Process" Environment 25 June, 1983.
5. Axelson, O., Josefson, H., Rehn, M., Sundell, L. "Svensk Pilotstudie Over Lungcancer Hos Gruvarbetare" (Swedish Pilot Study Concerning Lung Cancer Among Miners) Lakartidningen 68 (49):5687-5693. 1971.
6. Axelson, O., Sundell, L. "Mining, Lung Cancer and Smoking" Scandinavian Journal of the Work Environment and Health 4: 46-52. 1978.
7. Berry, G., Newhouse, M.L., Turok, M. "Combined Effect of Asbestos Exposure and Smoking on Mortality from Lung Cancer in Factory Workers" The Lancet: 476-479, September 2, 1972.
8. Blot, W.J., Harrington, J.M., Toledo, A., Hoover, R., Heath, C.W., Fraumeni, J.F. "Lung Cancer After Employment in Shipyards During World War II" The New England Journal of Medicine 299(12):620-624. 1978.
9. Blot, W.J., Morris, L.E., Stroube, R., Tagnon, I., Fraumeni, J.F. "Lung and Laryngeal Cancers in Relation to Shipyard Employment in Coastal Virginia" Journal of the National Cancer Institute 65(3):571-575. 1980.

10. Brett, G.Z., Benjamin, B. "Smoking Habits of Men Employed in Industry, and Mortality" British Medical Journal 3:82-85. 1968.
11. Breslow, L. Hoaglin, L., Rasmussen, G., Abrams, H.K. "Occupations and Cigarette Smoking as Factors in Lung Cancer" American Journal of Public Health 44:171-181. 1954.
12. Broad, W.J., "Researcher Denied Future U.S. Funds" Science 216:1081. 1982.
13. Con Edison Co. Inhouse Study on the Relationship of Gas Heating and Degree Days Con Edison Co., New York, 1980.
14. Dahlgren, E. "Lungcancer, Hjartkarlssjukdom Och Rokning Hos En Grupp Gruvarbetare" (Lungcancer, Cardiovascular Disease and Smoking in a Group of Mineworkers) Lakartidningen 76 (52):4811-4813. 1979.
15. Elmes, P.C., Simpson, M.J.C. "Insulation Workers in Belfast. 3. Mortality 1940-66" British Journal of Industrial Medicine 28:226-236. 1971.
16. Emerson, J.L., Thompson, D.J., Gerbig, C.G., Robinson, V.B. Results of Teratogenic Studies of 2,4,5-Trichlorophenoxyacetic Acid in the Rat" Report from the Human Health Research and Development Laboratories, the Dow Chemical Company, Zionsville, Indiana, March 1970.
17. Epstein, S.S. The Politics of Cancer Anchor Press/Doubleday, New York, New York. 1979.
18. Hammond, E.C., Selikoff, I.J. "Relation of Cigarette Smoking to Risk of Death of Asbestos-Associated Disease Among Insulation Workers in the United States" Biological Effects of Asbestos, Bogovski, P., et al, Editors, I.A.R.C. Scientific Publications #8, Lyon, France. pp. 312-317. 1973.
19. Jasanoff, S., Nelkin, D. "Science, Technology, and the Limits of Judicial Competence" Science 214:1211. 1981.
20. Kantrowitz, A., "Controlling Technology Democratically" American Scientist 63:505. 1975.
21. Kolonel, L.N. "Association of Cadmium with Renal Cancer" Cancer 37(4):1782-1787. 1976.
22. Kreyberg, L. "Lung Cancer in Workers in a Nickel Refinery" British Journal of Industrial Medicine 35(2):109-116. 1978.
23. Kuratsune, M., Tokudome, S., Shirakusa, T., Yoshida, M., Tokumitsu, Y. Hayano, T., Seita, M. "Occupational Lung Cancer Among Copper Smelters" International Journal of Cancer 13:552-558. 1974.
24. Levin, M.L., Kraus, A.S., Goldberg, I.D., Gerhardt, P.R. "Problems in the Study of Occupation and Smoking in Relation to Lung Cancer" Cancer 8:932-936. 1955.
25. Lundin, F.E., Lloyd, J.W., Smith, E.M., Archer, V.E., Holaday, D.A. "Mortality of Uranium Miners in Relation to Radiation Exposure, Hard-Rock Mining and Cigarette Smoking--1950 through September 1967" Health Physics 16:571-578. 1969.
26. Martishnig, K.M., Newell, D.J., Barnsley, W.C., Cowan, W.K., Feinmann, E.L., Oliver, E. "Unsuspected Exposure to Asbestos

and Bronchogenic Carcinoma" British Medical Journal 1:746-749. 1977.

27. McDonald, J.C., Liddell, F.D.K., Gibbs, G.W., Eyssen, G.E., McDonald, A.D. "Dust Exposure and Mortality in Chrysotile Mining, 1910-75" British Journal of Industrial Medicine 37(1):11-24. 1980.
28. Meurman, L.O., Kiviluoto, R., Hakama, M. "Combined Effect of Asbestos Exposure and Tobacco Smoking on Finnish Anthophyllite Miners and Millers" f 2>Annals of the New York Academy of Sciences 330:491-495. 1979.
29. Miller, C.T., Neutel, C.I., Nair, R.C., Marrett, L.D., Last, J.M., Collins, W.E. "Relative Importance of Risk Factors in Bladder Carcinogenesis" Journal of Chronic Diseases 31:51-56. 1978.
30. Morgan, R.W., Shettigara, P.T. "Occupational Asbestos Exposure, Smoking, and Laryngeal Carcinoma" Annals of the New York Academy of Sciences 271:308-310. 1976.
31. National Academy of Sciences. Indoor Air Pollution, Report of the N.A.S./N.R.C. Committee on Air Pollution, National Academy of Sciences. 1981.
32. Newhouse, M.L., Berry, G. "Patterns of Mortality in Asbestos Factory Workers in London" Annals of the New York Academy of Sciences 330:53-60. 1979.
33. Osburn, H.S. "Lung Cancer in a Mining District in Rhodesia" South African Medical Journal 43:1307-1312. 1969.
34. Pinto, S.S., Henderson, V., Enterline, P.E. "Mortality Experience of Arsenic-Exposed Workers" Archives of Environmental Health 33:325-331. 1978.
35. Reed, D., Lessard, R., Maheux, B. "Lung Cancer on a Nickel Smelting Island" (Meeting Abstract) American Journal of Epidemiology 108(3):233. 1978.
36. Rubino, G.F., Piolatto, G., Newhouse, M.L., Scansetti, G., Aresini, G.A., Murray, R. "Mortality of Chrysotile Asbestos Workers at the Balangero Mine, Northern Italy" British Journal of Industrial Medicine 36:187-194. 1979.
37. Rutstein, D.D., Mullan, M.D., Frazier, T.M., Halperin, W.E. "Sentinel Health Events (Occupational): A Basis for Physician Recognition and Public Health Surveillance" American Journal of Public Health 73:1054-1062. 1983.
38. Selikoff, I.J., Hammond, E.C., Chung, J. "Asbestos Exposure, Smoking, and Neoplasia" Journal of the American Medical Association 204(2):104-110. 1968.
39. Selikoff, I.J., Seidman, H., Hammond, E.C. "Mortality Effects of Cigarette Smoking Among Amosite Asbestos Factory Workers" Journal of the National Cancer Institute 65(3):507-513. 1980.
40. Shettigara, P.T., Morgan, R.W. "Asbestos, Smoking, and Laryngeal Carcinoma" Archives of Environmental Health 30:517-519. 1975.
41. Stell, P.M., McGill, T. "Asbestos and Laryngeal Carcinoma" Lancet 2:416-417. 1973.

42. Stell, P.M., McGill, T. "Exposure to Asbestos and Laryngeal Carcinoma" Journal of Larynogology and Otology 89(2):513-517. 1975.
43. Sterling, T.D. "Does Smoking Kill Workers or Working Kill Smokers?" International Journal of Health Services 8:437-452. 1978.
44. Sterling, T.D. "Statistical Problems in Multi-Variate (Etiological) Surveys" Canadian Journal of Statistics 7:205-215. 1979.
45. Sterling, T.D., Sterling, E. "CO Levels in Kitchens and Homes with Gas Cookers." Journal of the Air Pollution Control Association 29:238-244. 1979.
46. Sterling, T.D., Kobayashi, D. "Use of Gas Ranges for Cooking and Heating in Urban Dwellings." Journal of the Air Pollution Control Association 31:162-165. 1981.
47. Sundell, L. "Lung Cancer in Miners in Relation to Smoking Habits" European Journal of Respiratory Diseases 61(Suppl. 107):131-132. 1980.
48. U.S. Department of Health, Education, and Welfare, Public Health Service. Smoking and Health A Report of the Surgeon General. DHEW Publication No. (PHS) 79-50066. 1979.
49. U.S. Department of Health, Education, and Welfare, The Health Consequences of Smoking: Cancer A Report of the Surgeon General. United States Department of Health, Education, and Welfare, Public Health Service, Office of Smoking and Health, Rockville, Maryland, DHEW Publication No. (PHS) 82-50179. 1982.
50. Weiss, W. "Chloromethyl Ethers, Cigarettes, Cough and Cancer" Journal of Occupational Medicine 18(3):194-199. 1976.
51. Weiss, W. "The Cigarette Factor in Lung Cancer Due to Chloromethyl Ethers" Journal of Occupational Medicine 22(8):527-529. 1980.
52. Wolff, G., Lauter, J. "Zur Epidemiologie des Magenkrebses" Archiv fur Geschwulstforschung 46(1):1-14. 1976.
53. Wynder, E.L., Covey, L.S., Mabuchi, K., Mushinski, M. "Environmental Factors in Cancer of the Larynx, a Second Look" Cancer 38(4):1591-1601. 1976.
54. Wynder, E.L., Goldsmith, R. "The Epidemiology of Bladder Cancer, a Second Look" Cancer 40: 1246-1268. 1977.

FOOTNOTES

1. These examples are just that, examples. After some 25 years of experience in the fray of adversary encounters, one becomes somewhat cynical. Nevertheless, nothing said here should be viewed as a personal attack or criticism. This disclaimer is heartfelt. The examples here come from a genre of actions that appear to happen with sufficient frequency so as to set a standard of their own (as much as one may regret the existence of such a standard).

2. Parallel reasoning would deny that there has been a rise of unemployment, insisting instead that overemployment in previous years makes present unemployment rates look abnormally high, or that alcohol is not to blame for the increase in motor vehicle accidents, insisting instead that nondrinkers have too few accidents.
3. There are a number of computerized files on which such searches can be implemented at a small cost. Both MEDLINE and CANCERLIT were used.
4. Seventeen studies which discussed both smoking and occupation were not considered as the data on smoking was not analyzed in conjunction with the data on occupation. For example, data on smoking would be analyzed for an entire case-control group, and then each of several occupations within the case-control series would be analyzed separately. Results would be presented for both smoking in relation to cancer, and for occupation in relation to cancer, but the two sets of results would refer to two separate populations. In addition, 10 studies were not considered as smoking data was used only to standardize for smoking during the analysis of occupation.

PROTECTING WORKERS, PROTECTING PUBLICS:

THE ETHICS OF DIFFERENTIAL PROTECTION

Patrick Derr
Robert Goble
Roger E. Kasperson
Robert W. Kates

Center for Technology, Environment, and Development
Clark University
Worcester, MA 01610

When talking about risk in the private sector, it is appropriate that attention be directed to the host of issues currently besieging those charged with health and safety responsibility in industry, labor, and the federal government. These issues include such topics as: What role should science and values play in risk assessment? How should product liability and insurance best contribute to risk management in the private sector? How may the new developments in science and technology best be incorporated into occupational health and safety policies? What new institutional initiatives should be taken by the various components of society?

In response to these new challenges, it is essential that we not overlook other very fundamental problems which may be deeply entrenched in societal attitudes and practices, and thus less apparent as policy or health problems. One such case is the long-standing differential which has existed in the level of protection afforded to workers and members of the public from the same hazards of technology. Even in hunting and gathering societies, work was undeniably often risky. But the industrial revolution and the development of the factory system during the 18th and 19th centuries, and the rapid growth of science and technology during the 20th century have, as indicated most recently by the past and future toll from asbestos exposure, led to concentrations of new hazards in the workplace.

In this paper we report on a two-year investigation of this problem whose results have been provided in a series of 10 articles in

Environment (see Appendix 1). We summarize our findings on the following four objectives:

- To examine the scope and extent of differential protection of workers and the public as it exists in law and practice
- To estimate the health toll associated with differential protection
- To examine how differential protection is justified and the validity of those arguments
- To propose means by which society may best respond

DIFFERENTIAL PROTECTION IN LAW AND PRACTICE

To examine the scope and extent of differential protection, we have examined the Clark University data base of some 93 technological hazards, the differences in protection for some 10 hazards where both EPA and OSHA standards exist, and compared regulatory standards in the U.S. with those that exist in other countries.

The Center for Technology, Environment, and Development at Clark University has compiled information on 93 technological hazards. The list is diverse, including such hazards as skateboards, liquefied natural gas, saccharine, collapsing dams, microwave ovens, and contraceptives. We have used this hazard inventory and related data to answer two questions about the differential exposure of workers and the general public to technological hazards:

- Are most hazards in the inventory mainly occupational, mainly environmental, or a combination of both?
- For those hazards that threaten both workers and public, are there systematic differences in the exposure or risk?

For the 40 hazards (Table 1) for which we can compare occupational and public exposures, we have applied five measures of hazard exposure: the magnitude of the population at risk, the maximum number of people who might be killed in one incident, the estimated annual mortality due to the hazard, the maximum concentration of material or energy to which a person might be exposed in an incident, and average annual exposure to the hazardous material or energy (annual dose). For each measure we have asked whether the exposure of workers is greater than 10 times that of the public, less than a tenth that of

TABLE I. FORTY HAZARDS WITH BOTH WORKER AND PUBLIC EXPOSURE

ENERGY HAZARDS	MATERIALS HAZARDS
Nuclear warfare-blast	Automobiles-CO pollution
Handguns-shooting	Recombinant DNA-harmful release
Commercial aviation-noise	Alcohol-accidents
SST-noise	Smoking-chronic disease
Motor vehicles-noise	Pesticides-ingestion
Smoking-fire	Nuclear reactors-high level radiation
Trains-collisions	Nerve gases-accidential release
Microwave ovens-radiation	PCB-dispersion
Automobiles-collisions	Cadmium-dispersion
LNG-fire and explosions	Automobile-air-borne lead
Medical x-rays-radiation	Mercury usage-dispersion
Home appliances-fire	Radwaste-accidental release
General aviation-crashes	Mirex (pesticide)-dispersion
Commercial aviation-crashes	PVC-air emissions
Power mowers-accidents	DDT usage-dispersion
Chainsaws-accidents	Nuclear war-radiation
Elevators-falling	Nerve gas-wartime use
Motor vehicle racing-accidents	Chlorination of drinking water
Skyscrapers-fires	2,4,5 T - toxic effects
	Taconite mining-ingestion, breathing
	Spray asbestos-inhalation

the general public, or equal, within a factor of 10, to the exposure of the public. We consider only substantial differences in exposure, greater than 10-fold or less than one-tenth.

The results, expressed as a percentage of the 40 hazards falling into each range, show that, for most hazards, the size of a catastrophe, the annual mortality, and the maximum one-time exposure for workers and the public are equal within an order of magnitude. However, most hazards expose many more members of the general public than workers, while workers are usually exposed to a much higher dose of the hazard.

It is possible to estimate the difference in annual risk between workers and the public for the 40 hazards by dividing the estimates of annual mortality by the population at risk. These are shown in Figure 1, with the probability of death in any given year for workers shown on the horizontal axis and for the public on the vertical axis. Almost all of the hazards fall below the diagonal of equal risk,

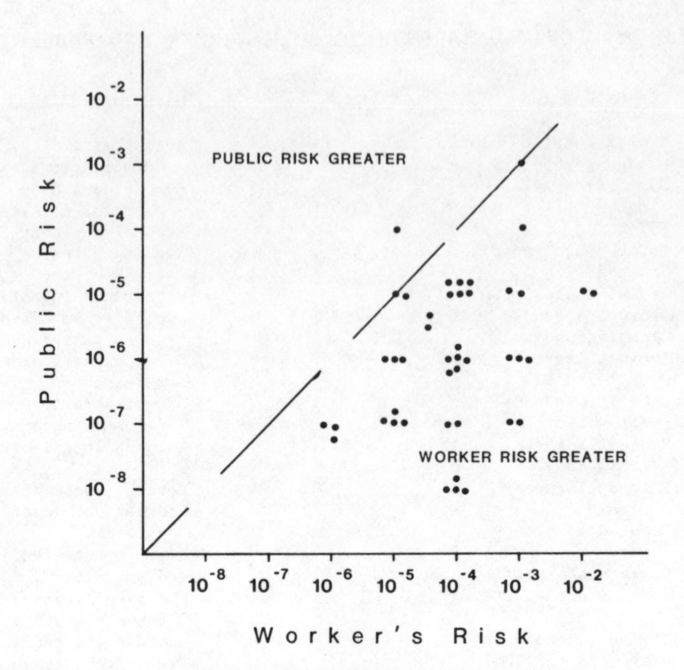

Figure 1. Worker's Risk

clearly demonstrating the tendency for greater worker risk and posing significant questions of equity.

Our second exercise was to examine the practice of regulatory agencies in the United States. Comparing EPA ambient standards with OSHA standards has, of course, its problems: the time periods over which concentrations are to be averaged are frequently different, and uncertainties in dose/response information make it difficult to determine whether a short exposure to high concentrations is better or worse than a long exposure to lower concentrations.

Despite these difficulties, Figure 2 shows the substances for which we can compare limits on concentrations, the current EPA and OSHA standards, and information on medical effects and background level. For each hazard, the ratio of the environmental standard to the occupational standard is shown. Except for ozone and carbon monoxide, for which there are only short-term standards and for which the primary health concerns are short-term stresses, the comparison is based on the limits the standard imposes on the cumulative annual dose.

It is apparent in Figure 2 that regulators afford a greater measure of protection to members of the general public than to workers. While one might expect that standards for both groups would be set at or below the point at which measured harm occurs, this is not the

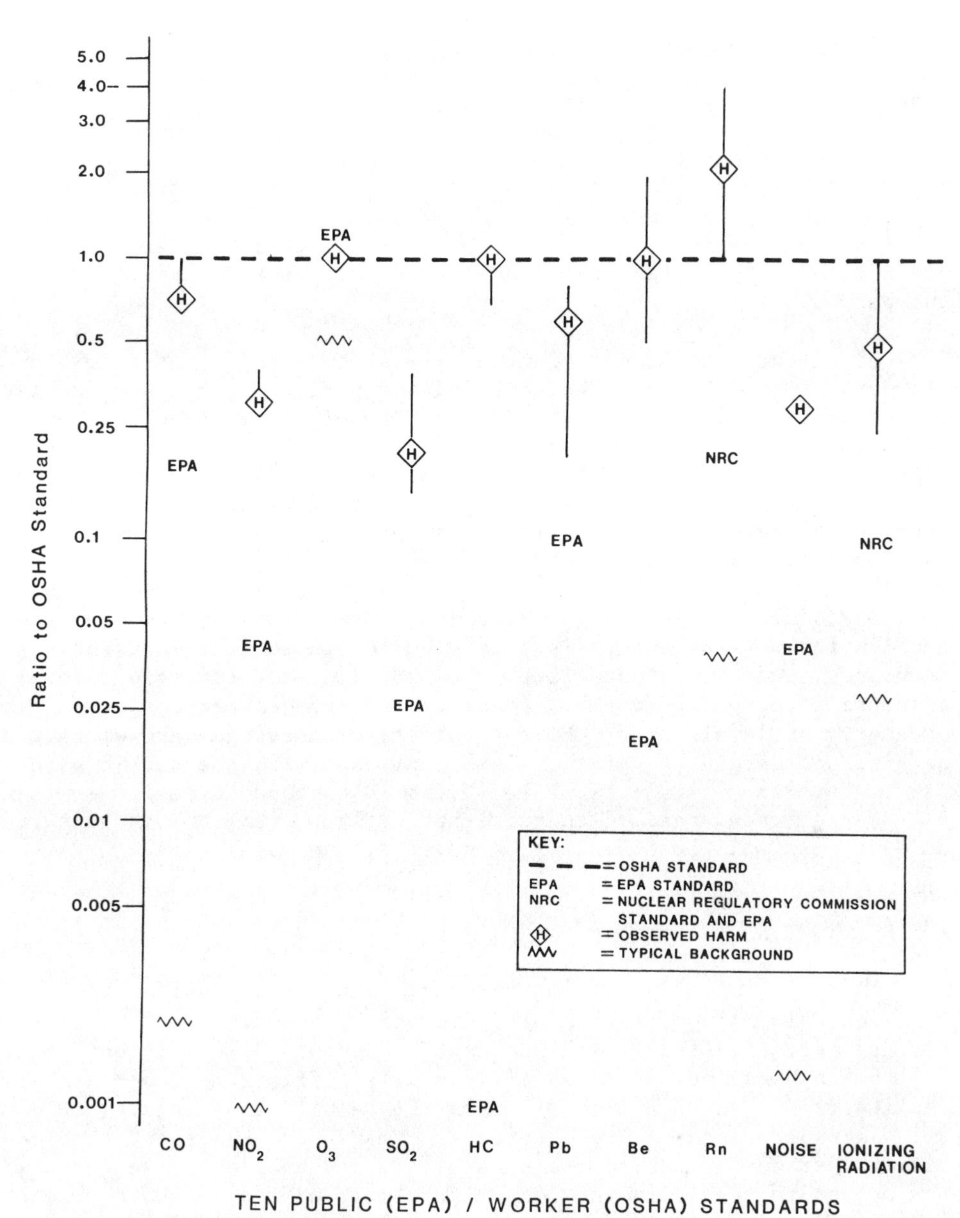

Figure 2. Ten Public (EPA)/Worker (OSHA) Standards

case. While general environmental standards are set below levels of measured harm, workplace standards tend to be set above that level.

To determine whether this pattern of differential protection is a uniquely American phenomenom, we next placed these results in the context of international experience. Specifically, we examined occupational standards for some 12 airborne toxic substances in 13 industrialized countries. That analysis revealed a general East-West dichotomy, with the Soviet Union and East European countries exhibiting the greatest stringency in standards and the United States the greatest permissiveness. While most of the differences are within an order of magnitude, factors of 20 or more appear for heptachlor, lead, and parathion.

Given these major differences, it is remarkable that a double standard of protection of workers and the public is tolerated no less in the Soviet Union than in the United States. Comparing some nine technological hazards, we find that differential protection is the rule in both countries, with permissible exposure levels for the public and for workers varying by factors of five to 200. Nor is the level of differential protection in a socialist society less than that for a capitalist society.

To conclude, there is a universal differential in societal protection for workers and the public embodied in law, administrative standards, and current practice. The baseline for the occupational standard is to permit workers to be exposed to deleterious materials or energy at levels at or above the level of observed harm without a significant margin of safety. This contrasts with the public standard, where permissible exposure levels 10 to 1,000 times lower than the occupational standard characteristically prevail. This pattern holds across political systems, economies, and ideologies.

THE HEALTH TOLL

There are now 220 million persons living in the United States. Of these, almost two million will die this year--mostly from heart disease, cancer, stroke, and accidents. Seventy-five million people will be injured, and 25 to 30 million will suffer some long-term disability from a chronic illness or as the result of an injury. Almost all will suffer some acute conditions: more than two such acute conditions per person per year are reported, and half of these are serious enough to result in the loss of at least one day of work or other activity.

Of these same 220 million Americans, about 97 million will be employed this year for money; 7 million more will be looking for work; and about 18 million will be retired. What proportion of the burden of death, disease, injury, and pain noted above originates in the work experience?

Unfortunately, this question cannot be answered authoritatively or with any precision. In fact, the public health impact of working experience is a hotly debated subject, in which injuries are relatively well-defined but illnesses, and especially cancer incidence, are poorly understood. Elsewhere (Derr, Goble, Kasperson, and Kates 1981), we have suggested that the annual toll, while impossible to pinpoint, is nonetheless large--10 million injuries, 30 million separate bouts of illness, 20,000 to 75,000 deaths, and cumulatively, two million disabled.

What fraction of this toll is attributable to differential protection can only be hinted at by our work. We conducted three case studies--radiation exposure at power plants, the pesticide parathion, and lead--in which we looked carefully at the health price of differential protection. In the case of radiation, the extra occupational toll is small but not negligible (five to nine premature deaths/year). The use of parathion involved a relatively comparable health toll: some 22 job-related deaths, 71 public fatalities, and accidental exposures of 1300 workers and 300 members of the public between 1966 and 1972. The toll from lead is much more serious: 40,000 to 50,000 workers with blood levels indicating physiological damage, including life-shortening effects to the neurological system and to organs such as the kidneys.

Meanwhile, the substantially lower Swedish mortality rates for workers (as compared with the U.S. rates) hint that more determined effort to reduce the differential might reduce the toll from occupational hazards (Kasperson 1983).

JUSTIFICATIONS FOR DIFFERENTIAL PROTECTION

The double standard for protecting workers and the public from particular technological hazards is not, in itself, necessarily unjust or inequitable. There can be compelling justifications for such differentials in particular cases. But neither should differential protection, in general, be presumed to be acceptable. Each case requires careful analysis of the moral argument, social context, and empirical facts. Some differentials may well prove just; others likely will not.

There are, in our view, four major moral justifications that may apply to particular cases of the double standard: differential protection maximizes benefits to society as a whole; workers are better able than members of the public to cope with hazardous exposures; workers are compensated for the risks they bear; and workers voluntarily consent to higher risk as a condition of employment. These are stated formally as principles of equity which could support different protection in Figure 3.

UTILITY: An allocation is just if, and only, if it maximizes the summed welfare of all members of the morally relevant community. If "summed welfare" is understood collectively, the roots of this principle can be traced to the earliest documents of our civilization. If "summed welfare" is understood distributively (as simply adding up individual welfares), the principle takes its classical formulation from the work of the Utilitarians, Bentham and Mill.

ABILITY: An allocation of risks is just if, and only if, it is based upon the ability of persons to bear those risks. Since "need for protection" mirrors "ability to bear risk," this principle is simply a special case of the more general claim that allocations are just if, and only if, they treat people according to their needs.

COMPENSATION: An allocation of risks is just if, and only if, those assuming the allocated risks are rewarded (compensated) accordingly. This principle is derived from the somewhat more general one that an allocation is just if, and only if, it is made according to the actually productive contributions of persons.

CONSENT: An allocation of risks is just if, and only if, it has the consent of those upon whom the risks are imposed. Typical formulations of the principle are found in the Nuremberg Code and in guidelines for experimentation on human subjects.

Figure 3. Four Equity Principles for Differential Protection

Utility

The principle of utility suggests that the discrepancy in protection may be justified on the grounds that the benefit to society outweighs the cost to workers. This is certainly plausible, since the high concentration of hazardous material in the U.S. occupational setting and the comparatively small number of people exposed suggest that there will be differences in the most efficient management of hazards in the two areas.

The evidence from our case studies however, shows that the particular existing discrepancies do not in fact maximize social welfare. The "spreading" of risk to temporary workers in nuclear power plants in preference to more effective exposure reduction management programs and broader use of remote control maintenance have contributed to a growing total radiation burden for society and to reduced incentive to employ cost-effective measures.

In the case of lead, where we were able to directly compare the imposition of controls based on human health effects, the imposed incremental cost-per-health effect on the margin was significantly lower for occupational standards than for environmental standards. It appears likely that the level of parathion exposures for field workers is not justified by any utility calculus.

Ability

Considerations of differential ability to bear hazards can justify differential protection in particular cases. The differential protection in lead and radiation standards, for example, can be partly justified by consideration of the specially vulnerable publics (e.g., infants, pregnant women) who are excluded from, or receive special protection, in employment. But there are other cases--and our work would suggest that they are more typical--in which considerations of ability do not justify current practices and standards.

Our case studies suggest that the regulatory agencies (EPA and OSHA) took seriously the need to identify sensitive populations. Yet, their treatments of sensitivity in the standard-setting process differed. The discrepancies in differential protection for workers and the public, it should be noted, cannot be accounted for by differential sensitivity. In the case of lead, for example, the public standard was based on the characteristics of children, the most sensitive subgroup; OSHA, by contrast, identified workers of reproductive age (both male and female) as the most sensitive group at risk but concluded that it was not feasible to set standards that would protect their potential offspring.

We have also considered a second issue with regard to differential sensitivity within the population of workers. The use of screening programs to identify and remove people from the workforce who are at greater risk of adverse health effects is increasingly common and carries the potential for abuse and unintended consequences (Lavine 1982). In some cases, the means by which less risk-tolerant workers are protected are themselves unjust, when all potentially fertile female employees, regardless of family plans, are excluded from workplaces posing possible teratogenic hazards.

In other cases, it is the _differential consideration_ of such ability that is unjust, as when blacks with hemoglobin defects are "protected" from military flight duty while white officers with recessive genes for similar hemoglobin defects receive no such similar "protection." In still others, the groups placed at _most_ risk are in fact _least_ able to bear the risks imposed, as in the use of the elderly, children, or the malnourished for agricultural work in pesticide-treated fields.

Compensation

Explicit compensation through wages for risk rarely occurs. Although a few jobs do appear to compensate occupational risk through specific increments in wages, these are the same parts of the labor market that are already best-off in other ways (Graham and Shakow 1983). Thus, policemen, who are at far less occupational risk than cab drivers, are explicitly compensated for risk, whereas cab drivers--already much less well paid than policemen--are not.

Temporary nuclear power plant workers hired for specific tasks in high radiation environments receive no specific compensation for risk.

The protracted legal debate over the "medical-removal" provision of the occupational lead standard never considered risk premiums in wages. Ruckelshaus did not count the increased payments that pesticide applicators and farm workers ought to receive as an additional cost of changing from DDT to parathion.

Of course, compensation for hazard exposure in wages need not be explicit. Our detailed analysis of the factors (including health risks) affecting worker earnings (Graham and Shakow 1983) concluded that some workers in the major unionized manufacturing sectors, the primary segment, may receive an implicit wage premium for hazard exposure. Most workers in the secondary labor segment, by contrast, do not receive any such increment to their wages even though they experience equal or greater risk and their actuarial mortality is higher.

Consent

An ethically adequate consent to specific occupational risks would require at minimum that it be both free and informed. Our work suggests that these criteria are rarely met in the workplace. Rather, a consistent pattern emerges that (1) workers are primarily provided information directed toward telling them what they should do to control their exposures once they are on the job; they are not provided information with the expectation that they will choose whether or not to accept the exposure; and (2) workers do not generally feel free to accept or reject exposure; the prospect of losing one's job is considered more serious than even the possibility of quite severe health effects (Melville 1981).

Because of workers' fear that severe lead poisoning could lead to dismissal, when OSHA established a medical-reproval provision for the lead standard, one-and-a-half years' job security and wages protection were offered so that employees would not refuse to have blood-lead measurements taken. Most temporary workers in nuclear power plants and most agricultural field workers are not in a position to refuse employment in an economy where the unemployment rate is running higher than 10 percent. It is largely for these reasons that the Swedish approach to occupational health protection assumes that free choice of employment by workers is impossible and that information concerning risk should be geared toward reduction programs (Kasperson 1983).

We conclude from this scrutiny of the four major justifications that they are rarely publicly discussed and are honored at best only in very limited ways. Some workers, but not most, are partly informed of risks and tolerate them, but only with hindsight after they have accepted employment. Only under rare conditions are risk premiums in wages directly paid; when they are, these compensations tend to correspond with social class rather than the level of risk experienced by the individual worker. Some workers demonstrate greater ability to tolerate hazardous exposure either because the least appropriate among them have been screened out of employment, because they are inherently healthier and are thus able to survive as the fittest, or because they have had training and experience to cope with or reduce their exposure. Additionally, there is the widespread belief (to be found both in opinion and law) and some evidence (as in our nuclear and parathion studies) that the overall aggregate social benefit can be increased by selective use of differentials in exposure.

HOW SHOULD SOCIETY RESPOND?

In light of our findings concerning the broad scope of the double standard, the significant associated health toll, and the intrinsic injustice to workers embodied in differential protection, we believe that society should act to rectify this situation. There are two major responses society can make: it can take measures to decrease differential protection for workers and publics, or it can increase the application of other means of redress to make the differential more morally acceptable. We recommend the following guidelines (Figure 4) for response.

Step 1: In all cases of differential protection for workers and publics, it should first be determined that the discrepancy carries significant benefits to society as a whole. If such benefits do not exist, the level of risk presented to workers should be reduced to that which prevails for the public.

Step 2: Even if differential protection carries significant benefits to society, action should be taken to reduce the risk to as many workers as possible and to reduce it to as close to the level of protection afforded to the public as can reasonably be achieved, where "reasonableness" is determined according to the viability of the industry and net benefits to society as a whole. The argument for such action is not only an equity one; the widely recognized ethical principle of nonmaleficence calls for the avoidance of harm wherever possible as a hallmark of decent and responsible behavior. Care should be taken that the risk is actually reduced and not simply relocated and that other equivalent new risks are not substituted in its place.

Step 3: For those workers for whom the risk cannot reasonably be reduced to the level of public protection, action should be taken to increase the degree of consent (through increased information and enlarged choice in employment) and compensation (as through insurance or risk premiums in wages). Increased information will also, of course, better equip workers to enter into negotiation with employers.

Step 4: As a last step, a determination should be made as to whether residual risk remains for certain groups of workers due to differential sensitivity to hazards. If so, special educational and protective measures should be undertaken to achieve as much equality

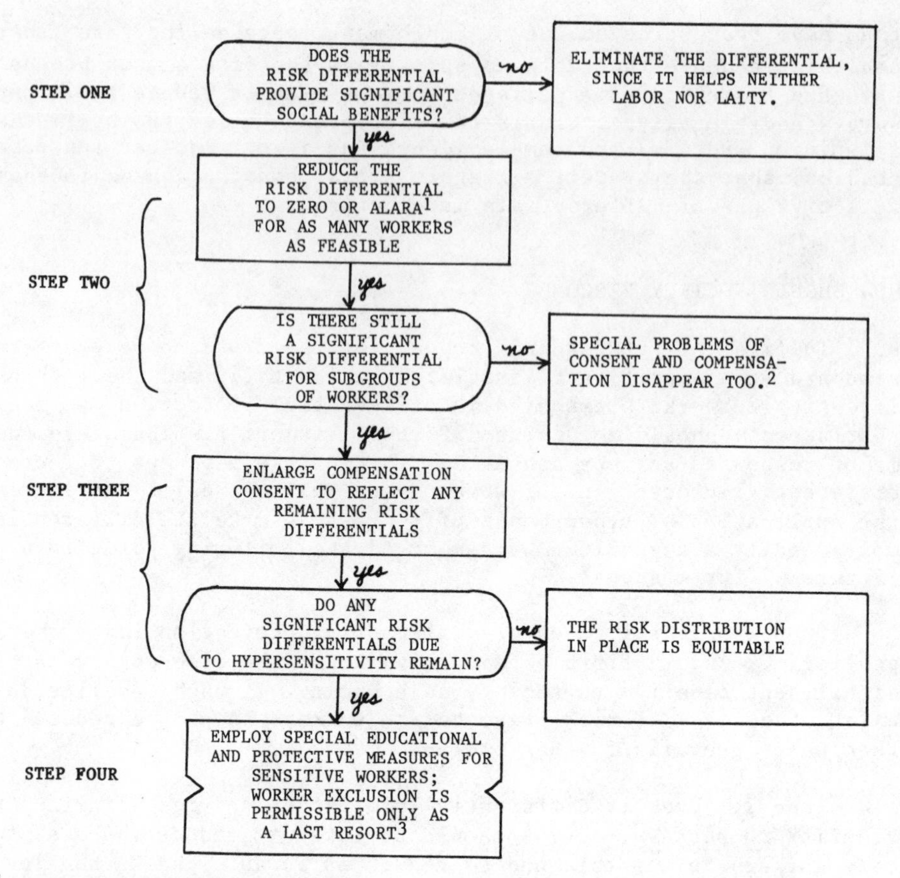

[1] as low as is reasonably achievable

[2] because, of course, there is no excess risk which requires any special consent or compensation

[3] because the avoidance of bodily harm in this way carries the potential for substituting other serious harms associated with exclusion and discrimination. Identifying and responding to differential sensitivity to hazards poses a complicated set of scientific and ethical problems, which we have begun to explore in new research effort.

Figure 4. Schematic Diagram of Guidelines for Responding to Inequities in Risk

in risk as is reasonable. Exclusion of groups of workers from exposure to the hazards should occur only as a last resort with transfer to equivalent jobs (as measured by various social goods) being the preferred strategy. Since this action carries the danger of abuse and other unintended consequences, it should be particularly avoided where the groups involved are traditional victims of economic or social discrimination.

These guidelines are our preferred approach. But recognizing that they will be employed only by those already committed to narrowing differential protection, we recommend that the Occupational Safety and Health Administration institute a comprehensive review of its existing standards for the protection of worker health and safety. Where OSHA finds that its standard exceeds (in equivalent terms) the level of protection afforded to members of the public in comparable standards and regulations enacted by other government agencies, we propose that the Secretary of Labor institute procedures to adopt the more protective standard.

APPENDIX I

A List of Articles Published in the Clark University Project:

Derr, Patrick, Robert Goble, Roger E. Kasperson, and Robert W. Kates (1981). "Worker/Public Protection: The Double Standard," Environment, 23 (September), 6-15, 31-36.
Derr, Patrick, Robert Goble, Roger E. Kasperson, and Robert W. Kates (1983). "Responding to the Double Standard of Worker/Public Protection," Environment, 25 (July/August), 6-11, 35-36.
Graham, Julie and Don Shakow (1981). "Risk and Reward: Hazard Pay for Workers," Environment, 23 (October), 14-20, 44-45.
Graham, Julie, Don M. Shakow, and Christopher Cyr (1983). "Risk Compensation--In Theory and in Practice," Environment, 25 (January/February), 14-20, 39-40.
Hattis, Dale R., Robert Goble, and Nicholas Ashford (1982). "Airborne Lead: A Clearcut Case of Differential Protection," Environment, 24 (January/February), 14-20, 33-42.
Johnson, Kirsten (1982). "Equity in Hazard Management: Publics, Workers, and Parathion," Environment, 24 (November), 28-38.
Kasperson, Roger E. (1983). "Worker Participation in Protection: The Swedish Alternative," Environment, 25 (May), 13-20, 40-43.
Kasperson, Roger E. and John Lundblad (1982). "Closing the Protection Gap: Setting Health Standards for Nuclear Power Workers," Environment, 24 (December), 14-20, 33-38.
Lavine, Mary P. (1982). "Industrial Screening Programs for Workers," Environment, 24 (June), 26-38.
Melville, Mary (1981). "Risks on the Job: The Worker's Right to Know," Environment, 23 (November), 12-20, 42-45.

ACKNOWLEDGMENT

The authors wish to express their appreciation to their colleagues in the Center for Technology, Environment, and Development of Clark University. The research for this article and for the series was supported by the National Science Foundation under grant number OSS 79-24516. Any opinions, findings, or recommendations expressed herein are those of the authors and do not necessarily reflect the views of the National Science Foundation.

THE CHEMICAL INDUSTRY'S VIEW OF RISK ASSESSMENT

Geraldine V. Cox
Timothy F. O'Leary
Gordon D. Strickland

Chemical Manufacturers Association
2501 M Street, N.W.
Washington, DC 20037

ABSTRACT

The Chemical Manufacturers Association (CMA) supports and encourages the use of risk analysis in the regulatory and legislative process. Only through dispassionate analysis of the facts, and careful risk assessment can this nation be assured that the decisions effecting environmental and health issues are made rationally. We can no longer afford to throw money away to solve nonproblems. Rather, we must use risk analysis as a support to the policymaker.

CMA examined key issues in risk analysis and risk management and adopted a policy for regulatory impact analysis of health, safety, and environmental regulations. The association recommends that:

- Regulations should be adopted when (1) a need for regulation has been demonstrated, (2) costs bear a reasonable relationship to the benefits, and (3) the most cost-effective approach is adopted.

- Regulatory impact analysis should not include quantification of intangibles in monetary terms.

- Regulatory agencies should use "good science" in defining both the need for a regulation and the benefits in terms of risk reduction it will provide.

- Regulatory agencies should evaluate alternative approaches.

CMA established a sepcial committee to investigate the issues of the use of risk analysis in regulatory decisionmaking. This paper will discuss the structure of CMA's effort and the history of our program.

KEY WORDS: Chemical Industry, Regulatory Analysis, Risk Assessment.

INTRODUCTION

Scientists and technicians of the Chemical Manufacturers Association (CMA) and its member companies believe that risk assessment is and always has been an important tool for making logical business and scientific decisions. Granted, risk assessment is a crude tool for making many decisions because many variables may be ill-defined or unknown. It is, however, a scientific process and one with proven value.

Properly developed risk assessment models have allowed researchers to march safely through such perilous scientific frontiers as space travel and nuclear research. Each step of these programs was subjected to risk analysis, and each decision was based on the safest path at each decision junction. The question is not whether risk assessment has value, but whether for a specific application, the developed model is too crude a tool for the job at hand.

Historically, all tools are crude when they are first invented, but if the job to be done is important, or if the job must be done on a grand scale, it is worthwhile to improve the tools. CMA contends that legislation and regulation are important and that they exist on a grand scale. Additionally, we believe that both can be improved through the risk assessment process: through assessing the impact of each alternative and selecting the best one to achieve our goals. CMA believes that while the tools for the job may be crude, the job justifies refining them until they are sufficient for the task. CMA also contends that our health, our environment, and our economic well being can no longer be entrusted to political expediency. We apply better science to packaging products and developing hot tubs than to developing laws and regulations that guard our health, environment, and economy. In an era when over 90 percent of the scientists who ever lived are still alive, there can be no excuse for abandoning good science.

William Ruckelshaus, Administrator of the Environmental Protection Agency, discussed risk assessment at a meeting at the National Academy of Sciences. He called upon this nation's scientists to help find better solutions to today's health and environmental problems.

The scientists and technicians of the chemical industry have been working toward that goal for some time by building better tools for legislative and regulatory impact assessment. Clearly, we are not alone. Most, if not all of you are working on the same project, but we are all working almost independently.

The public has a fear of risk analysis, and it is rooted to a basic fear of science and technology. We are a high technology society whose general population is not well versed in science. Most of our legislators and policymakers have had little more than high school science in their formal education, not to mention the legal profession that often is forced to determine scientific issues in a court of law. This is not a rational approach.

Critics of the risk assessment process love to ask if we can wait for dead bodies to pile up while we do the risk analysis. They also like to reference the one in a million as being a real victim. When we are dealing with numbers this low, we will never be able to measure this incidence in a general population with a natural disease at levels well above one in a million. Yet, these values are thrown at us as if they were real. Engineers have used risk assessment for many years when they design highway curves. They have an established bank in the curve that is designed to protect a car moving at a certain maximum speed. This practice is older than super highways, yet, there is not a major outcry when someone exceeds that speed and is injured.

The general public feels that the world is becoming a more dangerous place. Surveys conducted in 1980 by Cambridge Reports, Inc., Lou Harris and Associates, Inc., and Yankelovich, Skelly and White, Inc., painted a disturbing picture of public attitudes toward risk. Table I indicates that over two thirds of the public believed that the world was becoming more dangerous, and the proportion of people who

TABLE I. THE WORLD IS BECOMING A MORE DANGEROUS PLACE

In general, do you think the dangers to personal health and safety the average person is exposed to today are greater or less than the danger an average person was exposed to 50 years ago?

	1978	1979	1980
Greater	66%	67%	68%
Less	21	17	14
About the same	8	11	14
Don't Know	4	5	4

Cambridge Reports - 3078

believed that the world was less dangerous fell sharply between 1978 and 1980.

Surprisingly, the threat of nuclear war was not the major reason people felt less secure. In fact, the atomic bomb ranked very low among the reasons people gave for their feelings. Table II shows that pollution, other environmental problems, automobiles, and crime were the main reasons people felt that the world was becoming more dangerous.

TABLE II. WHY DO YOU FEEL THAT WAY?

Greater	
Pollution, environment	17%
More cars, vehicles traffic	13
More crime	11
Pace, complexity of world	10
Increased industry, machinery, technology	10
Air pollution, atmospheric chemicals	7
More population	5
Additives in food	4
Vehicular pollution	4
More accidents	3
More diseases, cancer causes	3
Radiation, nuclear waste	2
Pesticides, herbicides	1
Atomic Bomb	1
Other	6
Don't know	2

Cambridge Report - 3078

Chemicals were perceived as a major influence in increasing risk. Table III illustrates that over four fifths of the people believed that chemicals posed more of a risk in 1980 than they did 20 years earlier.

Table IV shows that almost nine out of ten people felt that all chemicals posed some risk. Nevertheless, nearly three quarters of those questioned also believed that the benefits of chemicals outweighed the risks.

The public fear of chemical risks seems to be the result of their exposure to one-sided news reporting rather than real exposure to chemicals. Scarcely a week goes by without newspapers, radio, or television reporting that yet something else was suspected of causing cancer. Often, such reports were and still are, based on preliminary

TABLE III. MORE OR LESS RISK THAN IN THE PAST?

For each of the following items, tell me if compared to 20 years ago you feel society today is exposed to more or less risk.

More Risk	Public
Crime and personal safety	85%
The chemicals we use	81
International political stability	73
Energy sources	71
Methods of travel	62
Domestic and political stability	61
Water supply	60
Personal financial stability	59
Food supply	56
Household products	52
Medical care	32
Occupational safety	31

Risk in a Complex Society
Lou Harris, 1980

findings or studies which are later found to be seriously flawed. Remember this laundry list:

Saccharin

BHT

PCBs

Nitrites

BHA

ad nauseam

TABLE IV. CHEMICALS AND RISKS

Do you agree or disagree with these statements about chemicals?

	1978	1979	1980
All chemicals pose some risk:			
Agree	85%	86%	88%
Disagree	12	12	10
Don't Know	3	2	2
Most chemicals made by the chemical industry are beneficial and it would be hard to live without them:			
Agree	67	75	73
Disagree	30	22	25
Don't know	3	3	2
In the case of most chemicals now on the market, the benefits outweigh the risks:			
Agree	-	-	70
Disagree	-	-	26
Don't know	-	-	4

Corporate Priorities, 1980

The announcement that nitrites were suspected carcinogens, for instance, sparked a near panic among the American consumers. Nitrites were widely used to preserve meat. They are still widely used because they do not pose a significant risk and provide great benefits in preventing spoilage.

Unfortunately, the truth never seems to catch up with the initial scare story. The story about another suspected source of cancer invariably appears on the front page of the newspaper. Six months or a year later, the story reporting that the danger is nonexistent or far less than first supposed, might be a short filler item--if it appears at all.

There is little wonder, then, at the public's fear of chemicals. The wonder is that the people still recognize the benefits that chemicals provide, even though that side of the story seldom appears.

This force of public opinion is bound to pressure the public policymaker into making rapid decisions to solve a problem that the public perceives--whether it is a real problem or not. The dispassionate use of risk analysis will avoid such stupidity as the political buy out of Times Beach based on levels of dioxin that were 15,000 times less than Seveso, Italy, where no chronic effects have been measured.

RISK ANALYSIS IN THE GOVERNMENT

The public's perception of danger lurking everywhere has been codified in laws and court decisions that attempt to lessen risk or even eliminate it. The "Delaney Clause" is a prime example of such an effort.

The clause inserted into the Food, Drug, and Cosmetic Act forbids any substance found to cause cancer from being added to food in any amount. In effect, it was an attempt to eliminate risk.

New agencies such as the Environmental Protection Agency, the Consumer Product Safety Commission, and the Occupational Safety and Health Administration were formed in an attempt to reduce risks associated with the environment, consumer products, and the workplace.

In pursuit of such laudable goals, however, the government has frequently ignored the effect it has on other equally worthy objectives. Improved living standards, a healthy economy, and the freedom to make personal choices have often been forgotten in the rush to reduce risk.

The public, however, has long recognized that society must balance risks and benefits. As Table V shows, a substantial majority of

Americans believed that it would be a mistake to ban everything that might cause cancer.

The realization that risk cannot be eliminated entirely has failed to penetrate very far into the legal or judicial system. On one hand, the "Cotton Dust" decision, the American Textile Manufacturers Institute, Inc., et al. v. Donovan, Secretary of Labor, et al., concluded that costs need not be considered in setting OSHA exposure standards for toxic materials or harmful physical agents. On the other hand, the "Benzene" decision, Industrial Union Department, AFL-CIO v. American Petroleum Institute, determined that costs should bear a reasonable relation to the benefits of a standard. At best, the results of recent court decisions are mixed.

Such a result is not surprising, given the piecemeal nature of consumer and environmental laws. Administrations generally have formed their policies in response to whatever crises that they faced at the moment of their decision. Their decisions lacked an underlying theme and legislative mandates were inconsistent. Government agencies had overlapping and sometimes conflicting responsibilities which lead to similar problems in the regulations they issued.

Shortly after President Reagan was sworn in, he mandated, in Executive Order 12291, that government agencies must perform regulatory impact analysis for major regulations.

Both Presidents Ford and Carter had previously sought to compel government agencies to consider the impact of regulations. President Ford ordered agencies to evaluate the inflationary effect of their proposals. President Carter went a step further. He ordered agencies to perform a regulatory analysis of significant regulations. Such analyses had to include a description of possible alternatives and an explanation of why one was chosen over all other possibilities.

TABLE V. CANCER AND THE DELANEY CLAUSE

Some products like birth control pills and certain food preservatives are very useful but increase the risk of getting cancer. Do you think that we should automatically ban the use of everything that may cause cancer?

	1978	1979	1980
Yes	22%	24%	26%
No	67	69	69
Don't know	11	7	9

Corporate Priorities, 1980

Nevertheless, President Reagan's order was a tremendous stride forward. It provided detailed instructions on what the agencies were required to do and gave the Presidential Task Force on Regulatory Relief and the Office of Management and Budget great authority in supervising the agencies' compliance with the Executive Order. Moreover, it was a tangible expression of the Administration's philosophy of issuing regulations only after they have been thoroughly evaluated. As such, it was a dramatic change in the government's attitude toward regulation.

The Executive Order requires that, in issuing new regulations and in reviewing old ones, the government agencies must:

- Demonstrate a need for the regulation and analyze its consequences
- Ensure that potential benefits outweigh potential costs
- Use the most cost-effective approach in setting objectives and ordering priorities

Regulatory impact analyses need to be carried out only for major rules. Such rules cost $100 million or more, significantly affect prices or adversely affect a range of other economic activities. Analyses must provide detailed information on a rule's possible effects. All costs and benefits must be described thoroughly, including who will bear the cost and who will reap the benefits. Moreover, an analysis must describe alternative approaches to achieve the regulatory goal.

The Executive Order has improved, a little, the quality of analyses done by government agencies, but much room for improvement remains. Most agencies still do not have detailed guidelines for carrying out the order and analyses show inconsistent assumptions, methods, and coverage of important points. The quality of some analyses reflects a lack of clear agency statements that these are important tools for decisionmakers.

ROLE OF THE CHEMICAL MANUFACTURING ASSOCIATION

CMA formed the Public Risk Analysis Special Committee (PRASC) late in 1980 to help develop a sound policy for analyzing and managing risks. PRASC was charged with developing policies on the value, limitation, and practice of public risk analysis. The committee was also responsible for determining public risk management techniques that CMA could advocate in legislative and regulatory arenas. Moreover, PRASC sought to communicate with the academic community, regulatory agencies, and others interested in risk analysis. A major goal was to let others learn from PRASC's experience.

At that time, and to a lesser extent even today, the interrelated concepts of risk, cost, and benefit were neither based on adequate identification of issues, or definition of terms; nor were they supported by a solid analytical or conceptual structure. The science of risk assessment was very primitive.

CMA structured PRASC to contain the many disciplines needed to analyze risks thoroughly. The group was composed of chemists, toxicologists, medical doctors, mathematicians, economists, lawyers, environmental specialists, industrial managers, and public affairs specialists. The multidisciplinary approach sought to ensure that all facets of risk were examined.

CMA believed that a sound, integrated natural policy for dealing with health and safety issues should rest on two basic concepts:

- Policy decisions should be based on sound science, reflecting all relevant scientific data.

- Policy decisions should be based on a reasoned balancing of risks, costs, and benefits as well as other special and economic issues. Such decisions should always seek to maintain personal freedom of choice.

In examining the types of risks to which the public is exposed and how to analyze them, CMA took as its starting point certain working hypotheses. Even three years later, the hypotheses still seem to provide a valid conceptual framework for analyzing health and safety issues.

The first hypothesis was that a risk-free society is not feasible. We can never eliminate all risk, even in a particular area.

Second, the degree of risk can vary widely from trivial to unacceptable. The degree of regulatory control, then, should vary according to the degree of risk involved.

Moreover, a sound regulatory policy would require different standards for qualitatively different risks. Those risks that people assume voluntarily should be treated differently from risks that are imposed involuntarily.

When risks were not "unacceptable" and assumed voluntarily, regulations should permit people to make an informed choice about the trade-off they prefer.

Finally, communicating too much information could be as confusing as communicating too little. The public could be overloaded with unstructured information or could be needlessly frightened of insignificant risks.

KEY ISSUES

The members of the Public Risk Analysis Special Committee began to consider the basic issues involved. PRASC felt that considering such issues might lead to a socially acceptable and scientifically sound framework for risk analysis.

One such issue was the relative roles of industry and government in protecting the public against risks. How far should government intrude into the decisionmaking process? What responsibility did industry have in this area?

"Zero Risk" was another issue. If "zero risk" was not feasible, what responsibilities did government and industry have to inform the public about risks? How could such information be provided without needlessly frightening the public or overloading it with information?

Risks differ in kind, degree, time, population at risk, type (voluntary or involuntary), and many other characteristics. Was it possible to rank risks by order of magnitude? If it were possible to rank such risks, should information be communicated to the public according to the magnitude of risk?

Different constituencies--business, labor unions, customers, environmentalists--viewed risks differently. Would it be possible to establish a mutually satisfactory framework for risk analysis?

What role should industry play in trying to develop such a framework and how far should industry go in trying to generate government and public support?

THE CHEMICAL INDUSTRY VIEW

Executive Order 12291 shifted the focus of the Chemical Manufacturers Association's activities dealing with risk analysis. The issue has become much broader and now encompasses the entire scope of regulatory analysis. Consequently, the Public Risk Analysis Special Committee within CMA developed, and the Association adopted, a policy for regulatory impact analysis of health, safety, and environmental regulations.

The policy states that regulatory agencies should perform regulatory impact analyses to make government decisionmaking processes more effective. CMA believes that improved analysis at the beginning of a regulatory proposal will allow workable and effective rules to be in place sooner. Scientific, technical, and economic issues should be examined before important decisions have been made.

The Chemical Manufacturers Association recommends the following guidelines for regulatory impact analysis:

o Regulations should be adopted when (1) a need for regulation has been demonstrated, (2) costs bear a reasonable relationship to benefits, and (3) the most cost-effective approach is adopted.

o Regulatory impact analysis should not include quantification of intangibles in monetary terms.

o Regulatory agencies should use "good science" in defining both the need for a regulation and the benefits in terms of risk reduction it will provide.

o Regulatory agencies should evaluate alternative approaches to regulation.

These guidelines both support Executive Order 12291 and add some new concepts.

CMA believe that regulations should be adopted only where they will significantly reduce risk. Regulations should not be used to induce small changes or to reduce already minor risks. Justification for a regulation should be based on scientific data that clearly identify the hazard to be reduced and show to what extent the regulation will reduce the hazard.

The anticipated cost of a regulation should include both the direct costs of complying with it and its direct costs throughout the economy. Similarly, the benefits to be included are the direct and indirect benefits of the regulation.

The Association believes that a regulation should take the least burdensome approach that will achieve its goals. Resources are wasted whenever regulation imposes requirements not directly related to its objectives.

While, Executive Order 12291 does not address the issue of quantifying intangibles, the Chemical Manufacturers Association has taken the position that regulatory agencies should not place a dollar value on human life, other health effects such as pain and suffering, or aesthetics. Such quantification is not meaningful to society and the use of a mechanistic cost-benefit ratio for decisionmaking would be unwise. Regulatory impact analysis should provide decisionmakers with as much information as practicable to ensure that regulations express human as well as economic values.

CMA has not accepted the notion that all values can be reduced to money. Quite apart from the ethical difficulties of considering human beings as just another commodity to be used and discarded, reliable techniques do not exist to place a price tag on the intrinsic value of a human life. Likewise, no reliable way yet exists to place a dollar value on aesthetics such as a blue sky or a scenic view. Analytical techniques, such as they are, have great conceptual problems and yield inconsistent and wildly divergent results. CMA believes that human lives and aesthetics should not be treated as commercial transactions, but that trade-offs should be social decisions expressed through elected and appointed officials.

Executive Order 12291 did not specifically address the need for scientific review, but CMA urges agencies to use quantitative risk assessments that are based on substantial evidence. Unsupported assumptions or seriously flawed scientific studies form a poor basis for regulation.

Analysis should be reviewed by scientists from industry, universities, and other interested groups as well as by agency scientists to ensure that the data are valid and that interpretations are correct.

CMA supports the concept that regulatory agencies should analyze the potential costs and benefits of reasonable alternatives for achieving regulatory goals. Nonregulatory approaches, such as economic incentives, can be more effective and less costly than regulations.

Agencies should also consider alternative methods of regulatory control. Such alternatives might include flexible compliance standards, variances, and exceptions.

CMA is following these policies in its studies of the impact of regulations. A new committee, the Regulatory Impact Special Committee (RISC), is reviewing the major issues of regulatory impact analysis and the ways that government agencies are carrying out Executive Order 12291.

RISC has defined major terms related to risk analysis and regulatory impact analysis, and has summarized the viewpoints and philosophies of industry, environmental groups, labor unions, government agencies, and others concerned with risk assessment or the analysis of regulations. In addition, RISC has developed a checklist for reviewing regulatory impact analysis and has outlined what can and cannot be quantified in such an analysis. The committee has made suggestions to government agencies on their proposed guidelines for analyzing regulations and has formally commented to EPA and its Science Advisory Board on the Agency's guidelines for exposure assessment.

Work along these lines will continue with an increased emphasis on examining the analyses that government agencies are doing to support proposed regulations. For instance, RISC and other CMA groups are now involved in reviewing EPA's proposed effluent guidelines for the organic chemicals, plastics, and synthetic fibers industries. Previously, the committee dissected the studies supporting OSHA's hazard communication rule and EPA's proposed rule on premanufacture notification. In most cases, agencies were found to use questionable data, assumptions, or analytical methods as the foundation for proposed regulations. CMA hopes that close scrutiny will encourage agencies to improve the quality of their analyses.

CONCLUSION

The Chemical Manufacturers Association will continue to examine the potential of regulatory impact analyses to improve the quality of the rules by which businesses and consumers must live. The chemical industry expects to contribute substantially to this growing area of study.

SECTION 4: RISK ANALYSIS AND RISK MANAGEMENT: ISSUES, METHODS, AND CASE STUDIES

RISK ANALYSIS AND RISK MANAGEMENT

Chauncey Starr

Electric Power Research Institute
New York, NY

HISTORICAL RISK ANALYSIS

It is interesting to note that humanity's basic attitudes to risks and their relation to the quality of life seem to be ageless and unchanging. To quote Horace,[1] the Roman poet of 2000 years ago:

"From hour to hour not one of us
Takes thought of his peculiar doom;
Bold sailors dread the Bosporus
Nor heed what other fate may loom..."

As Horace points out, the average person is not likely to develop a balanced perspective of life's spectrum of risks, or of the associated benefits of life's activities. Although risks are real, and often quantifiable in the aggregate--as with physical accidents--individual perceptions and attitudes usually are not derived from these realities. Thus, public concern with issues of risk may result in powerful popular movements, but these are rarely a useful guide for the most effective allocation of national resources to increase public health and safety. Providing a better and less subjective guide is the objective of risk analysis.

It is useful to recognize the important but nevertheless partial role that risk analysis plays in the societal actions that pragmatically determine the real risks to the exposed public. I have shown in Figure 1 the closed loop of the key decision functions which determine societal risk exposure. Of course, the practical processes involved in these functions depend on continuous and complex interactions among the implementing organizations, and these are not represented on this schematic.

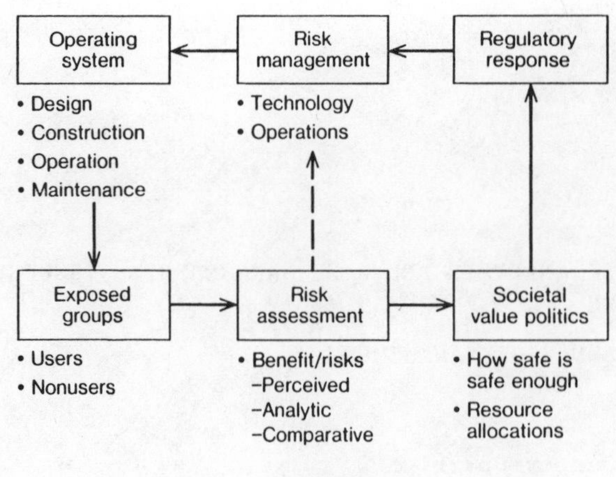

Figure 1

Risk Decision Functions

To those of us concerned with both the processes and the functions, the most difficult to place in an analytic context is the political interplay of group values, ideologies, and goals. For the analytic professional, the political process of compromise and accommodation to conflicting views often appears irrational, although there must be a political rationality in its outcomes. Most serious, however, are those occasions when the political outcome is unwittingly counterproductive for public health and safety because of the compromises and accommodations. We will discuss this more fully later.

The assessment and management of risks--both individual and group--have always been an implicit, and sometimes explicit (as in warfare), part of human activity. What has changed with time is the decisionmakers' perception of relative social values. What has generally not changed is the pervasive self-interest of decisionmakers, whether it be their physical survival, accretion or protection of wealth, political power, or some more abstruse agenda. The notion customarily promoted by national decisionmakers that government actions are taken to optimize the general welfare is rarely valid.

This is more than a superficial and cynical observation. Modern analytical assessment of societal benefits and risks, as it is attempted analytically in the industrial countries, is customarily targeted on the interests of society as a whole rather than on those of a narrow group. For this reason objective analytical assessments of societal benefits and risks are not likely to be directly influential in the political process, except as an information reference or as a basis for public education. As expressed in a Wall Street Journal editorial:[2]

> "...according to the rules and regulations of contemporary national politics,

> to wit: the origin and substantive details of the problem at hand are largely irrelevant; what's important is to frame the issue as essentially a battle for votes..."

On the other hand, these analytic assessments may be very influential in determining the choice of risk management techniques applied to operating systems, both by regulators and system producers and users. Further, the persistent introduction of quantitative risk analysis into public debate eventually creates a more realistic perception politically. In its absence the qualities of risk tend to overshadow the quantities (e.g., carcinogen issues).

The contribution to the reduction of public risks by modern risk assessments, involving quantification of system event probabilities and their consequences, is best understood by considering the accepted approaches to risk prior to the middle of this century. Civil engineering structures--buildings, bridges, dams, etc.--are classic historical examples. The historical design objective was to avoid failure of the structure, defined as collapse under expected usage. To provide such assurance, the designers applied a traditional "safety factor." For example, if a rope was tested to hold 100 pounds, a safety factor of 10 would be provided if the maximum load did not exceed 10 pounds. In practice, these safety factors traditionally ranged from a low of about three to as much as 40, depending on the designers' judgment and the tradition for each type of usage, i.e., steady state, cyclic stress, shock, corrosion, etc. Thus, the safety factor supplied a design umbrella large enough to cover all the areas of the designers' known range of ignorance, i.e., the "known unknowns." The system worked reasonably well, although an occasional structure collapsed because of an "unknown unknown"; for example, the Tacoma bridge collapse caused by unanticipated wind-induced oscillations.

The safety factor design approach was socially acceptable at that time. The engineering profession said, "trust us," and the public did. There were no probabilistic risk assessments involving off-design failure analyses, no environmental impact statements, nor any of the other modern trappings of project reviews. The designers' judgment on the choice of safety factors integrated all uncertainties without an explicit justification of the choices. The public risk was implicitly covered by the design objective of avoiding failure, but was never explicitly estimated. When the unforeseen occasionally occurred, it was usually accepted as an "act of God."

The historical approach to the risk management of a replaceable product which permitted experience feedback was one of empirical "trial and error," as, for example, with autos and airplanes. Operating experience was fed back to guide improvements, a process that continues today. The traditional "safety factor" was less important in such product designs, because the feedback process was sufficiently rapid (a few years) to permit improvements needed for achieving a performance target. The collective risk was initially low, because only a few individuals were involved in the early developmental stages, although individual risks were high. As usage grew, the public was asked by manufacturers to "trust us." Today, with a large fraction of the public using autos and airplanes, failures, and "recalls" are more serious. This has given rise to recent consideration of comprehensive assessments of both benefits and risks arising from technological changes in such fields.

It should be recognized that the "safety factor" and "trial and error" methodologies continue to be pragmatically useful, and are only slowly being replaced by modern risk assessment approaches in a limited number of publicly pervasive systems. The penetration of large-scale technologies has become much more rapid than decades ago, so the "trial and error" method can be very costly both in public health and cost. Further, some large-scale systems involve so many interdependent components, that the individual "safety factor" approach would be compounded to the point of making the system inoperable (e.g., air transport). Finally, very rare, but high consequence events may require decades or centuries to provide the feedback information for guiding decisions, and each such occurrence may be undesirably costly to public health and safety. It is these considerations that have encouraged the development of modern risk assessment approaches.

MODERN RISK ANALYSIS

The objective of modern public risk assessment is to provide a basis for actions to minimize the impairment of public health and safety arising from technical systems creating risks. This objective is not directly concerned with the ability of a technical system to perform its functions, in contrast to the historical "safety factor" or "trial and error" approaches. The risk assessment focus is on injury to the user and the public, with the technical equipment being considered a potential risk-creating source.

The analytical approach to estimating the probability of each event in such a system analysis utilizes empirical data when it is available, or experience with similar circumstances and equipment, or professional judgment based on a composite of experience. The same situation applies to consequence estimates. Thus, risk analysis generally embodies the heuristic approach of empirical learning, with large uncertainties in event probabilities and public consequences. Nevertheless, the central values of the final estimates do provide a "best knowledge" estimate of the relative importance of a risk. Further, the detailed analysis of the off-design and failure modes of the system provide a very useful disclosure of the key components or subsystems which have most influence on the public risk. Such insight guides redesign and operating and maintenance techniques tailored to reduce risk.

A noteworthy example is the recent use of Probabilistic Risk Assessment (PRA) for nuclear power plants. These have not only provided a better professional estimate of failure probabilities, but also have stimulated technical fixes and wiser off-design operational responses. Such studies also provide greater confidence to operators who must act in emergencies, because they understand more completely how the system will respond to their measures. The nuclear utilities are now undertaking PRAs voluntarily in recognition of these benefits.

Recognizing both the limitations and merits of such quantitative risk assessments, it is questionable whether they have any significant influence on the perceptions or attitudes of the nonprofessional public. As with the "safety factor" and "trial and error" methods, the public is again being asked to trust the experts. For that portion of the public that has a generic distrust of authority, the "trust us" posture is an incitement to argument. Unfortunately, public arguments on issues of health and risk can quickly become polarized advocacies. The public cannot discriminate between the experts and pseudo-experts, or recognize the misleading aspects of the selective truths often presented by advocates. I do not believe that public debate and political processes are a suitable mechanism for evaluating quantitative risk assessments. Unless everyone becomes an expert, the "trust us" issue is unavoidable, and I prefer to trust the professional process, with peer review and peer debate. The uncertainties of science and technology are easier to deal with when separated from the manipulated perceptions of the public and the objectives of politicians.

In a recent speech,[3] William D. Ruckelshaus, the current EPA Administrator, discussed the practical problems of societal value politics and the need for "a government-wide process for assessing and managing health, safety, and environmental risks." He suggested the following:

> "It is my strong belief that where EPA or OSHA or any of the social regulatory agencies is charged with protecting public health, safety, or the environment, we should be given, to the extent possible, a common statutory formula for accomplishing our tasks. This statutory formula may well weigh public health very heavily in the equation as the American people certainly do.
>
> The formula should be as precise as possible and should include a responsibility to assess the risk and to weigh that, not only against the benefits of the continued use of the substance

> under examination, but against the risks associated with substitute substances and the risks associated with the transfer of the substance from one environmental medium to another via pollution control practices."

I agree with Mr. Ruckelshaus. Quantified risk analysis is not only needed for good public administration, but it is also needed to remove risk as a political stratagem for other purposes. A risk not placed in perspective easily becomes a manipulated vehicle for creating public anxieties, distrust, and demands for institutional or social change.

As shown in Figure 1, the overall process of society's approach to risks involves a sequence of steps, each requiring an action by a group with a delegated responsibility. All societies, regardless of political organization, involve such decision steps. The issue is always to whom are the respective responsibilities for each step delegated and on what information they act.

SAFETY GOALS

The situation is substantially different when we address the societal question of "how safe is safe enough." The implicit end-result of a society's answer to that question is the allocation of the resources needed to achieve an agreed-upon safety goal. Unfortunately, in all societies such an allocation is part of a "zero sum" game, i.e., resources applied to one goal leave less available for other goals. Thus, this competition among social goals inevitably involves every special and group interest that influences a society's decisions.

It is interesting to observe that the setting of safety goals, either absolute or comparative, provokes intense debate both professionally and publicly. Nevertheless, such goals are usually secondary in practical importance to the many implicit and obscure decisions which allocate the resources to achieve the goals. Safety goals have great political currency, since they embody idealistic consensus views on health and safety. I doubt if anyone wants to be exposed to toxic substances, or to die by accident or from disease. Politicians do not get elected by voting for exposure to carcinogens. But, clearly, we are not spending unlimited sums to achieve health and safety goals. Every society has many social goals, and the competition among these limits the allocation of resources to health and safety. As a result, every society determines an acceptable "non-zero" risk level for each of its activities. The factors entering into a determination of an acceptable risk level broadly involve the societal benefits and costs and the available resources. These factors include both tangible and intangible aspects, and are weighted by social values and public perceptions. The common bureaucratic phrase, "benefit/cost ratio," sometimes applied to evaluations of risk management, is overly simplistic, and is useful only for narrow issues involving small costs. The broad social objective of risk analysis is, therefore, the most effective use of the resources allocated to public health and safety.

An illustrative example of the interaction of these factors is the continuing political debate on the value of a human life when considering public investments in health and safety. Many authors have commented on the extremely wide range of implicit values of a life associated with our society's historical expenditure for reducing risks. It is evident that, while much talked about, an analytically quantified value of a life is not a determining parameter in the allocation of societal resources.

What is significant is the aggregate societal expenditure on reducing life-taking risks and their distribution. That aggregate expenditure is eventually distributed by the political processes among the risky activities. In principle, the effective value of a life saved would then be the ratio of the aggregate expenditure to the number of lives saved. In a wealthy society this would be much more than in a poor one. Such a calculation would, of course, depend on the estimated number of lives saved. For those activities where a body of empirical data exists (e.g., autos), a reasonable estimate may perhaps be made, but for most activities (e.g., air pollution, etc.), the risks are speculative and estimates are very unreliable. The point is that such concepts as an absolute value of a life play no real role in the social policies that result in the allocation of resources to reduce specific risk. The pertinent consideration is the size of the allocation relative to the disposable resources of society, and the competition with many other needs.

If specific allocations were done with economic efficiency, the Pareto principle would be followed, i.e., the resource allocations would be such as to equalize their marginal life-saving effect among the activities. Thus, the total life-saving would be a maximum for the available aggregate expenditure. Of course, this is never done and would be impractical in such a complex society as ours. Investments in health and safety come from many sources--personal, group, local, regional, and national. Their aggregate can only be guessed at, and their distribution among activities is uncoordinated. There are, however, a few "invisible hands" that bring some balance to the situation, principally the budgetary limitations of each decision center, their desire to shift expenditures to each other, and the usually common nature of public concerns.

The complex politics of societal values is thus a dominant determiner of the subsequent behavior of regulatory bodies and their actions for risk management. In the political process the most powerful component appears to be public perceptions of risk. Others have written at length on the origins of public perceptions, their relevance to reality, and their occasionally counterproductive outcomes. The importance of public views and anxieties is often underestimated by many professionals involved in risk analysis, and this topic deserves special attention. It is my perception that public concerns are often misallocated relative to public risk. I am sure that we all believe some risks are overemphasized, and that some are unduly ignored.

RISK MANAGEMENT

Risk management is finally carried out always by individuals, companies, or other operating units--not by regulating agencies. The function of regulatory agencies concerned with public health and safety is to assure that risk management techniques are implemented in operating systems which involve the public, as shown in Figure 1. I will not discuss here the organization and operations of regulatory agencies, a much belabored subject. I will address some of the policy issues involved in risk management which may determine its effectiveness.

In principle, the objective of risk management of a specific activity is to minimize social losses arising from an existing or potential risk. The preceding political process in Figure 1 presumably should have considered the issues of societal benefits and national resources, and should have defined for the regulatory agency the criteria for imposing remedial costs upon society. Thus, ideally, the regulatory agency should not need to consider the benefits of the regulated activity or whether the Nation can afford the costs. Unfortunately, the situation is rarely that clear. In practice, an image-motivated political body may vaguely direct a regulatory body to minimize both the risk and social cost of doing so, thus transferring to the regulatory body the political chore of balancing societal benefits, costs, and resources under a mandate that the public be protected from unreasonable risks.

The regulatory techniques of risk management fall into two classes: (1) imposition of technical and operating criteria; and (2) encouragement of operating system self-management. In both cases rewards and penalties are used to enforce these objectives. Rewards include licensing (or the equivalent approval to sell) and support of public acceptability. Penalties include a range of punitive actions, liabilities, and, most importantly, a degradation of public acceptability. The effectiveness of these regulatory techniques has been much studied, debated, and reviewed. I will not discuss them further. However, it is useful to consider the basic limitations common to all such regulatory actions.

The effectiveness of risk management is constrained by the complexity of most risk situations and their uncontrollable factors. While frequently occurring risks (e.g., auto collisions) provide an empirical base for determining many of the parameters involved, this is not the case for rare occurrences or for statistically low-level risks obscured in a large aggregation of similar consequence events. Thus, the predictability of the outcome of a risk management action is often severely limited. Because most such actions involve significant resource costs, their unpredictable outcome tends to discourage all but the most obvious measures.

The infrequent but high-consequence events present special problems of predictability and risk management. Every accident is the end result of a chain of events starting with some small initiator. There are usually a very large number of such potential initiators, each starting a different chain of events. For high-frequency risks, the empirical data base usually discloses the most common consequence, and thus provides a useful risk management opportunity. For infrequent accidents, a very few sequences may have been observed, but managing these provides very little assurance that the potential spectrum of initiators and sequences has been significantly reduced.

The most extreme risk scenarios are predominantly based on hypothetical rare sequences (e.g., the risks of nuclear power). This leads to risk management approaches which concentrate on virtuous "good practices," such as frequent maintenance, component testing, meticulous supervision, operator training quality, sobriety, alertness, honesty, cleanliness, etc. Of course, technical modifications to existing systems are included in risk management actions to address perceived defects, but it is often controversial that they actually reduce potential risks. Such modifications were imposed by regulation in nuclear power plants after the Three Mile Island accident. As another example, earthquake safety codes have historically contained a changing mix of both empirical and theoretical improvements. The basic difficulty with rare event risk management is that the paucity of empirical information forces a dependence on unverifiable professional judgment in fields of great uncertainty. This is also the case for very low-level effects. Thus, public anxiety cannot be allayed by visible proof, and may, in fact, be enhanced by visible risk management. Such anxiety leads to a continually increasing political demand for further risk reductions, continuing public anxiety and social expenditures disproportionate to the real risk.

In almost all risk situations, human behavior plays a major role in every step leading to or flowing from risk management. As a system component, people cannot be regulated, although they can be persuaded. The population is exposed to risks of two limiting types: involuntary risks created by systems in which an exposed individual does not exercise control (e.g., air pollution); and voluntary risks which the individual does control (e.g., swimming or skiing). Of course, in many practical cases the situation is mixed (e.g., auto transport). The risk management approaches are substantially different for these two classes of exposures.

In voluntary exposures the individual has presumably weighed the benefit/risk/resource balance and chosen to undertake the activity--skiing, for example. Society recognizes a social cost (or a loss of a socially productive investment) inherent in the voluntary risk taking, and generally attempts to encourage self-management of the risk by the individual. This may be done by indoctrination, training, social pressures to act responsible (e.g., seat belt and antismoking campaigns, etc.), and intangible rewards for safe performance. Regulatory bodies may also try to reduce the risk by controlling the equipment involved; for example, cigarette tars, ski lifts, snowmobiles, or, as in the case of alcoholic drinks, by setting an age barrier. Such regulations constrain individual freedom of choice, and are therefore not always accepted, as with motorcycle helmets.

The involuntary exposures of the public represent a quite different domain of risk management. Here the individual bears the risk without any direct means of avoiding it, has little control over how it is managed, and usually does not directly or visibly pay for its control. The result is a public pressure to reduce involuntary risks to a much lower level than an individual might accept voluntarily. In one of my early papers,[4] I estimated that the disparity between voluntary and involuntary historically-accepted levels of risk ranged between 100 and 1,000 times. Thus, the regulation of involuntary exposures is under much political pressure, and the risk management techniques are more coercive than in the voluntary case.

As mentioned before, such risk management emphasizes control of the activity by allowing it to operate under mandated conditions and predetermined criteria. It also uses direct and indirect economic penalties. For example, in addition to punitive penalties imposed by a regulatory body, an industry may face public liability suits, or the massive economic loss of being shut down. These actions are philosophically based on the concept of "make the polluter pay." They may be effective in a narrow application or for a short term. The hidden aspect of this process of risk management is that eventually almost all the costs of regulation and risk reduction are passed through to the public in the increased costs of goods and services. There appears to be no way to avoid having the public pay these costs. This means that risk management actions should be based on a societal evaluation justifying the costs rather than on the narrow issue of the feasibility of an industry to absorb the costs.

ECONOMICS OF IRRESPONSIBILITY

I'd like to end this presentation on risk management by leaving with you some thoughts on the "economics of irresponsibility." As every risk involves a cost--and most imposed risk reductions involve a cost--the question of who bears these costs may have a substantial effect on the effectiveness of risk management. If the individual or group creating the risk could encompass all the economic values of the societal benefit/cost/resource mix, there would be a strong motivation to manage the situation to achieve a social cost minimum. If, however, any of the participants in the decision process avoids the responsibility for considering the costs involved, his actions are most likely to increase the final social costs. Thus, the total economic cost of risk management may depend strongly on the degree of cost irresponsibility permitted in the decision process.

Let us consider a few illustrative examples of such behavior. Insurance generally spreads the cost of a single occurrence among a large group. Does this avoidance of the full consequence of an accident reduce the motivation for effective individual risk management (i.e., the moral hazard issue)? Although it is a matter of degree, I believe it does. The extreme case of rewarding cost irresponsibility is the preplanned arson of an old building to collect the fire insurance. As another example, does flood insurance encourage more people to build homes in known flood plains? I believe it does.

The issue can be generalized. Before the great depression of the 1930s, during the era of the "safety factor," individuals usually blamed themselves for the consequences of a voluntary risk, and treated mid- and low-level involuntary exposures as "acts of God." A shift in social philosophy started during those depression years when government began to protect the public from the vicissitudes of life, and now has become a widespread public attitude that all involuntary and many voluntary risks are the fault of government, society, or other institutions. One rarely hears the phrase, "act of God," any more. The recent flurry of product liability suits resulting from an individual's use of purchased equipment in a manner far beyond the purpose for which it was sold represents such an extreme shifting of responsibility away from the individual.

I am not arguing the ethical merits of providing a societal umbrella over individual behavior. Perhaps society should be responsible for the acts of the drunk driver--either because of inadequate indoctrination or other societal circumstances. The question I am raising is whether this shifting of cost and responsibility creates in the decision processes an "economics of irresponsibility" which seriously distorts the most effective use of national resources to improve public health and safety. I believe that "moral hazard" is equally applicable to regulatory actions as it is to insurance, and should be considered in risk management decisions.

This review of some of the basic issues involved in risk analysis was not intended to be either comprehensive or detailed. My purpose was to highlight both the importance and limitations of risk analysis, and to stimulate your thoughts on these matters. Practitioners in risk analysis quickly develop humility when faced with practial issues, but inadequate as it may be, it is better than anything else available.

FOOTNOTES

1. Suggested by Willie Hammer, "Product Safety Management and Engineering," Prentice Hall, 1980, a comprehensive review.
2. Issue of June 29, 1983.
3. "Science, Risk, and Public Policy"; National Academy of Sciences, June 22, 1983.
4. "Social Benefit versus Technological Risk," Science, Vol. 165, p. 1232, September 19, 1969.

DIETARY CARCINOGENS AND ANTICARCINOGENS:

OXYGEN RADICALS AND DEGENERATIVE DISEASES

Bruce N. Ames

Department of Biochemistry
University of California
Berkeley, CA 94720

SUMMARY

The human diet contains a great variety of natural mutagens and carcinogens, as well as many natural antimutagens and anticarcinogens. Many of these mutagens and carcinogens may act through the generation of oxygen radicals. Oxygen radicals may also play a major role as endogenous initiators of degenerative processes, such as DNA damage and mutation (and promotion), that may be related to cancer, heart disease, and aging. Dietary intake of natural antioxidants could be an important aspect of the body's defense mechanism against these agents. Many antioxidants are being identified as anticarcinogens. Characterizing and optimizing such defense systems may be an important part of a strategy of minimizing cancer and other age-related diseases.

INTRODUCTION

Comparison of data from different countries reveals wide differences in the rates of many types of cancer. This leads to hope that each major type of cancer may be largely avoidable, as is the case for cancers due to tobacco, which constitute 30 percent of the cancer deaths in the United States and the United Kingdom.[1] Despite numerous suggestions to the contrary, there is no convincing evidence of any generalized increase in U.S. (or U.K.) cancer rates other than what could plausibly be ascribed to the delayed effects of previous increases in tobacco usage.[1-3] Thus, whether or not any recent changes in life style or pollution in industrialized countries will substantially affect future cancer risks, some important determinants of current risks remain to be discovered among long-established aspects of our way of life. Epidemiologic studies have indicated that dietary practices are the most promising area to explore.[1,4] These studies suggest that a general increase in consumption of fiber-rich cereals, vegetables, and fruits and decrease in consumption of fat-rich products and excessive alcohol would be prudent.[1,4] There is still a lack of definitive evidence about the dietary components that are critical for humans and about their mechanisms of action. Laboratory studies of natural foodstuffs and cooked food are beginning to uncover an extraordinary variety of mutagens and possible carcinogens and anticarcinogens. In this article I discuss dietary mutagens and carcinogens and anticarcinogens that seem of importance and speculate on relevant biochemical mechanisms, particularly the role of oxygen radicals and their inhibitors, in the fat-cancer relationship, promotion, anticarcinogenesis, and aging.

NATURAL MUTAGENS AND CARCINOGENS IN FOOD

Plant Material

Plants in nature synthesize toxic chemicals in large amounts, apparently as a primary defense against the hordes of bacterial, fungal, and insect and other animal predators.[5-40] Plants in the human diet are no exception. The variety of these toxic chemicals is so great that organic chemists have been characterizing them for over 100 years, and new plant chemicals are still being discovered.[12,24,25] However, toxicological studies have been completed for only a very small percentage of them. Recent widespread use of short-term tests for detecting mutagens[41,42] and the increased number of animal cancer tests on plant substances[6] have contributed to the identification of many natural mutagens, teratogens, and carcinogens in the human diet.[5-40] Sixteen examples are discussed below.

Safrole, estragole, methyleugenol, and related compounds are present in many edible plants.[5] Safrole, estragole, and methyleugenol are carcinogens in rodents, and several of their metabolites are mutagens.[5] Oil of sassafras, which had been used in "natural sassaparilla" root beer, is about 75 percent safrole. Black pepper contains small amounts of safrole and large amounts (close to 10 percent by weight) of the closely related compound piperine.[26] Extracts of black pepper cause tumors in mice at a variety of sites at a dose of extract equivalent to 4 mg of dried pepper per day (about 160 mg/kg per day) for three months; an estimate of the average human intake of black pepper is over 140 mg per day (about 2 mg/kg per day) for life.[26]

Most hydrazines that have been tested are carcinogens and mutagens, and large amounts of carcinogenic hydrazines are present in edible mushrooms. The widely eaten false morel (Gyromitra esculenta) contains 11 hydrazines, three of which are known carcinogens.[28] One of these, N-methyl-N-formylhydrazine, is present at a concentration of 50 mg per 100 g and causes lung tumors in mice at the extremely low dietary level of 20 μg per mouse per day.[28] The most common commercial mushroom, Agaricus bisporus, contains about 300 mg of agaritine, the δ-glutamyl derivative of the mutagen 4-hydroxymethylphenylhydrazine, per 100 g of mushrooms, as well as smaller amounts of the closely related carcinogen N-acetyl-4-hydroxymethylphenylhydrazine.[28] Some agaritine is metabolized by the mushroom to a diazonium derivative which is a very potent carcinogen (a single dose of 400 ng/g gave 30 percent of mice stomach tumors) and which is also present in the mushroom in smaller amounts.[28] Many hydrazine carcinogens may act by producing oxygen radicals.[43]

Linear furocoumarins such as psoralen derivatives are potent light-activated carcinogens and mutagens and are widespread in plants of the Umbelliferae family, such as celery, parsnips, figs, and parsley (for instance, four mg per 100 g of parsnip).[17,19,44] The level in celery (about 100 μg per 100 g) can increase about 100-fold if the celery is stressed or diseased.[19] Celery pickers and handlers commonly develop skin rashes on their arms when exposed to diseased celery.[19] Oil of bergamot, a citrus oil, is very rich in a psoralen and was used in the leading suntan lotion in France.[17] Psoralens, when activated by sunlight, damage DNA, and induce tanning more rapidly than the ultraviolet component of sunlight, which is also a carcinogen.[17] Psoralens (plus light) are also effective in producing oxygen radicals.[18]

The potato glycoalkaloids solanine and chaconine are strong cholinesterase inhibitors and possible teratogens and are present at about 15 mg per 200 g of potato.[12,13] When potatoes are diseased, bruised, or exposed to light, these and other (24) glycoalkaloids reach levels that can be lethal to humans.[12] Plants typically respond to damage by making more (and often different) toxic chemicals as a defense against insects and fungi.[19,24,25] The different cultivars of potatoes vary in the concentration of these toxic glycoalkaloids (the concentration is a major determinant of insect and disease resistance); one cultivar bred for insect resistance had to be withdrawn from use because of its toxicity to humans (40 mg of glycoalkaloids in a 200-g potato is considered to be a toxic level).[12]

Quercetin and several similar flavonoids are mutagens in a number of short-term test systems. Flavonoids are extremely widespread (daily levels close to one g) in the human diet.[8,16,20,21] There is evidence for the carcinogenicity of quercetin in two strains of rats,[8] although it was negative in other experiments.[21]

Quinones and their phenol precursors[9,14,16,23,45] are widespread in the human diet. Quinones are quite toxic as they can act as electrophiles or accept a single electron to yield the semiquinone radical, which can either react directly with DNA[14,46] or participate in a redox cycle of superoxide radical generation by transferring the electron to O_2.[47] The superoxide radical and its metabolic product H_2O_2 can, in turn, lead to the oxidation of fat in cellular membranes by a lipid peroxidation chain reaction, thus generating mutagens and carcinogens, as discussed below. A number of quinones and dietary phenols have been shown to be mutagens.[7,9,16,23,44] Mutagenic anthraquinone derivatives are found in plants such as rhubarb and in mold toxins.[7,16,48] Many dietary phenols can spontaneously autoxidize to quinones, generating hydrogen peroxide at the same time (examples are catechol derivatives such as the caffeic acid component of chlorogenic acid[9], which is present at about 250 mg per cup of coffee). The amounts of these phenols in human urine (and in the diet) are appreciable.[45] Catechol, for example, is excreted in human urine at about 10 mg per day and appears to be mainly derived from metabolism of plant substances.[45] Catechol is a potent promoter of carcinogenesis,[45] an inducer of DNA damage, a likely active metabolite of the carcinogen benzene,[46] and a toxic agent in cigarette smoke.[45] Carecholamine induction of cardiomyopathy is thought to occur through generation of oxygen radicals.[49]

Theobromine, a relative of caffeine, has been shown to be genotoxic in a variety of tests, to potentiate (as does caffeine) DNA damage by various carcinogens in human cells, and to cause testicular atrophy and spermatogenic cell abnormalities in rats.[27] Cocoa powder is about two percent theobromine, and therefore humans may consume hundreds of milligrams of theobromine a day from chocolate. Theobromine is also present in tea.

Pyrrolizidine alkaloids are carcinogenic, mutagenic, and teratogenic and are present in thousands of plant species (often at $>$ 1 percent by weight), some of which are ingested by humans, particularly in herbs and herbal teas and occasionally in honey.[7,29] Pyrrolizidine alkaloid poisonings in humans (as well as in other mammals) cause lung and liver lesions and are commonly misdiagnosed.[29]

The broad (fava) bean (Vicia faba), a common food of the Mediterranean region, contains the toxins vicine and convicine at a level of about two percent of the dry weight.[30] Pythagoras forbade his followers to eat the beans, presumably because he was one of the millions of Mediterranean people with a deficiency of glucose-6-phosphate dehydrogenase. This deficiency results in a low glutathione concentration in blood cells, which causes increased resistance to the malarial parasite, probably accounting for the widespread occurrence of the mutant gene in malarial regions. However, the low glutathione concentration also results in a marked sensitivity to agents that cause oxidative damage, such as the fava bean toxins and a variety of drugs and viruses. Sensitive individuals who ingest fava beans develop a severe hemolytic anemia caused by the enzymatic hydrolysis of vicine to its aglycone, divicine, which forms a quinone that generates oxygen radicals.[30]

Allyl isothiocyanate, a major flavor ingredient in oil of mustard and horseradish, is one of the main toxins of the mustard seed and has been shown to cause chromosome aberrations in hamster cells at low concentration[50] and to be a carcinogen in rats.[31]

Gossypol is a major toxin in cottonseed and accounts for about one percent of its dry weight.[32] Gossypol causes pathological changes in rat and human testes, abnormal sperm, and male sterility.[32,33] Genetic damage has been observed in embryos sired by gossypol-treated male rats: dominant lethal mutations in embryos were measured after males were taken off gossypol treatment and allowed to mate.[33] Gossypol appears to be a carcinogen as well: it has been reported to be a potent initiator and also a promoter of carcinogenesis in skin painting studies with mice.[34] Crude, unrefined cottonseed oil contains considerable amounts of gossypol (100 to 750 mg per 100 ml). Thus, human consumption may be appreciable in countries, such as Egypt, where fairly crude cottonseed oil is commonly used in cooking. Gossypol is being tested as a male contraceptive in over 10,000 people in China (at an oral dose of about 10 mg per person per day), as it is inexpensive and causes sterility during use.[33] Gossypol's mode of action as a spermicide may be through the production of oxygen radicals.[35]

Plant breeders have developed "glandless cotton," a new strain with low levels of gossypol, but seeds from this strain are much more susceptible to attack by the fungus, Aspergillus flavus, which produces the potent carcinogen aflatoxin.[36]

Sterculic acid and malvalic acid are widespread in the human diet. They are toxic cyclopropenoid fatty acids present in cottonseed oil and other oils from seeds of plants in the family Malvaceal (for instance, cotton, kapok, okra, and durian).[51] Another possible source of human exposure is consumption of fish, poultry, eggs, and milk from animals fed on cottonseed.[51] Cyclopropenoid fatty acids are carcinogens which markedly potentiate the carcinogenicity of aflatoxin in trout, cause atherosclerosis in rabbits, are mitogenic in rats, and have a variety of toxic effects in farm animals.[51] The toxicity of these fatty acids could be due to their ease of oxidation to form peroxides and radicals.[51]

Leguminous plants such as lupine contain very potent teratogens.[22] When cows and goats forage on these plants, their offspring may have severe teratogenic abnormalities; an example is the characteristic "crooked calf" abnormality due to the ingestion of anagyrine from lupine.[22] In addition, significant amounts of these teratogens are transferred to the animals' milk, so that drinking the milk during pregnancy is a serious teratogenic hazard.[22] In one rural California family, a baby boy, a litter of puppies, and goat kids all had "crooked" bone birth-defect abnormalities. The pregnant mother and the dog had both been drinking milk obtained from the family goats, which had been foraging on lupine (the main forage in winter).[22] It was at first mistakenly thought that the birth defects were caused by spraying of 2,4-D.

Sesquiterpene lactones are widespread in many plants,[37] although because they are bitter they are not eaten in large amounts. Some have been shown to be mutagenic.[37] They are a major toxin in the white sap of Lactuca virosa (poison lettuce), which has been used as a folk remedy. Plant breeders are now transferring genes from this species to commercial lettuce to increase insect resistance.[38]

The phorbol esters present in the Euphorbiacea, some of which are used as folk remedies or herb teas, are potent promoters of carcinogenesis and may have been a cause of nasopharyngeal cancer in China and esophageal cancer in Curacao.[39]

Alfalfa sprouts contain canavanine, a highly toxic arginine analog that is incorporated into protein in place of arginine. Canavanine, which occurs in alfalfa sprouts at about 1.5 percent of their dry weight,[40] appears to be the active agent in causing the severe lupus erythematosus-like syndrome seen when monkeys are fed alfalfa sprouts.[40] Lupus in man is characterized by a defect in the immune system which is associated with autoimmunity, antinuclear antibodies, chromosome breaks, and various types of pathology.[40] The chromosome breaks appear to be due to oxygen radicals as they are prevented by superoxide dismutase.[52] The canavanine-alfalfa sprout pathology could be due in part to the production of oxygen radicals during phagocytization of antibody complexes with canavanine-containing protein.

The 16 examples above, plus coffee (discussed below), illustrate that the human dietary intake of "nature's pesticides" is likely to be several grams per day--probably at least 10,000 times higher than the dietary intake of man-made pesticides.[53]

Levels of plant toxins that confer insect and fungal resistance are being increased or decreased by plant breeders.[38] There are health costs for the use of these natural pesticides, just as there are for man-made pesticides,[41,54] and these must be balanced against the costs of producing food. However, little information is available about the toxicology of most of the natural plant toxins in our diet, despite the large doses we are exposed to. Many, if not most, of these plant toxins may be "new" to humans in the sense that the human diet has changed drastically with historic times. By comparison, our knowledge of the toxicological effects of new man-made pesticides is extensive, and general exposure is exceedingly low.[53]

Plants also contain a variety of anticarcinogens[55], which are discussed below.

Alcohol

Alcohol has long been associated with cancer of the mouth, esophagus, pharynx, larynx, and, to a lesser extent, liver,[1,56] and it appears to be an important human teratogen, causing a variety of physical and mental defects in babies of mothers who drink.[57] Alcohol drinking causes sperm abnormalities in mice[57a] and is a synergist for chromosome damage in humans.[58] Alcohol metabolism generates acetaldehyde, which is a mutagen, and teratogen,[59] a cocarcinogen and possibly a carcinogen,[60] and also radicals that produce lipid hydroperoxides[61] and other mutagens and carcinogens (see below)[62]. In some epidemiologic studies on alcohol,[56] it has been suggested that dietary green vegetables are a modifyng factor in the reduction of cancer risk.

Mold Carcinogens

A variety of mold carcinogens and mutagens are present in mold-contaminated food such as corn, grain, nuts, peanut butter, bread, cheese, fruit, and apple juice.[15,63] Some of these, such as sterigmatocystin and aflatoxin, are among the most potent carcinogens and mutagens known.[15,63] Dietary glutathione has been reported to counteract aflatoxin carcinogenicity.

Nitrite, Nitrate, and Nitrosamines

A number of human cancers, such as stomach and esophageal cancer, may be related to nitrosamines and other nitroso compounds formed from nitrate and nitrite in the diet.[64,65] Beets, celery, lettuce, spinach, radishes, and rhubarb all contain about 200 mg of nitrate per 100-g portion.[65] Anticarcinogens in the diet may be important in this context as well.[66]

Fat and Cancer: Possible Oxidative Mechanisms

Epidemiologic studies of cancer in humans suggest, but do not prove, that high fat intake is associated with colon and breast cancer.[1,4,67] A number of animal studies have shown that high dietary fat is a promoter and a presumptive carcinogen.[4,67,68] Colon and breast cancer and lung cancer (which is almost entirely due to cigarette smoking) account for about half of all U.S. cancer deaths. In addition to the cyclopropenoid fatty acids already discussed, two other plausible mechanisms involving oxidative processes could account for the relation[69] between high fat and both cancer and heart disease.

Rancid Fat. Fat accounts for over 40 percent of the calories in the U.S. diet,[67] and the amount of ingested oxidized fat may be appreciable.[70,71] Unsaturated fatty acids and cholesterol in fat are easily oxidized, particularly during cooking.[70,71] The lipid peroxidation chain reaction (rancidity) yields a variety[71-73] of mutagens, promoters, and carcinogens such as fatty acid hydroperoxides,[62] cholesterol hydroperoxide,[74] endoperoxides, cholesterol and fatty acid epoxides,[74-77] enals and other aldehydes,[44,59,78] and alkoxy and hydroperoxy radicals.[44,72] Thus the colon and digestive tract are exposed to a variety of fat-derived carcinogens. Human breast fluid can contain enormous levels (up to 780 μM)[75] of cholesterol epoxide (an oxidation product of cholesterol), which could originate from either ingested oxidized fat or oxidative processes in body lipids. Rodent feeding studies with oxidized fat[79] have not yielded definitive results.

Peroxisomes. Peroxisomes oxidize an appreciable percentage of dietary fatty acids, and removal of each two-carbon unit generates one molecule of hydrogen peroxide (a mutagen, promoter, and carcinogen).[80,81] Some hydrogen peroxide escapes the catalase in the peroxisome,[80,82,83] thus contributing to the supply of oxygen radicals, which also come from other metabolic sources.[72,83-85] Hydroperoxides generate oxygen radicals in the presence of iron-containing compounds in the cell.[72] Oxygen radicals, in turn, can damage DNA and can start the rancidity chain reaction which leads to the production of the mutagens and carcinogens listed above.[72] Drugs such as clofibrate, which cause lowering of serum lipids and proliferation of peroxisomes in rodents, result in age pigment (lipofuscin) accumulation (a sign of lipid peroxidation in tissues) and liver tumors in animals.[80] Some fatty acids, such as $C_{22:1}$ and certain trans fatty acids, appear to cause peroxisomal proliferation because they are poorly oxidized in mitochondria and are preferentially oxidized in the peroxisomes, although they may be selective for heart or liver.[86] There has been controversy about the role of trans fatty acids in cancer and heart disease, and recent evidence suggests that trans fatty acids might not be a risk factor for atherosclerosis in experimental animals.[87] Americans consume about 12 g of trans fatty acids a day[87] and a similar amount of unnatural cis isomers (which need further study[88]), mainly from hydrogenated vegetable fats. Dietary $C_{22:1}$ fatty acids are also obtained from rapeseed oil and fish oils.[86] Thus oxidation of certain fatty acids might generate grams of hydrogen peroxide per day within the peroxisome.[86] Another source of fat toxicity could be perturbations in the mitochondrial or peroxisomal membranes caused by abnormal fatty acids, yielding an increased flux of superoxide and hydrogen peroxide. Mitochondrial structure is altered when rats are fed some abnormal fatty acids from partially hydrogenated fish oil.[89] Dietary $C_{22:1}$ fatty acids and clofibrate also induce ornithine decarboxylase,[86] a common attribute of promoters.

A recent National Academy of Sciences committee report suggests that a reduction of fat consumption in the American diet would be prudent,[4] although other scientists argue that, until we know more about the mechanism of the fat-cancer relation and about which types of fat are dangerous, it is premature to recommend dietary changes.[90]

COOKED FOOD AS A SOURCE OF INGESTED BURNT AND BROWNED MATERIAL

Work of Sugimura and others has indicated that the burnt and browned material from heating protein during cooking is highly mutagenic.[21,91] Several chemicals isolated on the basis of their mutagenicity from heated protein or pyrolyzed amino acids were found to be carcinogenic when fed to rodents.[21] In addition, the browning reaction products from the caramelization of sugars or the reaction of amino acids and sugars during cooking (for instance, the brown material on bread crusts and toasted bread) contain a large variety of DNA-damaging agents and presumptive carcinogens.[23,38,92] The amount of burnt and browned material in the human diet may be several grams per day. By comparison, about 500 mg of burnt material is inhaled each day by a smoker using two packs of cigarettes (at 20 mg of tar per cigarette) a day. Smokers have more easily detectable levels of mutagens in their urine than nonsmokers,[93] but so do people who have consumed a meal of fried pork or bacon.[94] In the evaluation of risk from burnt material it may be useful (in addition to carrying out epidemiologic studies) to compare the activity of cigarette tar to that of the burnt material from cooked food (or polluted air) in short-term tests and animal carcinogenicity tests involving relevant routes of exposure. Route of exposure and composition of the burnt material are critical variables. The risk from inhaled cigarette smoke can be one reference standard: an average life shortening of about eight years for a two-pack-a-day smoker. The amount of burnt material inhaled from severely polluted city air, on the other hand, is relatively small: it would be necessary to breathe smoggy Los Angeles air (111 $\mu g/m^3$ total particulates; 31 $\mu g/m^3$ soluble organic matter) for one to two weeks to equal the soluble organic matter of the particulates or the mutagenicity from one cigarette (20 mg of tar).[95] Epidemiologic studies have not shown significant risks from city air pollution alone.[1,96] Air in the houses of smokers is considerably more polluted than city air outside.[97]

Coffee, which contains a considerable amount of burnt material, including the mutagenic pyrolysis product methylglyoxal, is mutagenic.[21,98] However, one cup of coffee also contains about 250 mg of the natural mutagen chlorogenic acid[9] (which is also an antinitrosating agent[66]), highly toxic atractylosides,[10] the glutathione transferase inducers kahweal palmitate and cafestol palmitate,[11] and about 100 mg of caffeine (which inhibits a DNA-repair system and can increase tumor yield[99] and cause birth defects at high levels in several experimental species).[100] There is preliminary, but not conclusive, epidemiologic evidence that heavy coffee drinking is associated with cancer of the ovary, bladder, pancreas, and large bowel.[101]

Cooking also accelerates the rancidity reaction of cooking oils and fat in meat,[70,71] thus increasing consumption of mutagens and carcinogens.

ANTICARCINOGENS

We have many defense mechanisms to protect ourselves against mutagens and carcinogens, including continuous shedding of the surface layer of our skin, stomach, cornea, intestines, and colon.[102] Understanding these mechanisms should be a major goal of cancer, heart, and aging research. Among the most important defenses may be those against oxygen radicals and lipid peroxidation if, as discussed here, these agents are major contributors to DNA damage.[103] Major sources of endogenous oxygen radicals are hydrogen peroxide[83] and superoxide[72,104] generated as side products of metabolism, and the oxygen radical burst from phagocytosis after viral or bacterial infection or the inflammatory reaction.[105] A variety of environmental agents could also contribute to the oxygen radical load, as discussed here and in recent reviews.[72,106] Many enzymes protect cells from oxidative damage;[107] examples are superoxide dismutase,[104] glutathione peroxidase, DT-diaphorase,[108] and the glutathione transferases.[109] In addition, a variety of small molecules in our diet are required for antioxidative mechanisms and appear to be anticarcinogens; some of these are discussed below.

Vitamin E (tocopherol) is the major radical trap in lipid membranes[72] and has been used clinically in a variety of oxidation-related diseases.[110] Vitamin E ameliorates both the cardiac damage and carcinogenicity of the quinones adriamycin and daunomycin, which are mutagenic, carcinogenic, cause cardiac damage, and appear to be toxic because of free radical generation.[111] Protective effects of tocopherols against radiation-induced DNA damage and mutation and dimethylhydrazine-induced carcinogenesis have also been observed.[112] Vitamin E markedly increases the endurance of rats during heavy exercise, which causes extensive oxygen radical damage to tissues.[113]

β-Carotene is another antioxidant in the diet that could be important in protecting body fat and lipid membranes against oxidation. Carotenoids are free-radical traps and remarkably efficient quenchers of singlet oxygen.[114] Singlet oxygen is a very reactive form of oxygen which is mutagenic and particularly effective at causing lipid peroxidation.[114] It can be generated by pigment-mediated transfer of the energy of light to oxygen, or by lipid peroxidation, although the latter is somewhat controversial. β-Carotene and similar polyprenes are present in carrots and in all food that contains chlorophyll, and they appear to be the plants' main defense against singlet oxygen generated as a by-product from the interaction of light and chlorophyll.[115] Carotenoids have been shown to be anticarcinogens in rats and mice.[116] Carotenoids (in green and yellow vegetables) may be anticarcinogens in humans.[1,56,117] Their protective effects in smokers might be related to the high level of oxidants in both cigarette smoke and tar.[45,118] Carotenoids have been used medically in the treatment for some genetic diseases, such as porphyrias, where a marked photosensitivity is presumably due to singlet oxygen formation.[119]

Selenium is another important dietary anticarcinogen. Dietary selenium (usually selenite) significantly inhibits the induction of skin, liver, colon, and mammary tumors in experimental animals by a number of different carcinogens, as well as the induction of mammary tumors by viruses.[120] It also inhibits transformation of mouse mammary cells.[121] Low selenium concentrations may be a risk factor in human cancer.[122] A particular type of heart disease in young people in the Keshan area of China has been traced to a selenium deficiency, and low selenium has been associated with cardiovascular death in Finland.[123] Selenium is in the active site of glutathione peroxidase, an enzyme essential for destroying lipid hydroperoxides and endogenous hydrogen peroxide and thus helping to prevent oxygen radical-induced lipid peroxidation,[107] although not all of the effects of selenium may be accounted for by this enzyme.[120] Several heavy-metal toxins, such as Cd^{2+} (a known carcinogen) and Hg^{2+}, lower glutathione peroxidase activity by interacting with selenium.[107] Selenite (and vitamin E) has been shown to counter the oxidative toxicity of mercuric salts.[124]

Glutathione is present in food and is one of the major antioxidants and antimutagens in the soluble fraction of cells. The glutathione transferases (some of which have peroxidase activity) are major defenses against oxidative and alkylating carcinogens.[109] The concentration of glutathione may be influenced by dietary sulfur amino acids.[125,126] N-Acetylcysteine, a source of cysteine, raises glutathione concentrations and reduces the oxidative cardiotoxicity of adriamycin and the skin reaction to radiation.[127] Glutathione concentrations are raised even more efficiently by L-2-oxothiazolidine-4-carboxylate, which is an effective antagonist of acetaminophen-caused liver damage.[126] Acetaminophen is thought to be toxic through radical and quinone oxidizing metabolites.[128] Dietary glutathione may be an effective anticarcinogen against aflatoxin.[129]

Dietary ascorbic acid is also important as an antioxidant. It was shown to be anticarcinogenic in rodents treated with ultraviolet radiation, benzo[a]pyrene, and nitrite (forming nitroso carcinogens),[64,65,130] and it may be inversely associated with human uterine cervical dysplasia (although this is not proof of a cause-effect relationship).[131] It was recently hypothesized that ascorbic acid may have been supplemented and perhaps partially replaced in humans by uric acid during primate evolution.[132]

Uric acid is a strong antioxidant present in high concentrations in the blood of humans.[132] The concentration of uric acid in the blood can be increased by dietary purines; however, too much causes gout. Uric acid is also present in high concentrations in human saliva[132] and may play a role in defense there as well, in conjunction with lactoperoxidase. A low uric acid level in blood may possibly be a risk factor in cigarette-caused lung cancer in humans.[133]

Edible plants and a variety of substances in them, such as phenols, have been reported to inhibit (cabbage) or to enhance (beets) carcinogenesis[11,55,134] or mutagenesis[23,66,92,135] in experimental animals. Some of these substances appear to inhibit by inducing cytochrome P-450 and other metabolic enzymes (134; see also 11), although on balance it is not completely clear whether it is generally helpful or harmful for humans to ingest these inducing substances.

The hypothesis that as much as 80 percent of cancer could be due to environmental factors was based on geographic differences in cancer rates and studies of migrants.[136] These differences in cancer rates were thought to be mainly due to life-style factors, such as smoking and dietary carcinogens and promoters,[136] but they also may be due in good part (see also 1) to less than optimum amounts of anticarcinogens and protective factors in the diet.

The optimum levels of dietary antioxidants, which may vary among individuals, remain to be determined; however, at least for selenium,[120] it is important to emphasize the possibility of deleterious side effects at high doses.

OXYGEN RADICALS AND DEGENERATIVE DISEASES ASSOCIATED WITH AGING

A plausible theory of aging holds that the major cause is damage to DNA[102,137] and other macromolecules and that a major source of this damage is oxygen radicals and lipid peroxidation.[43,84,103,138-141] Cancer and other degenerative diseases, such as heart disease,[102] are likely to be due in good part to this same fundamental destructive process. Age pigment (lipofuscin) accumulates during aging in all mammalian species and has been associated with lipid peroxidation.[73,84,138,139] The fluorescent products in age pigment are thought to be formed by malondialdehyde (a mutagen and carcinogen and a major end product of rancidity) cross-linking protein and lipids.[138] Metabolic rate is directly correlated with the rate of lipofuscin formation (and inversely correlated with longevity).[139]

Cancer increases with about the fourth power of age, both in short-lived species such as rats and mice (about 30 percent of rodents have cancer by the end of their two- to three-year life-span) and in long-lived species such as humans (about 30 percent of people have cancer by the end of their 85-year life-span).[142] Thus, the marked increase in life-span that has occurred in 60 million years of primate evolution has been accompanied by a marked decrease in age-specific cancer rates; that is, in contrast to rodents, 30 percent of humans do not have cancer by the age of three.[142] One important factor in longevity appears to be basal metabolic rate,[139,141] which is much lower in man than in rodents and could markedly affect the level of endogenous oxygen radicals.

Animals have many antioxidant defenses against oxygen radicals. Increased levels of these antioxidants, as well as new antioxidants, may also be a factor in the evolution of man from short-lived prosimians.[143] It has been suggested that an increase in superoxide dismutase is correlated (after the basal metabolic rate is taken into account) with increased longevity during primate evolution, although this has been disputed.[141] Ames et al. proposed[132] that as uric acid was an antioxidant and was present in much higher concentrations in the blood of humans than in other mammals, it may have been one of the innovations enabling the marked increase in life span and consequent marked decrease in age-specific cancer rates which occurred during primate evolution. The ability to synthesize ascorbic acid may have been lost at about the same time in primate evolution as uric levels began to increase.[144]

Cancer and Promotion

Both DNA-damaging agents (initiating mutagens)[21,41,42] and promoters[145] appear to play an important role in carcinogenesis.[21,146] It has been postulated that certain promoters of carcinogenesis act by generation of oxygen radicals and resultant lipid peroxidation.[73], [146-149] Lipid peroxidation cross-links proteins[43,150] and affects all aspects of cell organization,[72] including membrane and surface structure, and the mitotic apparatus. A common property of promoters may be their ability to produce oxygen radicals. Some examples are fat and hydrogen peroxide (which may be among the most important promoters),[67,68,81] TCDD,[151] lead and cadmium,[152] phorbol esters,[147,149,153] wounding of tissues,[154] asbestos,[155] peroxides,[156] catechol,[45] (see quinones above), mezerein and teleocidin B,[147] phenobarbital,[157] and radiation.[72,158] Inflammatory reactions involve the production of oxygen radicals by phagocytes,[105] and this could be the basis of promotion for asbestos[155] or wounding.[154] Some of the antioxidant anticarcinogens (discussed above) are also antipromoters,[73,121,146,159,160] and phorbol ester-induced chromosome damage[149] or promotion of transformation[159] is suppressed by superoxide dismutase, as would be expected if promoters were working through oxidative mechanisms.[73,161] Many "complete" carcinogens cause the production of oxygen radicals; examples are nitroso compounds, hydrazines, quinones, polycyclic hydrocarbons (through quinones), cadmium and lead salts, nitro compounds, and radiation. A good part of the toxic effects of ionizing radiation damage to DNA and cells is thought to be due to generation of oxygen radicals,[103,162] although only a tiny part of the oxygen radical load in humans is likely to be from this source.

Recent studies give some clues as to how promoters might act. Promoters disrupt the mitotic apparatus, causing hemizygosity and expression of recessive genes.[163] Phorbol esters generate oxygen radicals, which cause chromosome breaks[164] and increase gene copy number.[165] Promoters also cause formation of the peroxide hormones of the prostaglandin and leukotriene family by oxidation of arachidonic acid and other C_{20} polyenoic fatty acids, and inhibitors of this process appear to be antipromoters.[160] These hormones are intimately involved in cell division, differentiation, and tumor growth[166] and could have arisen in evolution as signal molecules warning the cell of oxidative damage. Effects on the cell membrane have also been suggested as the important factor in promotion, causing inhibition of intercellular communication[167] or protein kinase activation.[167a]

Heart Disease

It has been postulated that atherosclerotic lesions, which are derived from single cells, are similar to benign tumors and are of somatic mutational origin.[102,168] Fat appears to be one major risk factor for heart disease as well as for colon and breast cancer.[69] In agreement with this, a strong correlation has been observed between the frequency of atherosclerotic lesions and adenomatous polyps of the colon.[69] Thus, the same oxidative processes involving fat may contribute to both diseases. Oxidized forms of cholesterol have been implicated in heart disease,[169] and atherosclerotic-like lesions have been produced by injecting rabbits with lipid hydroperoxide or oxidized cholesterol.[169] The anticarcinogens discussed above could be anti-heart disease agents as well. As pointed out in the preceding section, vitamin E ameliorates both the cardiac damage and carcinogenicity of the free-radical-generating quinones adriamycin and daunomycin; N-acetylcysteine reduces the cardiotoxicity of adriamycin; and selenium is an anti-risk factor for one type of heart disease.

Other Diseases.

The brain uses 20 percent of the oxygen consumed by man and contains an appreciable amount of unsaturated fat. Lipid peroxidation (with consequent age pigment) is known to occur readily in the brain,[72] and possible consequences could be senile dementia or other brain abnormalities.[84] Several inherited progressive diseases of the central nervous system, such as Batten's disease, are associated with lipofuscin accumulation and may be due to a lipid peroxidation caused by a high concentration of unbound iron.[170] Mental retardation is one consequence of an inherited defective DNA repair system (XP complementation group D) for depurinated sites in DNA.[171]

Senile cataracts have been associated with light-induced oxidative damage.[172] The retina and an associated layer of cells, the pigment epithelium, are extremely sensitive to degeneration in vitamin E and selenium deficiency.[173] The pigment epithelium accumulates massive amounts of lipofuscin in aging and dietary antioxidant deficiency.[173] The eye is well known to be particularly rich in antioxidants.

The testes are quite prone to lipid peroxidation and to the accumulation of age pigment. A number of agents, such as gossypol, which cause genetic birth defects (dominant lethals) may be active by this mechanism. The various agents know to cause cancer by oxidative mechanisms are prospective mutagenic agents for the germ line. Thus, vitamin E, which was discovered 60 years ago as a fertility factor,[72] and other antioxidants such as selenium,[174] may help both to engender and to protect the next generation.

RISKS

There are large numbers of mutagens and carcinogens in every meal, all perfectly natural and traditional (see also 21,23). Nature is not benign. It should be emphasized that no human diet can be entirely free of mutagens and carcinogens and that the foods mentioned are only representative examples. To identify a substance, whether natural or man-made, as a mutagen or a carcinogen, is just a first step. Beyond this, it is necessary to consider the risks for alternative courses of action and to quantify the approximate magnitude of the risk, although the quantification of risk poses a major challenge. Carcinogens differ in their potency in rodents by more than a millionfold,[175] and the levels of particular carcinogens to which humans are exposed can vary more than a billionfold. Extrapolation of risk from rodents to humans is difficult for many reasons, including the longevity difference, anti-oxidant factors, and the probable multicausal nature of most human cancer.

Tobacco smoking is, without doubt, a major and well-understood risk, causing about 30 percent of cancer deaths and 25 percent of fatal heart attacks (as well as other degenerative diseases) in the United States.[1] These percentages may increase even more in the near future as the health effects of the large increase in women smokers become apparent.[1] Diet, which provides both carcinogens and anticarcinogens, is extremely likely to be another major risk factor. Excessive alcohol consumption is another risk, although it does not seem to be of the same general importance as smoking and diet. Certain other high-dose exposures might also turn out be important for particular groups of people--for instance, certain drugs, where consumption can reach hundreds of milligrams per day; particular cosmetics; and certain occupational exposures,[2] where workers inhale dusts or solvents at high concentration. We must also be prudent about environmental pollution.[41,54] Despite all of these risks, it should be emphasized that the overall trend in life expectancy in the United States is continuing steadily upward.[176]

The understanding of cancer and degenerative disease mechanisms is being aided by the rapid progress of science and technology, and this should help to dispel confusion about how important health risks can be identified among the vast number of minor risks. We have many methods of attacking the problem of environmental carcinogens (and anticarcinogens), including human epidemiology,[1] short-term tests,[41,42,177] and animal cancer tests.[175] Powerful new methods are being developed (for instance, see 58,177) for measuring DNA damage or other pertinent factors with great sensitivity in individuals. These methods, which are often noninvasive as they can be done on blood or urine (even after storage), can be combined with epidemiology to determine whether particular factors are predictive of disease. Thus, more powerful tools will be available for optimizing antioxidants and other dietary anti-risk factors, for identifying human genetic variants at high risk, and for identifying significant health risks.

REFERENCES

1. R. Doll and R. Peto, J. Natl. Cancer Inst. 66:1192 (1981).
2. R. Peto and M. Schneiderman, eds., Banbury Report 9. Quantification of Occupational Cancer, Cold Spring Harbor Laboratory, Cold Spring Harbor, NY (1981).
3. Cancer Facts and Figures, 1983, American Cancer Society, New York (1982).
4. National Research Council, Diet, Nutrition and Cancer, National Academy Press, Washington, DC (1982).
5. E.C. Miller, J.A. Miller, I. Hirono, T. Sugimura, S. Takayama, eds., Naturally Occurring Carcinogens-Mutagens and Modulators of Carcinogensis, Japan Scientific Societies Press and University Park Press, Tokyo and Baltimore (1979); E.C. Miller, et al., Cancer Res. 43:1124 (1983); C. Ioannides, M. Delaforge, D.V. Parke, Food Cosmet. Toxicol. 19:657 (1981).
6. G. J. Kapadia, ed., Oncology Overview on Naturally Occuring Dietary Carcinogens of Plant Origin, International Cancer Research Data Bank Program, National Cancer Institute, Bethesda, Maryland (1982).
7. A.M. Clark, in Environmental Mutagenesis, Carcinogenesis, and Plant Biology, E.J. Klekowski, Jr., ed., Praeger, New York (1982), vol. 1, pp. 97-132.
8. A. M. Pamukcu, S. Yalciner, J. F. Hatcher, G. T. Bryan, Cancer Res. 40:3468 (1980); J. F. Hatcher, A. M. Pamukcu, E. Erturk, G. T. Bryan, Fed. Proc. Fed. Am. Soc. Exp. Biol. 42:786 (1983).
9. H. F. Stich, M. P. Rosin, C. H. Wu, W. D. Powrie, Mutat. Res. 90:201 (1981); A. A. Avery'yanov, Biokhimiya 46:256 (1981); A. F. Hanham, B. P. Dunn, H. F. Stich, Mutat. Res. 116:333 (1983).
10. K. H. Pegel, Chem. Eng. News 59:4 (20 July 1981).
11. L. K. T. Lam, V. L. Sparnins, L. W. Wattenberg, Cancer Res. 42:1193 (1982).
12. S. J. Jadhav, R. P. Sharma, D. K. Salunkhe, CRC Crit. Rev. Toxicol, 9:21 (1981).
13. R. L. Hall, Nutr. Cancer 1 (No. 2):27 (1979).
14. H. W. Moore and R. Czerniak, Med. Res. Rev. 1:249 (1981).
15. I. Hirono, CRC Crit. Rev. Toxicol. 8:235 (1981).
16. J. P. Brown, Mutat. Res. 75:243 (1980).
17. M. J. Ashwood-Smith and G. A. Poulton, ibid. 85:389 (1981).
18. A. Ya. Potapenko, M. V. Moshnin, A. A. Krasnovsky, Jr., V. L. Sukhorukov, Z. Naturforsch. 37:70 (1982).
19. G. W. Ivie, D. L. Holt, M. C. Ivey, Science 213:909 (1981); R. C. Beier and E. H. Oertli, Phytochemistry, in press; R. C. Beier, G. W. Ivie, E. H. Oertli, in "Xenobiotics in Foods and Feeds," ACS Symp. Ser., in press; ________, D. L. Holt, Food Chem. Toxicol. 21:163 (1983).
20. G. Tamura, C. Gold, A. Ferro-Luzzi, B. N. Ames, Proc. Natl. Acad. Sci. U.S.A. 77:4961 (1980).

21. T. Sugimura and S. Sata, Cancer Res. (Suppl.) 43:2415s (1983); T. Sugimura and M. Nagao, in Mutagenicity: New Horizons in Genetic Toxicology, J. A. Heddle, ed., Academic Press, New York (1982), pp. 73-88.
22. W. W. Kilgore, D. G. Crosby, A. L. Craigmill, N. K. Poppen, Calif. Agric. 35 (No. 11) (November 1981); D. G. Crosby, Chem. Eng. News 61:37 (11 April 1983); C. D. Warren, ibid. (13 June 1983), p. 3.
23. H. F. Stich, M. P. Rosin, C. H. Wu, W. D. Powrie, in Mutagenicity: New Horizons in Genetic Toxicology, J. A. Heddle, ed., Academic Press, New York (1982), pp. 117-142; ____, W. D. Powrie, Cancer Lett. 14:251 (1981).
24. N. Katsui, F. Yagihashi, A. Murai, T. Masamune, Bull. Chem. Soc. Jpn. 55:2424 (1982); ______, ibid., p. 2428; R. M. Bostock, R. A. Laine, J. A. Kuc, Plant Physiol. 70:1417 (1982).
25. H. Griesebach and J. Ebel, Angew Chem. Int. Ed. Engl. 17:635 (1978).
26. J. M. Concon, D. S. Newburg, T. W. Swerczek, Nutr. Cancer 1 (No. 3):22 (1979).
27. H. W. Renner and R. Munzner, Mutat. Res. 103:275 (1982); H. W. Renner, Experientia 38:600 (1982); D. Mourelatos, J. Dozi-Vassiliades, A. Granitsas, Mutat. Res. 104:243 (1982); J. H. Gans, Toxicol. Appl. Pharmacol. 63: 312 (1982).
28. B. Toth, in Naturally Occurring Carcinogens-Mutagens and Modulators of Carcinogenesis, E. C. Miller, J. A. Miller, I. Hirono, T. Sugimura, S. Takayama, eds., Japan Scientific Societies Press and University Park Press, Tokyo and Baltimore (1979), pp. 57-65; A. E. Ross, D. L. Nagel, B. Toth, J. Agric. Food Chem. 30:521 (1982); B. Toth and K. Patil, Mycopathologia 78:11 (1982); B. Toth, D. Nagel, A. Ross, Br. J. Cancer 46:417 (1982).
29. R. Schoental, Toxicol. Lett. 10:323 (1982); R. J. Huxtable, Perspect. Biol. Med. 24:1 (1980); H. Niwa, H. Ishiwata, K. Yamada, J. Chromatogr. 257:146 (1983).
30. M. Chevion and T. Navok, Anal. Biochem. 128:152 (1983); V. Lattanzio, V. V. Bianco, D. Lafiandra, Experientia 38:789 (1982); V. L. Flohe, G. Niebch, H. Reiber, Z. Klin. Chem. Klin. Biochem., 9:431 (1971); J. Mager, M. Chevion, G. Glaser, in Toxic Constituents of Plant Foodstuffs, I. E. Liener, ed., Academic Press, New York (1980), pp. 265-294.
31. J. K. Dunnick, et al., Fundam. Appl. Toxicol. 2:114 (1982).
32. L. C. Berardi and L. A. Goldblatt, in Toxic Constituents of Plant Foodstuffs, I. E. Liener, ed., Academic Press, ed. 2, New York (1980), pp. 183-237.
33. S. P. Xue, in Proceedings, Symposium on Recent Advances in Fertility Regulation, Beijing (2 to 5 September 1980), p. 122.
34. R. K. Haroz and J. Thomasson, Toxicol. Lett. Suppl. 6:72 (1980).
35. M. Coburn, P. Sinsheimer, S. Segal, M. Burgos, W. Troll, Biol. Bull. (Woods Hole, Mass.) 159:468 (1980).

36. C. Campbell, personal communication.
37. G. D. Manners, G. W. Ivie, J. T. MacGregor, Toxicol. Appl. Pharmacol. 45:629 (1978); G. W. Ivie and D. A. Witzel, in Plant Toxins, vol. 1, Encyclopedic Handbook of Natural Toxins, A. T. Tu and R. F. Keeler, eds., Dekker, New York (in press).
38. J. C. M. Van der Hoeven, et al., in Mutagens in Our Environment, M. Sorsa and H. Vainio, eds., Liss, New York, (1982), pp. 327-338; J. C. M. van der Hoeven, W. J. Lagerweij, I. M. Bruggeman, F. G. Voragen, J. H. Koeman, J. Agric. Food Chem., in press.
39. T. Hirayama and Y. Ito, Prev Med. 10:614 (1981); E. Hecker, J. Cancer Res. Clin. Oncol. 99:103 (1981).
40. M. R. Malinow, E. J. Bardana, Jr., B. Pirofsky, S. Craig, P. McLaughlin, Science 216:415 (1982).
41. B. N. Ames, ibid. 204:587 (1979). "Mutagen" will be used in its broad sense to include clastogens and other DNA-damaging agents.
42. H. F. Stich and R. H. C. San, eds., Short-Term Tests for Chemical Carcinogens, Springer-Verlag, New York (1981).
43. P. Hochstein and S. K. Jain, Fed. Proc. Fed. Am. Soc. Exp. Biol. 40:183 (1981).
44. D. E. Levin, M. Hollstein, M. F. Christman, E. Schwiers, B. N. Ames, Proc. Natl. Acad. Sci. U.S.A. 79:7445 (1982). Many additional quinones and aldehydes have now been shown to be mutagenic.
45. S. G. Carmell, E. J. LaVoie, S. S. Hecht, Food Chem. Toxicol. 20:587 (1982).
46. K. Morimoto, S. Wolff, A. Koizumi, Mutat. Res. Lett. 119:355 (1983); T. Sawahata and R. A. Neal, Mol. Pharmacol. 23:453 (1983).
47. H. Kappus and H. Sies, Experientia 37:1233 (1981).
48. L. Tikkanen, T. Matsushima, S. Natori, Mutat. Res. 116:297 (1983).
49. P. K. Singal, N. Kapur, K. S. Dhillon, R. E. Beamish, N. S. Dhalla, Can. J. Physiol. Pharmacol. 60:1390 (1982).
50. A. Kasamaki, et al., Mutat. Res. 105:387 (1982).
51. J. D. Hendricks, R. O. Sinnhuber, P. M. Loveland, N. E. Pawlowski, J. E. Nixon, Science 208:309 (1980); R. A. Phelps, F. S. Shenstone, A. R. Kemmerer, R. J. Evans, Poult. Sci. 44:358 (1964); N. E. Pawlowski, personal communication.
52. I. Emerit, A. M. Michelson, A. Levy, J. P. Camus, J. Emerit, Hum. Genet. 55:341 (1980).
53. FDA Compliance Program Report of Findings. FY79 Total Diet Studies--Adult (No. 7305.002); available from National Technical Information Service, Springfield, VA). It is estimated that the daily dietary intake of synthetic organic pesticides and herbicides is about 60 g, with chlorpropham, malathion, and DDE accounting for about three-fourths of this. An estimate of 150 μg of daily exposure in Finland to

pesticide residues has been made by K. Hemmimki, H. Vainio, M. Sorsa, S. Salminen [J. Environ. Sci. Health Cl (No. 1), 55 (1983)].

54. N. K. Hooper, B. N. Ames, M. A. Saleh, J. E. Casida, Science 205:591 (1979).

55. L. W. Wattenberg, Cancer Res. (Suppl.) 43:2448s (1983).

56. J. Hoey, C. Montvernay, R. Lambert, Am. J. Epidemiol. 113:668 (1981); A. J. Tuyns, G. Pequignot, M. Gignoux, A. Valla, Int. J. Cancer 30:9 (1982); A. Tuyns, in Cancer Epidemiology and Prevention, D. Schottenfeld and J. F. Fraumeni, Jr., eds., Saunders, Philadelphia (1982), pp. 293-303; R. G. Ziegler, et al., J. Natl. Cancer Inst. 67:1199 (1981); W. D. Flanders and K. J. Rothman, Am. J. Epidemiol. 115:371 (1982).

57. E. L. Abel, Hum. Biol. 54:421 (1982); H. L. Rosset, L. Weiner, A. Lee, B. Zuckerman, E. Dooling, E. Oppenheimer, Obstet. Gynecol. 61:539 (1983).

57a. R. A. Anderson, Jr., B. R. Willis, C. Oswald, L. J. D. Zaneveld, J. Pharmacol. Exp. Ther. 225:479 (1983).

58. H. F. Stich and M. P. Rosin, Int. J. Cancer 31:305 (1983).

59. R. P. Bird, H. H. Draper, P. K. Basrur, Mutat. Res. 101:237 (1982); M. A. Campbell and A. G. Fantel, Life Sci. 32:2641 (1983).

60. V. J. Feron, A. Kruysse, R. A. Woutersen, Eur. J. Cancer Clin. Oncol. 18:13 (1982).

61. T. Suematsu, et al., Alcoholism: Clin. Exp. Res. 5:427 (1981); G. W. Winston and A. I. Cederbaum, Biochem. Pharmacol. 31:2301 (1982); L. A. Videla, V. Fernandez, A. de Marinis, N. Fernandez, A. Valenzuela, Biochem. Biophys. Res. Commun. 104:965 (1982); T. E. Stege, Res. Commun. Chem. Pathol. Pharmacol. 36:287 (1982).

62. M. G. Cutler and R. Schneider, Food Cosmet. Toxicol. 12:451 (1974).

63. Y. Tazima, in Environmental Mutagenesis, Carcinogenesis and Plant Biology, E. J. Klekowski, Jr., ed., Praeger, New York (1982), vol. 1, pp. 68-95.

64. P. N. Magee, ed., Banbury Report 12. Nitrosamines and Human Cancer, Cold Spring Harbor Laboratory, Cold Spring Harbor, NY (1982); P. E. Hartman, in Chemical Mutagens, F. J. de Serres and A. Hollaender, eds., Plenum, New York (1982), vol. 7, pp. 211-294; P. E. Hartman, Environ. Mutagen 5:111 (1983).

65. Committee on Nitrite and Alternative Curing Agents in Food, Assembly of Life Sciences, National Academy of Sciences, The Health Effects of Nitrate, Nitrite, and N-Nitroso Compounds, National Academy Press, Washington, DC (1981).

66. H. F. Stich, P. K. L. Chan, M. P. Rosin, Int. J. Cancer 30:719 (1982); H. F. Stick and M. P. Rosin, in Nutritional and Metabolic Aspects of Food Safety, M. Friedman, ed., Plenum, New York (in press).

67. L. J. Kinlen, Br. Med. J. 286:1081 (1983); D. J. Fink and D. Kritchevsky, Cancer Res. 41:3677 (1981).
68. C. W. Welsch and C. F. Aylsworth, J. Natl. Cancer Inst. 70:215 (1983).
69. P. Correa, J. P. Strong, W. D. Johnson, P. Pizzolato, W. Haenszel, J. Chronic Dis. 35:313 (1982).
70. F. B. Shorland, et al., J. Agric. Food Chem. 29:863 (1981).
71. M. G. Simic and M. Karel, eds., Autoxidation in Food and Biological Systems, Plenum, New York (1980).
72. W. A. Pryor, ed., Free Radicals in Biology, Academic Press, New York (1976 to 1982), vols. 1 to 5.
73. H. B. Demopoulos, D. D. Pietronigro, E. S. Flamm, M. L. Seligman, J. Environ. Pathol. Toxicol. 3:273 (1980).
74. F. Bischoff, Adv. Lipid Res. 7:165 (1969).
75. N. L. Petrakis, L. D. Gruenke, J. C. Craig, Cancer Res. 41:2563 (1981).
76. H. S. Black and D. R. Douglas, ibid. 32:2630 (1972).
77. H. Imai, N. T. Werthessen, V. Subramanyam, P. W. LeQuesne, A. H. Soloway, M. Kanisawa, Science 207:651 (1980).
78. M. Ferrali, R. Fulceri, A. Benedetti, M. Comporti, Res. Commun. Chem. Pathol. Pharmacol. 30:99 (1980).
79. N. R. Artman, Adv.Lipid Res. 7:245 (1969).
80. J. K. Reddy, J. R. Warren, M. K. Reddy, N. D. Lalwani, Ann. N.Y. Acad. Sci. 386:81 (1982); J. K. Reddy and N. D. Lalwani, CRC Crit. Rev. Toxicol., in press.
81. H. L. Plaine, Genetics 40:268 (1955); A. Ito, M. Naito, Y. Naito, H. Watanabe, Gann 73:315 (1982); G. Speit, W. Vogel, M.Wolf, Environ. Mutagen, 4:135 (1982): H. Tsuda, Jpn. J. Genet 56:1 (1981); N. Hirota and T. Yokoyama Gann 72:811 (1981).
82. S. Horie, H. Ishii, T. Suga, J. Biochem. (Tokyo) 90:1691 (1981); D. P. Jones, L. Eklow, H. Thor, S. Orrenius, Arch. Biochem. Biophys. 210:505 (1981).
83. B. Chance, H. Sies, A. Boveris, Physiol. Rev. 59:527 (1981).
84. D. Harman, Proc. Natl. Acad. Sci. U.S.A. 78:7124 (1981), in Free Radicals in Biology, W. A. Pryor, ed., Academic Press, New York (1982), vol. 5, pp. 255-275.
85. I. Emerit, M. Keck, A. Levy, J. Feingold, A. M. Michelson, Mutat. Res. 103:165 (1982).
86. C. E. Neat, M. S. Thomassen, H. Osmundsen, Biochem. J. 196:149 (1981); J. Bremer and K. R. Norum, J. Lipid Res. 23:243 (1982); M. S. Thomassen, E. N. Christiansen, K. R. Norum, Biochem J. 206:195 (1982); H. Osmundsen, Int. J. Biochem. 14:905 (1982); J. Norseth and M. S. Thomassen, Biochem. Biophys. Acta, in press.
87. M. G. Enig, R. J. Munn, M. Keeney, Fed. Proc. Fed. Am. Soc. Exp. Biol. 37:2215 (1978); J. E. Hunter, J. Natl. Cancer Inst. 69:319 (1982); A. B. Awad, ibid., p. 320; H. Ruttenberg, L. M. Davidson, N. A. Little, D. M. Klurfeld, D. Kritchevsky, J. Nutr. 113:835 (1983).
88. R. Wood, Lipids 14:975 (1979).

89. E. N. Christiansen, T. Flatmark, H. Kryvi, Eur. J. Cell Biol. 26:11 (1981).
90. Council for Agricultural Science and Technology, Diet, Nutrition, and Cancer: A Critique (Special Publication 13), Council for Agricultural Science and Technology, Ames, Iowa (1982).
91. L. F. Bjeldanes, et al., Food Chem. Toxicol. 20:357 (1982); M. W. Pariza, L. J. Loretz, J. M. Storkson, N. C. Holland, Cancer Res. (Suppl.) 43:2444s (1983).
92. H. F. Stich, W. Stich, M. P. Rosin, W. D. Powrie, Mutat. Res. 91:129 (1981); M. P. Rosin, H. F. Stich, W. D. Powrie, C. H. Wu, ibid. 101:189 (1982); C.-I. Wei, K. Kitamura, T. Shibamoto, Food Cosmet. toxicol. 19:749 (1981).
93. E. Yamasaki and B. N. Ames, Proc. Natl. Acad. Sci. U.S.A. 74:3555 (1977).
94. R. Baker, A. Arlauskas, A. Bonin, D. Angus, Cancer Lett. 16:81 (1982).
95. D. Schuetzle, D. Cronn, A. L. Crittenden, R. J. Charlson, Environ. Sci. Technol. 9:838 (1975); G. Gartreil and S. K. Friedlander, Atmos. Environ. 9:279 (1975); L. D. Kier, E. Yamasaki, B. N. Ames, Proc. Natl. Acad. Sci. U.S.A. 71:4159 (1974); J. N. Pitts, Jr., Environ. Health Perspect. 47:115 (1983).
96. J. E. Vena, Am. J. Epidemiol. 116:42 (1982); R. Cederiof, R. Doll, B. Fowler, Environ. Health Perspect. 22:1 (1978); F. E. Speizer, ibid. 47:33 (1983).
97. B. Brunekreef and J. S. M. Boleij, Int. Arch. Occup. Environ. Health 50:299 (1982).
98. H. Kasai, et al., Gann 73:681 (1982).
99. V. Armuth and I. Berenblum, Carcinogenesis 2:977 (1981).
100. S. Frabro, Reprod. Toxicol. 1:2 (1982).
101. D. Trichopoulos, M. Papapostolou, A. Polychronopoulou, Int. J. Cancer 28:691 (1981); P. Hartge, L. P. Lesher, L. McGowan, R. Hoover, ibid. 30:531 (1982); B. MacMahon, Cancer (Brussels) 50:2676 (1982); H. S. Cuckle and L. J. Kinlen, Br. J. Cancer 44:760 (1981); R. L. Phillips and D. A. Snowdon, Cancer Res. (Suppl.) 43:2403s (1983); L. D. Marrett, S. D. Walter, J. W. Meigs, Am. J. Epidemiol. 117:113 (1983); D. M. Weinberg, R. K. Ross, T. M. Mack, A. Paganini-Hill, B. E. Henderson, Cancer (Brussels) 51:675 (1983).
102. P. E. Hartman, Environ. Mutagen., in press.
103. J. R. Totter, Proc. Natl. Acad. Sci. U.S.A. 77:1763 (1980).
104. I. Fridovich, in Pathology of Oxygen, A. Autor, ed., Academic Press, New York (1982), pp. 1-19; L. W. Oberley, T. D. Oberley, G. R. Buettner, Med. Hypotheses 6:249 (1980).
105. B. Halliwell, Cell. Biol. Int. Rep. 6:529 (1982); A. I. Tauber, Trends Biochem. Sci. 7:411 (1982); A. B. Weitberg, S. A. Weitzman, M. Destrempes, S. A. Latt, T. P. Stossel, N. Engl. J. Med. 308:26 (1983). Neutrophils also produce HOCl, which is both a chlorinating and oxidizing agent.

106. M. A. Trush, E. G. Mimnaugh, T. E. Gram, Biochem. Pharmacol. 31:3335 (1982).
107. L. Flohe, in Free Radicals in Biology, W. A. Pryor, ed., Academic Press, New York (1982), vol. 5 pp. 223-254.
108. C. Lind, P. Hochstein, L. Ernster, Arch. Biochem. Biophys. 216:178 (1982).
109. M. Warholm, C. Guthenberg, B. Mannervik, C. von Bahr, Biochem. Biophys. Res. Commun. 98:512 (1981).
110. J. G. Bieri, L. Corash, V. S. Hubbard, N. Eng. J. Med. 308:1063 (1983).
111. Y. M. Wang, et al., in Molecular Interrelations of Nutrition and Cancer, M. S. Arnott, J. van Eys, Y.-M. Wang, eds., Raven, New York (1982), pp. 369-379.
112. C. Beckman, R. M. Roy, A. Sproule, Mutat. Res. 105:73 (1982); M. G. Cook and P. McNamara, Cancer Res. 40:1329 (1980).
113. K. J. A. Davies, A. T. Quintanilha, G. A. Brooks, L. Packer, Biochem. Biophys. Res. Commun. 107:1198 (1982).
114. C. S. Foote, in Pathology of Oxygen, A. Autor, ed., Academic Press, New York (1982), pp. 21-44; J. E. Packer, J. S. Mahood, V. O. Mora-Arellano, T. F. Slater, R. L. Willson, B. S. Wolfenden, Biochem. Biophys. Res. Commun. 98:901 (1981); W. Bors, C. Michel, M. Saran, Bull. Eur. Physiopathol. Resp. 17 (Suppl.):13, (1981).
115. N. I. Krinsky and S. M. Deneke, J. Natl. Cancer Inst. 69:205 (1982); J. A. Turner and J. N. Prebble, J. Gen. Microbiol. 119:133 (1980); K. L. Simpson and C. O. Chichester, Annu. Rev. Nutr. 1:351 (1981).
116. G. Rettura, C. Dattagupta, P. Listowsky, S. M. Levenson, E. Seifter, Fed. Proc. Fed. Am. Soc. Exp. Biol. 42:786 (1983); M. M. Mathews-Roth, Oncology 39:33 (1982).
117. R. Peto, R. Doll, J. D. Buckley, M. B. Sporn, Nature (London) 290:201 (1981); R. B. Shekelle, et al., Lancet, 1981-II, 1185 (1981); T. Hirayama, Nutr. Cancer, 1:67 (1979); G. Kvale, E. Bjelke, J. J. Gart, Int. J. Cancer 31:397 (1983).
118. W. A. Pryor, M. Tamura, M. M. Dooley, P. I. Premovic, D. F. Church, in Oxy-Radicals and Their Scavenger Systems: Cellular and Medical Aspects, G. Cohen and R. Greenwald, eds., Elsevier, Amsterdam (1983), vol. 2, pp. 185-192; W. A. Pryor, B. J. Hales, P. I. Premovic, D. F. Church, Science 220:425 (1983).
119. M. M. Mathews-Roth, J. Natl. Cancer Inst. 69:279 (1982).
120. A. C. Griffin, in Molecular Interrelations of Nutrition and Cancer, M. S. Arnott, J. Vaneys, Y. M. Wang, eds., Raven, New York (1982), pp. 401-408; D. Medina, H. W. Lane, C. M. Tracey, Cancer Res. (Suppl.) 43:2460s (1983); M. M. Jacobs, Cancer Res. 43:1646 (1983); H. J. Thompson, L. D. Meeker, P. J. Becci, S. Kokoska, ibid. 42:4954 (1982); D. F. Birt, T. A. Lawson, A. D. Julius, C. E. Runice, S. Salmasi, ibid., p. 4455; C. Witting, U. Witting, V. Krieg, J. Cancer Res. Clin. Oncol. 104:109 (1982).

121. M. Chatterjee and M. R. Banerjee, Cancer Lett. 17:187 (1982).
122. W. C. Willett, et al., Lancet, 1983-II, 130 (1983).
123. J. T. Salonen, G. Alfthan, J. Pikkarainen, J. K. Huttunen, P. Puska, ibid., 1982-II, 175 (1982).
124. M. Yonaha, E. Itoh. Y. Ohbayashi, M. Uchiyama, Res. Commun. Chem. Pathol. Pharmacol. 28:105 (1980); L. J. Kling and J. H. Soares, Jr., Nutr. Rep. Int. 24:29 (1981).
125. N. Tateishi, T. Higashi, A. Naruse, K. Hikita, Y. Sakamoto, J. Biochem. (Tokyo) 90:1603 (1981).
126. J. M. Williamson, B. Boettcher, A. Meister, Proc. Natl. Acad. Sci. U.S.A. 79:6246 (1982).
127. CME Symposium on "N-Acetylcysteine (NAC): A Significant Chemoprotective Adjunct," Sem. Oncol. 10 (Suppl. 1):1 (1983).
128. J. A. Hinson, L. R. Pohl, T. J. Monks, J. R. Gillette, Life Sci. 29:107 (1981).
129. A. M. Novi, Science 212:541 (1981).
130. W. B. Dunham, et al., Proc. Natl. Acad. Sci. U.S.A. 79:7532 (1982); G. Kallistratos and E. Fasske, J. Cancer Res. Clin. Oncol. 97:91 (1980).
131. S. Wassertheil-Smoller, et al., Am. J. Epidemiol. 114:714 (1981).
132. B. N. Ames, R. Cathcart, E. Schwiers, P. Hochstein, Proc. Natl. Acad. Sci. U.S.A. 78:6858 (1981).
133. A. Nomura, L. K. Heilbrun, G. N. Stemmermann, in preparation.
134. J. N. Boyd, J. G. Babish, G. S. Stoewsand, Food Chem. Toxicol. 20:47 (1982).
135. A. W. Wood, et al., Proc. Natl. Acad. Sci. U.S.A. 79:5513 (1982).
136. T. H. Maugh II, Science 205:1363 (1979) (interview with John Higginson).
137. H. L. Gensler and H. Bernstein, Q. Rev. Biol. 6:279 (1981).
138. A. L. Tappel, in Free Radicals in Biology, W. A. Pryor, ed., Academic Press, New York (1980), vol. 4, pp. 1-47.
139. R. S. Sohal, in Age Pigments, R. S. Sohal, ed., Elsevier/North-Holland, Amsterdam (1981), pp. 303-316.
140. J. E. Fleming, J. Miquel, S. F. Cottrell, L. S. Yengovan, A. C. Economos, Gerontology 28:44 (1982).
141. J. M. Tolmasoff, T. Ono, R. G. Cutler, Proc. Natl. Acad. Sci. U.S.A. 77:2777 (1980); R. G. Cutler, Gerontology, in press; J. L. Sullivan, ibid. 28:242 (1982).
142. R. Peto, Proc. R. Soc. London Ser. B 205:111 (1979); D. Dix, P. Cohen, J. Flannery, J. Theor. Biol. 83:163 (1980).
143. R. G. Cutler, in Testing the Theories of Aging, R. Adelman and G. Roth, eds., CRC Press, Boca Raton, FL, (in press).
144. D. Hersh, R. G. Cutler, B. N. Ames, in preparation.
145. E. Boyland, in Health Risk Analysis, Franklin Institute Press, Philadelphia (1980), pp. 181-193; E. Boyland, in Cancer Campaign, vol. 6, Cancer Epidemiology, E. Grundmann, ed., Fischer, Stuttgart (1982), pp. 125-128.
146. J. L. Marx, Science 219:158 (1983).
147. B. D. Goldstein, G. Witz, M. Amoruso, D. S. Stone, W. Troll,

Cancer Lett. 11:257 (1981); W. Troll, in Environmental Mutagens and Carcinogens, T. Sugimura, S. Kondo, H. Takebe, eds., Univ. of Tokyo Press, Tokyo and Liss, New York (1982), pp. 217-222.
148. B. N. Ames, M. C. Hollstein, R. Cathcart, in Lipid Peroxide in Biology and Medicine, K. Yagi, ed., Academic Press, New York (1982), pp. 339-351.
149. I. Emerit and P. A. Cerutti, Proc. Natl. Acad. Sci. U.S.A. 79:7509 (1982); Nature (London) 293:144 (1981); P. A. Cerutti, I. Emerit, P. Amstad, in Genes and Proteins in Oncogenesis, I. B. Weinstein and H. Vogel, eds., Academic Press, New York (in press); I. Emerit, A. Levy, P. Cerutti, Mutat. Res. 110:327 (1983).
150. J. Funes and M. Karel, Lipids 16:347 (1981).
151. S. J. Stohs, M. Q. Hassan, W. J. Murray, Biochem. Biophys. Res. Commun. 111:854 (1983).
152. C. C. Reddy, R. W. Scholz, E. J. Massaro, Toxicol. Appl. Pharmacol. 61:460 (1981).
153. H. Nagasawa and J. B. Little, Carcinogenesis 2:601 (1981); V. Solanki, R. S. Rana, T. J. Slaga, ibid., p. 1141; T. W. Kensler and M. A. Trush, Cancer Res. 41:216 (1981).
154. R. H. Simon, C. H. Scoggin, D. Patterson, J. Biol. Chem. 256:7181 (1981); T. S. Argyris and T. J. Slaga, Cancer Res. 41:5193 (1981).
155. G. E. Hatch, D. E. Gardner, D. B. Menzel, Environ. Res. 23:121 (1980).
156. A. J. P. Klein-Szanto and T. J. Slaga, J. Invest. Dermatol. 79:30 (1982).
157. C. C. Weddle, K. R. Hornbrook, P. B. McCay, J. Biol. Chem. 251:4973 (1976).
158. A. G. Lurie and L. S. Cutler, J. Natl. Cancer Inst. 63:147 (1979).
159. C. Borek and W. Troll, Proc. Natl. Acad. Sci. U.S.A. 80:1304 (1983); C. Borek, in Molecular Interrelations of Nutriton and Cancer, M. S. Arnott, J. van Eys, Y.-M. Wang, eds., Raven, New York (1982), pp. 337-350.
160. T. J. Slaga, et al., in Carcinogenesis: A Comprehensive Treatise, Raven, New York (1982), vol. 7, pp. 19-34; K. Ohuchi and L. Levine, Biochem. Biophys. Acta 619:11 (1980); S. M. Fischer, G. D. Mills, T. J. Slaga, Carcinogenesis 3:1243 (1982).
161. R. P. Mason, in Free Radicals in Biology, W. A. Pryor, ed., Academic Press, New York (1982), vol. 5, pp. 161-222.
162. G. McLennan, L. W. Oberley, A. P. Autor, Radiat. Res. 84:122 (1980).
163. J. M. Parry, E. M. Parry, J. C. Barrett, Nature (London) 294:263 (1981); A. R. Kinsella, Carcinogenesis 3:499 (1982).
164. H. C. Birnboim, Can. J. Physiol. Pharmacol. 60:1359 (1982).
165. A. Varshavsky, Cell 25:561 (1981).
166. T. J. Powles, et al., eds., Prostaglandins and Cancer: First International Conference, Liss, New York (1982).

167. J. E. Trosko, C.-C. Chang, A. Medcalf, Cancer Invest., in press.
167a. I. B. Weinstein, Nature (London) 302:750 (1983).
168. J. A. Bond, A. M. Gown, H. L. Yang, E. P. Benditt, M. R. Juchau, J. Toxicol. Environ. Health 7:327 (1981).
169. Editorial, Lancet, 1980-I, 964, (1980); K. Yagi, H. Ohkawa, N. Ohishi, M. Yamashita, T. Nakashima, J. Appl. Biochem. 3:58 (1981).
170. J. M. C. Gutteridge, B. Halliwell, D. A. Rowley, T. Westermarck, Lancet, 1982-II, 459 (1982).
171. J. E. Cleaver, in Metabolic Basis of Inherited Disease, J. B. Stanbury, J. B. Wyngaarden, D. S. Fredrickson, J. L. Goldstein, eds., McGraw-Hill, ed. 5, New York (1983), pp. 1227-1250.
172. K. C. Bhuyan, D. K. Bhuyan, S. M. Podos, IRCS Med. Sci. 9:126 (1981); A. Spector, R. Scotto, H. Weissbach, N. Brot, Biochem. Biophys. Res. Commun. 108:429 (1982); S. D. Varma, N. A. Beachy, R. D. Richards, Photochem. Photobiol. 36:623 (1982).
173. M. L. Katz, K. R. Parker, G. J. Handelman, T. L. Bramel, E. A. Dratz, Exp. Eye Res. 34:339 (1982).
174. D. Behne, T. Hofer, R. von Berswordt-Wallrabe, W. Elger, J. Nutr. 112:1682 (1982).
175. B. N. Ames, L. S. Gold, C. B. Sawyer, W. Havender, in Environmental Mutagens and Carcinogens, T. Sugimura, S. Kondo, H. Takebe, eds., Univ. of Tokyo Press, Tokyo, and Liss, New York (1982), pp. 663-670.
176. National Center for Health Statistics, Advance Report, Final Mortality Statistics, 1979; Monthly Vital Statistics Report 31, No. 6, suppl. [DHHS publication (PHS) 82-1120, Public Health Service, Hyattsville, MD, 1982]; Metropolitan Life Insurance Company Actuarial Tables, April 1983.
177. B. A. Bridges, B. E. Butterworth, I. B. Weinstein, eds., Banbury Report 13. Indicators of Genotoxic Exposure, Cold Spring Harbor Laboratory, Cold Spring Harbor, NY (1982); R. Montesano, M. F. Rajewsky, A. E. Pegg, E. Miller, Cancer Res. 42:5236 (1982); H. F. Stich, R. H. C. San, M. P. Rosin, Ann. N.Y. Acad. Sci., in press; I. B. Weinstein, Annu. Rev. Public Health 4:409 (1983).
178. I am indebted to G. Ferro-Luzzi Ames, A. Blum, L. Gold, P. Hartman, W. Havender, N. K. Hooper, G. W. Ivie, J. McCann, J. Mead, R. Olson, R. Peto, A. Tappel, and numerous other colleagues for their criticisms. This work was supported by DOE contract DE-AT03-76EV70156 to B.N.A. and by National Institute of Environmental Health Sciences Center Grant ES01896. This article has been expanded from a talk presented at the 12th European Environmental Mutagen Society Conference, Espoo, Finland, June 1982 [in Mutagens in Our Environment, M. Sorsa and H. Vainio, eds., Liss, New York (1982)]. I wish to dedicate this article to the memory of Philip Handler, pioneer in the field of oxygen radicals.

INSTITUTIONAL INITIATIVES FOR RISK MANAGEMENT

Thomas H. Moss

Case Western Reserve University
Cleveland, OH 44106

The issue of institutional initiatives for risk management is a crucial topic for discussion, as it confronts an important next step in the evolution of risk assessment as a tool of public policy. We know that risk analysis has become a defined discipline with well--known techniques and methodologies. We're well beyond using it in public policy on a "risk only" basis, in which we trigger actions regardless of the gradation of the risk. We're also beyond the naive technocratic notion that we can use it in public policy in a purely mathematical way, computing a hierarchy of risks and expecting public policymakers to let the numbers dominate whatever other values or philosophies they have. Similarly, the notion of a "science court" has dimmed in attractiveness, as we have grasped the complexity of non-technical factors which must be weighed in a risk decision. But, having turned away from the most simplistic applications, we must now face the question of devising institutions which can credibly handle the full array of uncertain and ambigious technical and non-technical issues in risk management decisionmaking.

Clearly, experiments are going on. If we look at compilations of risk analysis applications in legislation, in the courts, or in the federal and state regulatory agencies, we see a wide range of ideas being tried out.[1] In one perspective we are in a horrifying mess of inconsistencies in the manner in which risk ideas have crept into law and regulation. In another view, however, we are seeing a very health and pragmatic process of determining empirically the concepts which can win public acceptance and be technically and politically workable. It is a kind of "social Darwinistic" struggle among ideas and arrangements, to see which can stand the test of time.

The federal action with respect to saccharin was a perfect example of social pragmatism dominating legal or administrative purity. A clear legal test exists regarding carcinogens in food: the Delaney clause of the Federal Food, Drug, and Cosmetic Act. Though the provision is unpopular in scientific circles, it is unambiguous. Many in Congress were concerned that if applied to enforce a ban on saccharin, the negative economic and health consequences would far outweigh the small cancer risk saccharin seemed to represent.

The normal recourse in such cases where existing law and policy seem counter-productive is simply to repeal the provision. In fact, the Congress let the basic law stand, but overrode its application in this particular case. This was an institutional experiment: to maintain a rigid and extremely conservative standard in law but override it from the legislature in particular cases where its rigidity seems to defeat the overall goal of protecting human health.

This, however, is just one of a host of other ongoing legal and institutional arrangements which are being tried out in handling risk in public policy. Special review boards, reliance on National Academy studies, public participation mechanisms, judicial "confidence proceedings," and other ideas are all being used to meet the special demands of bringing risk considerations into regulatory law and institutions.

Though my scientific sense of order sometimes causes me to view these diverse and often internally contradictory experiments with alarm, my sense of politics does lead me to see value in the diversity and trial and error approach. However, as we propose and discuss institutional experiments today or at any other time, I believe there are some constraints we must respect.

The institutional arrangements must be built on a foundation of reasonable public understanding of the issues. The most elegant and efficient system in the world will have no acceptability if it has to rely on blind public acceptance. Technically defined procedures and standards which cannot be related to ordinary experience and the comprehension of a sizeable fraction of the lay public are doomed to failure as public policy. Indeed, this constraint implies that those of us who would like to see a wider and more sophisticated use of risk analysis in public policy have an inseparable interest in promoting any measures that boost the level of scientific understanding in the public. That public understanding is the limiting factor on the ability of the political system to responsibly fine-tune the use of risk concepts in public policy.

Any institutional arrangements must give a prominent and respectful place to values entirely outside of the realm of scientific and quantitative analysis. These may include values which lead to inconsistencies or contradicitons in public policy priorities when measured against a scientifically determined hierarchy of risks. Handling these values will require a kind of "split personality" in policy-making. That is, an ability is needed to understand and use the best of scientifically determined risk assessments, while at the same time maintaininig a consciousness of the legitimate place for other non-technical values in forming public policy.

The institutional arrangements must be designed to clamp down the shrill adversarial atmosphere that has characterized many recent public policy debates involving risk. Major business, environmental, consumer, and even scientific organizations have been tempted by short-term political expediency to approach public policy debates on risk with a win-lose, rather than a win-win, problem-solving attitude. In some instances this has seemed productive for both sides in reaching their overall goals. Strong posturing, extreme statements, speaking in partial truths and ridiculing opponents' arguments all seemed like good political tactics. However, in the long run it is rapidly breeding a kind of public cynicism about risk debates that will make truly useful applications or risk notions in public policy impossible. Lack of public sophistication in basic scientific notions, of course, breeds and sustains these adversarial tactics of half-truth and exaggeration.

A key goal of workable institutional arrangements for effectively bringing risk considerations into public policy will be to achieve a problem solving atmosphere and framework, not an arena to determine winners and losers. The public is nearly fed up with the adversary games being played in risk debates by technical opponents who should know better. Very soon it is likely to invoke a plague on the entire risk community unless we can improve the dialogue.

FOOTNOTES

1. T. H. Moss and B. Lubin, Risk Analysis, a Legislative Prospective, Chapter 4 in Health Risk Analysis, C. R. Richmond, P. J. Walsh, E. D. Copenhaver. Franklin Institute Press, 1981.

RISK ASSESSMENT: THE REALITY OF UNCERTAINTY

Edwin L. Behrens

Public Affairs Committee
American Industrial Health Council

ABSTRACT

Three realities about uncertainty need to be recognized if we are to progress as a nation in the management of risk, particularly where federal regulations affecting health and the environment are concerned. First, uncertainty and risk are inevitable. Second, public policy has not been well served in the long run when efforts have been made in the past to shortcut substantive approaches to risk management. There seems to be generally increasing acceptance that "zero risk" is not an adequate basis for the development of most regulatory policies. Last, risk and uncertainty must be addressed, and there must be convergence as to how they are to be managed. Risk assessment is increasingly accepted as a legitimate and necessary component of the regulatory process, and its use is even expected by federal courts despite the need for further improvements in the underlying science. Public policies incorporating risk assessment can be improved: (1) by implementing recommendations of the recent NAS report on Risk Assessment in the Federal Government: Managing the Process; (2) by continuing with research to improve scientific understanding relevant to risk assessment (with particular focus on exposure and epidemiology); and (3) by increasing public understanding of risk assessment and its role in public policy development. All are relevant to the interests and involvement of the Society for Risk Analysis.

KEYWORDS: Risk Assessment, Risk Management, Regulations, Regulatory Policy Development, Zero Risk

INTRODUCTION

My objective in this paper is to review some of the American Industrial Health Council's (AIHC) thoughts about "initiatives" to "institutionalize" or advance concepts to help ensure the soundness of that science which serves as a basis for federal regulatory actions. Risk assessment, of course, is a necessary component for virtually any effort to interpret and apply scientific data in the development of federal regulations where health and the environment are concerned.

I believe there are three realities about uncertainty to be recognized in public policy development that are related to risk assessment. First, there is the reality that uncertainty and risk are inevitable. Second, there is the reality that one cannot adequately address risk by taking shortcuts in the development of public policy where health and the environment are concerned. Finally, there remains the reality that risks and uncertainties must be addressed if we are to progress in the development of public policies that provide a sound basis for public administration. As a society, we must begin to achieve some convergence in how they are managed.

I would like to comment on the inevitability of risk and a change in perception that I believe has occurred. I cannot substantiate this observation, perhaps, but it is a perception that I have.

While the evolution of risk assessment as a tool for regulatory decisionmaking has not occurred without its share of public debate, the focus of that debate seems to have shifted in the very recent past. The shift that I believe has occurred seems to be a shift in focus away from questioning the legitimacy of conducting risk assessments, per se. The shift has been to when, where, and how they should be utilized in helping to decide the terribly complicated, and oftentimes contentious and critical decisions that must be made where chronic health concerns are involved. It is a shift that seems to signal an increasing acceptance that risks and uncertainty are an inevitable part of regulatory decisionmaking.

I do not believe those who were opposed to the use of risk assessment were opposed to risk assessment, per se. But its use was opposed by some because such use implied that public policy would be predicated on the basis of accommodating to some level of risk, rather than on the basis of allowing no risk at all. Obviously, in a "zero risk" world, there is no need to ask, "How much?"

A tempering of that opposition has occurred, I believe, because we are fundamentally a nation of pragmatic people who have always been able to separate reality from the ideal; accepting the one, on the one hand, to progress without sacrificing the ideality of the other.

This shift, if it has occurred, is most appropriate, because it places the focus of debate on risk management, which is really where policy debate belongs. It is in this context that questions relating to whether there should be any allowance for risk or not, when, and, if so, how much, should be addressed. Also involved are questions related to who should decide and on what basis the decisions should be made.

Clearly, risk assessment, which is merely a tool (albeit an important tool) to assist a decisionmaker, is not in itself a proper focus for major public policy debate. (This is not to say, of course, that the appropriate role for risk assessment and the relevance to be placed on it in any given instance are not legitimate items for debate. They are and will be!)

But the consequence of this shift should augur well for risk assessment. Legitimate concerns have always been raised, even within the risk assessment community itself, about the quality of data available for risk assessment purposes (e.g., exposure), about validity (e.g., statistical models), or about how to incorporate certain scientific considerations into risk assessment (or whether they should be considered or not).

These are issues that will continue to warrant scientific debate, but scientific dialogue can generally proceed best when it is not center-stage and under the spotlight as the focus of a heated public debate.

It is from conventional scientific questioning, and from the results of substantial research, that progress in the science of risk assessment has been made. Because of this progress, utilization of risk assessment has increased and evolved to the acceptance level and use that it has today. (Note that I say "has," not "enjoys," as I do not believe that risk assessment has yet quite acquired that level of grace in the minds of many legislators or others involved in the public policy development process.) This acceptance is not based on risk assessment as an established scientific truth. It has not even been established that the various components of risk assessment all satisfy conventional concepts of validity. Rather, risk assessment is accepted because of simple utility!

Risk assessments serve certain needs for decisionmakers that cannot be (or have not been) met any other way, and their use (despite their imperfections) will be increasingly recognized by public policymakers as long as they are not misapplied. I believe it is this sense of acceptance and utility that G. Box was attempting to convey when he said, "All models are wrong, but some are useful."[1]

That is certainly a far more optimistic assessment than Grossman's misquote of H. L. Mencken from Murphy's Law (Book III), "Complex problems have simple, easy-to-understand wrong answers." Risk assessment has already progressed well beyond Grossman, and if the discipline continues to perfect the science upon which it is based, it may eventually acquire the status and recognition enjoyed by Mencken! Before that time arrives, however, I believe the other two realities will need to be addressed.

The second recognizes the difficulties that have been encountered where federal policymakers have attempted to obviate the need to dimension risk. Such efforts generally seem to have the same singular effect. That is, they discount the role of scientific understanding as one, very critical element in the public decisionmaking process.

For example, in reading the testimony offered during the recent Food Safety hearings before the Senate Committee on Labor and Human Resources, two conclusions seem to stand out. The first is that regardless of whether one agrees that the Delaney Amendment should be changed or not, where such an absolute prohibition on risk is introduced into law it can never be viewed as a simple solution to a difficult problem. Second, it should never be imposed without an intensive and, importantly, prospective evaluation of the practical ramifications of such an approach.

C. W. McMillan, Assistant Secretary for Marketing and Inspection Services for the Department of Agriculture, said, "Serious thought should be given to modifying the Delaney clause so as to allow FDA to use risk assessment to reach sensible results consistent with public health. Such a statutory scheme would permit FDA to use the best science available to ascertain what the real risks presented by a substance are (risk assessment) and then determine how to manage that risk."[2]

FDA Commissioner, Arthur Hull Hayes, Jr., M.D., said, "It is clear that our ability to make balanced regulatory judgments is being stretched to the limit." He said that, while the Delaney clause made sense when enacted in 1958, "scientific methodology was less sophisticated--this approach has led to more and more problems as scientific methodology has advanced."[3]

President of Stanford University, Dr. Donald A. Kennedy, a former Commissioner of the FDA himself, said about Delaney, "First, it leaves no headroom for scientific progress. Second, by increasingly requiring agency action under conditions where risks appear to be trivial, it invites public ridicule and disregard for the food safety laws generally."[4]

Dr. Jere E. Goyan, Dean of the School of Pharmacy at the University of California and also a former Commissioner of the FDA, agreed that the Delaney clause "is not sensitive to advances that have been made in technology."[5]

Importantly, testimony affirms that the Delaney clause compromises FDA's ability to benefit from the increased scientific knowledge and understanding that have been acquired since 1958. Use of these advances in knowledge could help interpret data resulting from the corresponding progress that has been made by analytical chemists in their ability to detect the presence of potentially toxic materials at remarkably low levels.[6] In addition, Delaney precludes FDA from considering any other factors that may be relevant to public health in deciding how to deal with a food-additive for which the Delaney provisions have been found to apply.

Without debating whether the application of "zero risk" is appropriate in the context of food additives or not, the issues raised in these hearings validate the practical difficulties in dealing with such an approach. They clearly indicate that it is not an adequate basis to serve as the general or preferred approach by the Federal Government in its efforts to manage risk. Rather, the experience with Delaney merely confirms the inevitability of having to develop federal policies that attempt to achieve some consistency in their approach to risk management. Risk assessment must become a common element in any federal approach to the rational management of risk.

It should be emphasized that promoting the increased use of risk assessment, regulatory analyses, or any other attempts to quantify the regulatory decisionmaking process will never preempt the need for subjective judgment. Risk assessments and the use of other comparative judgments do not force a regulatory decision, but they can lead to a better decision.[7]

As an example of an agency's efforts to manage risk without basing agency actions upon a risk assessment, one merely has to look to OSHA's generic carcinogen proposal of 1977.[8] In that approach, OSHA proposed to establish generic criteria for the identification and classification of potential carcinogens. Identification and classification were based on defined criteria developed for administrative or regulatory expediency or, in the terminology of the recently issued NAS report on _Risk Assessment in the Federal Government: Managing the Process_, on "Hazard Identification."[9] Regulatory action was to proceed according to the prescribed classification in a relatively straightforward manner. Using the NAS report's own words in describing the OSHA approach: "The final rule, for example, did not address exposure assessment and rejected the use of dose-response assessment for any purpose except priority setting.... For reasons of efficiency, the guidelines were written in language that permitted little deviation from the judgments embodied in them."[10]

As a practical matter, the OSHA regulations, which were promulgated in 1980 and revised in 1981 in an attempt to accommodate the Supreme Court's ruling on benzene, have never served as the basis for regulatory action.

The benzene ruling required that OSHA find: (1) that a risk be significant and (2) that the risk would be significantly reduced by the proposed standard. More recently, federal policy interests have been further affected by the Fifth Circuit Court of Appeals' decision in the formaldehyde case. That decision has significant implications regarding risk assessment methodologies used throughout the Federal Government for regulatory purposes.

The formaldehyde decision set aside a CPSC ban of urea formaldehyde foam insulation, for the following reasons:

- o The finding of unreasonable chronic risk of cancer was not supported by substantial evidence
- o The finding of acute hazard was not supported by substantial evidence
- o The CPSC should have acted under the Federal Hazardous Substances Act rather than the Consumer Product Safety Act.[11]

Most importantly, the Court concluded that the finding of unreasonable risk of cancer was unsupported, based on a critical analysis and rejection of the risk assessment methodology followed by CPSC.

As our Chairman, Dr. Thomas Moss, identified in a seminar at Brookings Institution earlier this year, of 34 pieces of major regulatory legislation, 32 require a consideration of risk, while two are based on a "best available technology" approach.[12] Because of the scope of this legislation, the Fifth Circuit's decision seems likely to require changes in the use of risk assessment, not just by CPSC, but by other agencies as well. (Importantly, that decision looked beyond toxicology, and also addressed deficiencies in exposure data.)

Pending liability legislation is generally described as the "Lawyers' Relief Act" for any given year, and Regulatory Reform measures are generally similarly ascribed for economists. The two court decisions have the impact, however, not just of providing relief for a beleaguered risk assessment profession, but for establishing risk assessment as a legitimate and expected component in the regulatory decisionmaking process.

The Fifth Circuit Court decision said something else, however, that has received less attention. That statement will lead us to our third reality. What the court said is:

> "Regulatory agencies such as these were created primarily to protect the public from latent risk. For them to be effective, the public must be confident in their ability to determine which products are unsafe, which products are dangerous,

and which substances are carcinogenic. Interagency disagreement undermines such confidence."

There is a need to achieve more consistent, comprehensive approaches to the development of regulatory policies associated with chronic health concerns (e.g., carcinogenicity, mutagenicity, etc.) by the Federal Government. That is advocated by the NAS report and that is an objective that has been encouraged by the American Industrial Health Council (AIHC) in both this and the prior Administration. What is needed to proceed?

AIHC believes that the NAS report on risk assessment provides an opportunity for all parties interested in improving the regulatory decisionmaking process to work together. It provides a sound basis for continuing to perfect regulatory processes in the interest of sound science, consistent with protecting health and the environment, and suggests how more uniformity in the Federal Government's approach to risk assessment might be achieved. I believe that three recommendations are relevant to the Society's interests and involvement.

Congressional Action

Legislative opportunities offered by the NAS report should be pursued. Such legislation should not necessarily seek to implement every provision of the NAS report as written; and, since politics is the "art of the possible," it is unlikely that this would ever occur. Support of legislation by the NAS Committee members and other interested parties may require compromise, but it need not involve the sacrifice of principle.

Second, Congress should incorporate the NAS concept for providing quality assurance of the scientific data evaluations by establishing Scientific Advisory Panels for purposes of providing independent and expert peer review prior to major federal regulatory actions, where their use is not already explicitly recognized in law. (Panelists, of course, should be selected solely on the basis of their scientific and technical expertise.)

As a personal comment, I do not view risk assessment as necessarily an end itself warranting direct Congressional action to incorporate it as a component in federal regulatory decisionmaking. Risk assessment is only a tool. But Congress, in recognizing the inevitability of having to allow for some flexibility in accommodating to risk and uncertainty in the development of public policy, should provide for organizational mechanisms that ensure an appropriate separation between the scientific and regulatory decisionmaking functions within a federal agency. By this, and by requiring the independent peer review of that science which is to serve as the basis for major regulatory actions, Congress will be creating the environment in which sound risk assessments can be conducted and confidently utilized in a manner consistent with the protection of public health and the environment, and as increasingly expected by the federal courts.

The NAS report has now made these issues legitimate for the Congress to consider, and its issuance should signal and stimulate the start of national debate as to how improvements in policy development can be best implemented and how the fundamental objectives of the NAS report can be achieved.

Research

The Society should continue to identify and seek support for research consistent with the need to continue to improve the validity, reliability, and utility of risk assessments. Certain needs for this research seem clear and are discussed below.

Exposure. This seems to be an area in critical need of perfection, if the quality of risk assessments overall is to keep pace with the advances being made in toxicology and the other sciences that are involved in hazard identification and the evaluation of risk. Epidemiology is another.

Other Areas of Research. I refer you, for example, to the August, 1982, report of the Federal Task Force on Environmental Cancer and Heart and Lung Disease.[13] It includes an excellent articulation of research priorities for the determination of the relationship between environment and human disease.

I would like to conclude this section with a comment from Senator Kennedy, who said:

> "In the long term research on risk [and benefit] analysis will probably prove more important than any other single step we can take to improve our capacity to protect the public health through regulation."[14]

Communications

The public needs to better understand risk assessment and its role in the public policy development process. Although risk assessment is regarded as something new, some form of risk assessment has been practiced ever since the early attempts to regulate steam boiler construction for riverboats in the mid-19th century.[15] For toxicological purposes, however, its use is more recent. It has just been ten years, for example, since FDA first proposed its use of quantitative risk assessment in regulation.[16]

Because of public fears about chronic health concerns, the role of risk assessment and how it can help improve federal decisionmaking and priority setting needs to be better understood.

EPA Administrator William Ruckelshaus addressed this need in his recent policy speech before the NAS, and it has also been addressed by Senator Kennedy in the past. Kennedy said, "We must begin educating the public to the reality that there is no such thing as absolute safety. Regulation can never completely and totally protect the public. Large segments of the American public already accept this fact. But it is time for persons in positions of leadership to strengthen this understanding with more candid discussion of the limits of regulation."[17]

Obviously, Kennedy was not calling for tearing away the regulatory fabric that protects human health. What he was calling for is the public understanding necessary for good government.

Ruckelshaus recognized that scientists, and particularly the risk assessment community, will need to share in this communication process, using terminology that lay persons can understand.[18]

There are few areas of science today more critical to development of public policies associated with health and the environment, and few areas of science experiencing such rapid advancement and use. Much progress has been made, but there is much that remains to be done.

I would like to conclude with one observation regarding the likely success in seeing to the eventual implementation of more uniform approaches to risk assessment as a matter of national policy. It is from an unfinished speech that was being prepared by Franklin Delano Roosevelt at the time that he died. In that speech, he wrote this line, "The only limit to our realization of tomorrow will be our doubts of today."

FOOTNOTES

1. G. E. P. Box, Robustness in the Strategy of Scientific Model Building, Robustness in Statistics, 201-236 (1979).
2. C. W. McMillan, Testimony: Food Safety Hearings Before the Senate Committee on Labor and Human Resources (June 10, 1983).
3. Arthur Hull Hayes, Jr., M.D., Testimony: Food Safety Hearings Before the Senate Committee on Labor and Human Resources (June 10, 1983).
4. Donald Kennedy, Ph.D., Daily Excerpts of Testimony: Food Safety Hearings Before the Senate Committee on Labor and Human Resources (June 8, 1983).
5. Jere E. Goyan, Ph.D., Testimony: Food Safety Hearings Before the Senate Committee on Labor and Human Resources (June 8, 1983).

6. Edwin L. Behrens, Regulatory Impacts and the Need for Reform, Analytical Chemistry, 775A (June 1980).
7. Edwin L. Behrens, Comparative Risk Assessment, Hearing before the Subcommittee on Science, Research and Technology of the Committee on Science and Technology, U.S. House of Representatives, 349 (May 14, 1980).
8. National Academy of Sciences, Risk Assessment in the Federal Government: Managing the Process, 96-97 (1983).
9. Ibid., Page 3.
10. Ibid., Page 59.
11. R. C. Barnard and K. L. Rhyne, The Implication of the Formaldehyde Decision Regarding Risk Assessment (April 19, 1983).
12. T. H. Moss, Risk Assessment and the Legislative Process, Brookings Institution Workshop on Public Policy, Science and Environmental Risk, 3 (February 28, 1982).
13. U.S. Environmental Protection Agency; National Cancer Institute; National Heart, Lung, and Blood Institute; National Institute for Occupational Safety and Health; National Institute of Environmental Health Sciences; National Center for Health Statistics; Centers for Disease Control; Food and Drug Administration; Environmental Cancer and Heart and Lung Disease: Fifth Annual Report to Congress (August 1982).
14. Senator E. M. Kennedy, Remarks at the FDA Symposium on Risk/Benefit Decisions and the Public Health, 3 (February 15, 1978).
15. Subcommittee on Science, Research and Technology of the Committee on Science and Technology of the U.S. House of Representatives, A Review of Risk Assessment Methodologies, 10 (Congressional Research Service, Washington, DC, March 1983).
16. J. V. Rodricks, Risk Assessment in Perspective, 28 (September 1982).
17. Senator E. M. Kennedy, op. cit., Page 3 (February 15, 1978).
18. W. D. Ruckelshaus, Science, Risk and Public Policy, (June 22, 1983).

THE STATISTICAL CORRELATION MODEL FOR COMMON CAUSE FAILURES

J. A. Hartung
H. W. Ho
P. D. Rutherford
E. U. Vaughan

Atomics International Division
Energy Systems Group
Rockwell International
8900 De Soto Avenue
Canoga Park, CA 91304

ABSTRACT

The frequency of reactor accidents is greatly affected by the likelihood of dependent multiple failures such as common cause failures (CCFs). The ability to quantify and prevent such failures is therefore crucial to the assessment and management of nuclear risks. We recently proposed the statistical correlation model for CCFs with the hope that it could assist in their quantification and aid in the identification of improved defenses against them. Since that time, we have been working to transform the statistical correlation model from an abstract theoretical concept into a practical set of working tools to realize this potential. This paper reviews the theoretical basis for the statistical correlation model and summarizes the status of current work to develop improved methods and data to implement it in risk assessments and reliability analyses.

KEYWORDS: Statistical Correlation, Common Cause, Dependent Failures.

INTRODUCTION

The safety of nuclear power plants is based in large measure on the concept of defense in depth, wherein redundant systems and components and multiple fission product barriers are provided to guard against serious accidents. The probability of major accidents due to independent failures can be made very small by providing suitably redundant safety systems and fission product barriers. However, reliable accident prevention also requires avoidance of dependent multiple failures that may affect several systems, components, and/or fission product barriers at the same time.

Defenses against dependent failures have been designed into nuclear plants since early in the development of nuclear power. Segregation and protection of redundant plant protection system channels and siting of plants away from active earthquake faults are examples of early defenses. As the nuclear industry has matured, additional defenses have been identified and implemented, based largely on operating experience and engineering judgment. For example, following the Three Mile Island loss-of-coolant accident, the need was identified for improved instrumentation, control rooms, and operator training to help prevent operator confusion and errors that may lead to dependent failures. Other defenses have been identified by Epler,[1-3] Hayden,[4] Jolly and Wreathall,[5] Edwards and Watson,[6] and Bourne et al.,[7] to name a few.

In spite of this progress, dependent failures continue to be a problem for the nuclear industry as evidenced by their observed frequency of occurrence and potential consequences. The recent failure to scram incidents at the Salem plant served to highlight this problem area; however, the significance of dependent failures has been apparent for some time both in the operating experience record and in the results of recent probabilistic risk assessments. The highly redundant safety systems, components, and fission product barriers provided in nuclear power plants have apparently reduced the significance of independent failures to the point where dependent failures are at least a major, if not the dominant, contributor to risk.

DEFINITIONS AND BACKGROUND

There have been many definitions suggested in the literature for terms such as common cause and dependent failures.[1, 5-13] For purposes of this paper, common cause failures (CCFs) are defined as coexistent failures of two or more systems or components due to a common cause(s). A system or component is considered failed in this definition whenever it is unable to perform its intended function due to either internal failures (i.e., those that necessitate repair) or external faults (e.g., command faults or functional dependencies). The word "coexistent" implies that the subject items must be in failed states (i.e., unable to perform their intended functions) at the same time.

As defined here, CCFs need not be dependent, as it is theoretically possible for two items to fail independently due to the same cause. For example, two light bulbs may both be failed at the same time due to the same cause (say, wearout), and yet their failures may be independent of each other. However, evaluation of nuclear plant operating experience suggests that most CCFs are dependent, with independence being an idealized condition that is seldom if ever actually realized.

Conversely, most dependent failures can be traced back to a common cause(s). Sometimes, this common cause(s) is not the immediate cause of failure, but rather an earlier (root) cause. For example, in the Three Mile Island loss-of-coolant accident, misleading instrumentation and resulting operator confusion led to several operator errors, including failure to close the block valve downstream of the pilot-operated relief valve and excessive throttling of both safety injection systems. These failures were therefore CCFs by the above definition. Similarly, propagating or sequentially linked failures are also CCFs if they occur at the same time, since they have (by definition) a common initiating cause.

Common cause failures are difficult to quantify in risk assessments and reliability analyses. Some types of CCFs, such as those due to functional dependencies and natural phenomena, lend themselves naturally to explicit identification and quantification, albeit with large uncertainties. Other types of CCFs, such as those due to design and operational errors, management deficiencies, and manufacturing and construction flaws, are more difficult to explicitly foresee and quantify. Consequently, parametric techniques such as the geometric mean approach[14] and implicit models such as the β-factor,[15,16] binomial failure rate,[17] shock,[18] and advanced shock/binomial failure rate models[19-22] are often used to quantify such CCFs.

These implicit models have contributed greatly to our understanding and ability to quantify CCFs. However, they are not without their problems. For example, they do not readily lend themselves to the quantification of CCFs in diverse (as opposed to replicate) items and they do not adequately explain why observed common cause failure rates appear higher in accidents than during normal operation. Furthermore, these models are simply descriptive of failure rates and/or unavailabilities and therefore provide relatively few insights into the nature and causes of CCFs and potentially effective defenses against them.

STATISTICAL CORRELATION MODEL

We recently proposed the statistical correlation model with the hope that it could help to overcome at least some of these problems.[23] In this model, system and component unavailabilities are treated as potentially correlated random variables whose distribution encompasses both population (e.g., component-to-component) and temporal variability. The major unique features of this model are that system and component unavailabilities are assumed to be potentially correlated (rather than independent) and randomly variable with time (rather than constants or deterministic functions).

Figure 1 illustrates the statistical correlation model for a system of two redundant components whose availabilities are functions of time. This figure shows how a common cause may adversely affect (and correlate) component unavailabilities. Common cause failures may be introduced into the system in many ways, including functional dependencies, systems interactions, design and operational errors (including operator, maintenance, and/or testing errors), manufacturing and construction flaws, environmental effects, and external events. The possible existence of correlations amongst component unavailabilities introduces new terms into reliability equations. These new terms are treated explicitly in the statistical correlation model. The following equation shows the new (statistical correlation) term that applies to a system of two redundant components:

$$\overline{Q}(A \cdot B) = \overline{Q}_A \overline{Q}_B + r\sigma_A \sigma_B \tag{1}$$

where

$\overline{Q}(A \cdot B)$ = mean unavailability of a system of two redundant components (A and B)

$\overline{Q}_A$ & $\overline{Q}_B$ = mean unavailabilities of components A and B, respectively

σA & σB = standard deviations of the unavailabilities of components A and B, respectively

r = correlation coefficient between the unavailabilities of components A and B.

This form of equation is well known in statistics for correlated random variables; however, a derivation of it for component unavailabilities is presented in Reference 23. The first (conventional) term on the right-hand side of this equation accounts for independent failures, and the second (new) term is a correction to the first to account for possible statistical correlation effects. This second term can be interpreted as an implicit representation of the phenomena that are responsible for CCFs. This implicit representation is proposed as an alternative to explicit representation in which dependencies are resolved into the (almost limitless) conditionally independent events that are ultimately responsible for CCFs.

Figure 2 illustrates the statistical correlation model for a hypothetical system of two redundant components whose availabilities are

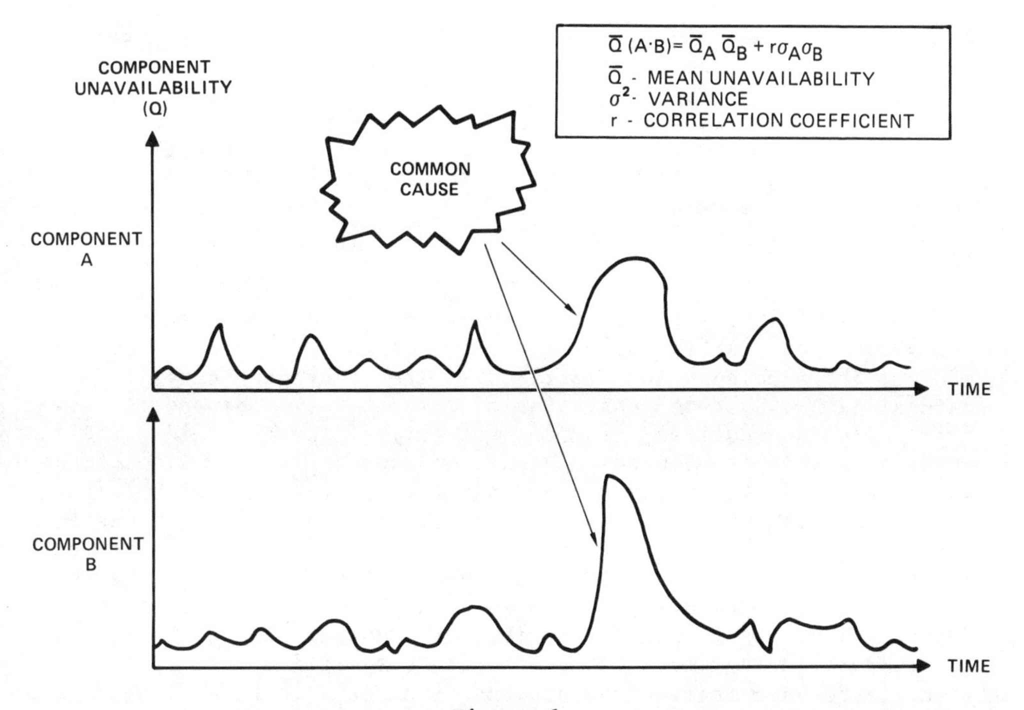

Figure 1

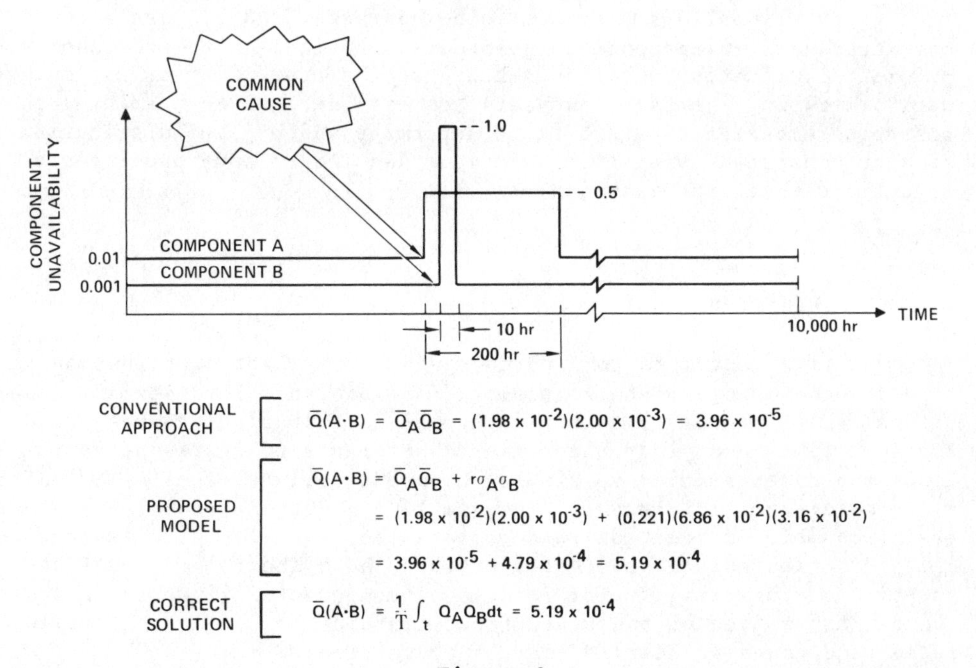

Figure 2

functions of time. This figure shows that the new statistical correlation) term introduced in Equation 1 is needed to obtain correct results. The conventional approach of multiplying mean unavailabilities together does not give correct results. In this example, the statistical correlation term exceeds the independent term by a large margin. Our evaluation of nuclear plant operating experience suggests that this is often the case, especially for systems involving three or more redundant components and accident sequences involving three or more failures.

The statistical correlation model is thought to be exact and independent of distribution for the two-component example discussed above. Its exactness and distribution free character can also be maintained for systems consisting of more than two components; however, to do so requires treatment not only of the two-fold correlation coefficients applicable to each pair of components, but also of the three-fold and higher order coefficients applicable to each triplet and larger combination of components. There are two approaches for handling these higher order coefficients: one is to treat them explicitly and the other is to treat them implicitly through the choice of an assumed joint distribution whose higher order correlations are determined by its marginal distributions and pair correlations. The first approach is theoretically better, but difficult to apply in practice because it requires so many parameters to be estimated. The second approach (which we are currently using is risk assessments and reliability analyses) is an approximation, but has the practical advantage that it allows a system to be characterized by just a few parameters and an appropriate joint distribution. Since the choice of a joint distribution determines how good the approximation is, much of our work during the last few years has been devoted to examining the operating experience data base to determine which joint distibution(s) is most consistent with it. This work had led to some preliminary selections that are discussed next.

JOINT DISTRIBUTIONS

Our first attempts to find an appropriate joint distribution centered on the multivariate lognormal distribution. This seemed to be a natural starting point, as the lognormal distribution has been used for some time in nuclear plant risk assessments to represent variations and uncertainties in single component failure rates. The multivariate lognormal distribution is simply a joint distribution of several (possible correlated) random variables, each of which is lognormally distributed. It therefore embodies both the conventional assumption of lognormality for individual component failure rates and the new assumption of the statistical correlation model, that failure rates (and unavailabilities) may be correlated.

However, there is one major difference between the conventional assumption of lognormality and the one proposed here. Whereas conventional analyses use the lognormal distribution to represent primarily uncertainties and sometimes population variability in time-averaged failure rates (and/or unavailabilities), we use it to encompass both temporal and population variability, as well as uncertainties. Consequently, our lognormal distributions have larger variances than those used in conventional analyses.

Having selected the multivariate lognormal distribution for initial investigation, the next step was to determine which parameter(s) should be represented by it. Our search led us to define a new parameter, which we call the "net hazard," that is defined as follows:

$$G(T) = \int_0^T g(t)dt \qquad (2)$$

where

$G(T)$ = net hazard at time T

$g(t)$ = net hazard rate at time t.

The net hazard rate [g(t)] is the hazard (or failure) rate of a component, adjusted to account for repair. As such, it can be either positive or negative, depending on whether failure or repair predominates. It is defined as the negative of the fractional time rate of change in the availability of a component. Component unavailability is therefore defined as follows:

$$Q(T) = 1-e^{-G(T)} \qquad (3)$$

where

$Q(T)$ = unavailability at time T

Although the net hazard rate [g(t)] can be either positive or negative, the net hazard [G(T)] is always greater than or equal to zero. Neither these parameters nor other similar parameters such as failure rate and unavailability can be measured directly, but must be inferred from observed successes and failures. Therefore, the choice of distributions to represent them must be based on inferences rather than direct measurements. Using a combination of theoretical arguments and semi-quantitative evaluation of operating experience, we hypothesized that the net hazard [G(T)] could be approximated by a multivariate lognormal distribution. This hypothesis plus equation 3 defines a truncated multivariate lognormal distribution for component unavailabilities.

To ascertain whether this hypothesis was reasonable, a computer code was developed to evaluate the unavailability of a system of components whose configuration and unavailabilities are represented by a Boolean expression and the desired truncated multivariate lognormal distribution, respectively. This code (COSAT) uses a covariance sampling technique to develop a sample of correlated variables, which are then used to evaluate the unavailability of the system via its Boolean expression.

The first test of this code and its underlying hypothesis was to determine whether it could reproduce the observed failure experience with replicate components. Much data has been collected on CCFs among replicate components,[6,15,19-22, & 24] so this seemed to be a natural starting point. Average failure rates (per demand) can be estimated from this data base as shown in Table I for diesel generators. The left-hand column in this table shows the number of replicate components in the sample, the second column indicates the number of demands on that size group of components, the third column shows the number of times that all components in the group failed, and the last column indicates the resultant estimated failure rate (per demand).

This data is shown graphically in Figure 3, along with a fit obtainable with COSAT. All data obtained so far for replicate components is similar in character to that shown in Figure 3, illustrating the diminishing rate of return for each successive component. Figure 4 illustrates some of this data for other components. Since the COSAT model for replicate components includes three parameters (a mean unavailability, the variance of that unavailability, and the correlation between unavailabilities), it is not surprising that three data points can be matched. However, what is more significant is that the remainder of the COSAT curve (for four or more components) is consistent with the data that we have examined to date. The goodness of this fit constitutes a test of the hypothesized multivariate lognormal distribution.

TABLE I

DIESEL GENERATOR FAILURE RATES

NUMBER OF COMPONENTS	NUMBER OF DEMANDS	NUMBER OF FAILURES	FAILURE RATE (PER DEMAND)
1	10,584	153	0.015
2	6,776	12	0.0018
3	2,408	2	0.0008
4	560	0	<0.0018

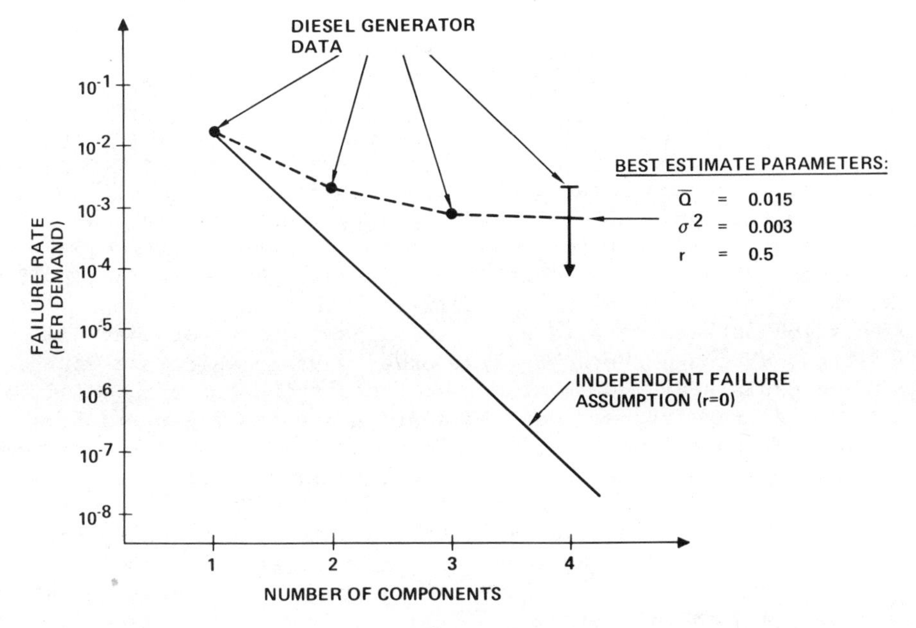

Figure 3

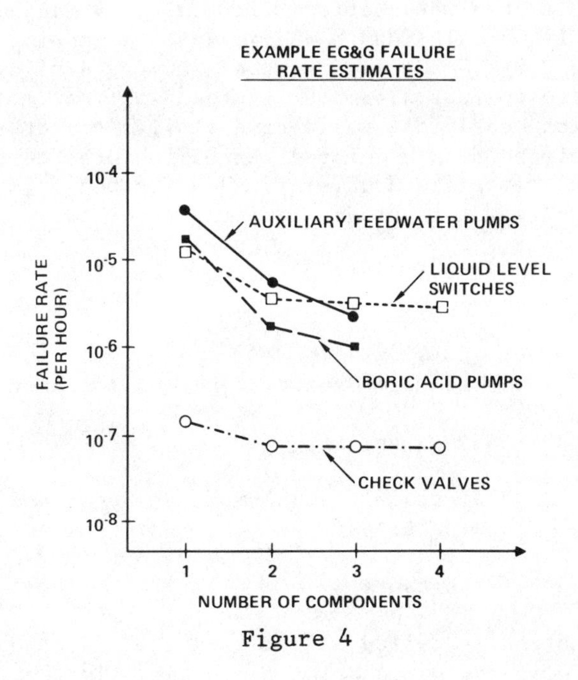

Figure 4

The next test of the COSAT model was to determine whether it could match the observed failure experience with diverse (as opposed to replicate) components. There has been relatively little work done to systematically collect and evaluate data on CCFs among diverse items. However, there have been many examples of such failures. For example, seven instances of failures among diverse pumps are noted in Reference 20. Pumps involved in these examples included those used for low-pressure and high-pressure coolant injection, containment spray, chemical and volume control, and residual heat removal. Also, Reference 25 notes that the conditional unavailability of diesel generators, given loss of offsite power, is higher by a factor of four than their unconditional unavailability. This suggests a possible correlation between offsite power and diesel generator unavailabilities, although there are also other ways in which the data base could be explained. Our evaluation of operating experience indicates that many of the most serious accidents and potential precursors of serious accidents have involved CCFs among diverse systems and components. Examples of events that involved such failures include the Three Mile Island loss of coolant accident, the Browns Ferry fire, the Oyster Creek low water level incident, and the Rancho Seco loss of instrumentation with steam generator dryout event, to name just a few. These and other multiple failure events are a rich source of data on CCFs among diverse items.

Problems were encountered when attempting to fit this data base with COSAT. Although a good fit (for diverse items) appeared theoretically possible, the mathematical difficulty of manipulating truncated multivariate lognormal distributions, and the resultant complexity and cost of using COSAT, made finding the fit very difficult. We, therefore, sought to develop a simpler approach to implement the statistical correlation model that would give similar or better results. The approach developed uses a joint distribution created for this purpose and named the "bivariate double-delta" distribution. Its marginal frequency density functions are defined as follows:

$$F(Q) = \delta a\,(1-Q) + (1-a)\delta(b-Q) \tag{4}$$

where

$F(Q)$ = frequency density function for unavailability

Q = unavailability

$\delta(1-Q)$ & $\delta(b-Q)$ = Dirac delta functions at $Q = 1$ and $Q = b$, respectively

a and b = constants

This function is called a "double-delta" function because it consists of two weighted Dirac delta functions. The constants a and b are uniquely determined by the mean value and variance of a component's unavailability, and the bivariate double-delta distribution is defined by a pair of double-delta functions and a correlation between them. Although this distribution may seem arbitrary, it was actually suggested by the critical properties of the truncated multivariate lognormal distribution.

The statistical correlation model can be implemented by applying Equation 1 in a step-wise fashion, using bivariate double-delta distributions to represent each pair of component unavailabilities and a mathematical procedure to account for the combined effect of multiple correlations. This approach has allowed us to match and explain operating experience data through a fairly simple and straightforward procedure. Figure 5 shows the match obtained for BWR relief valves as an example. As illustrated in this figure, the data are by no means conclusive but nevertheless generally tend to support the reasonableness of the model.

Because of the simplicity of this method, and its generally good agreement with operating experience data, this is the approach that we are currently using to implement the statistical correlation model in

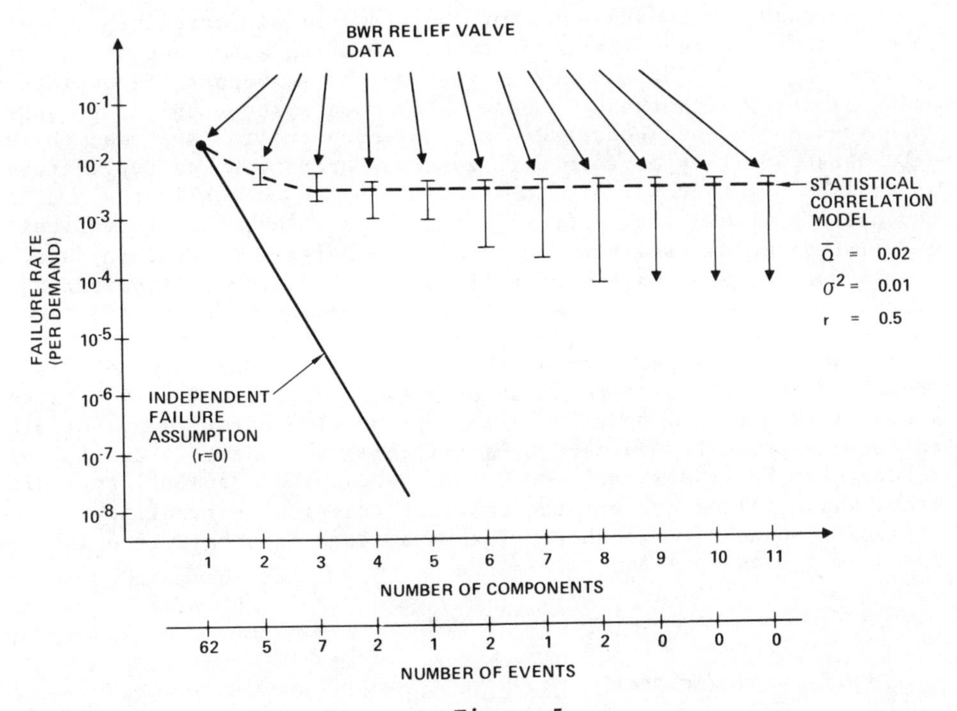

Figure 5

risk assessments and reliability analyses. It has been automated by developing a computer code (SCORE) that evaluates the unavailability, failure rate, and repair rate of a system based on input minimal cut sets, component data, and a correlation matrix. Although we are reasonably satisfied with the bivariate double-delta distribution approximation used in SCORE, it does have its limitations, and we are consequently continuing to search for improved joint distributions to represent component unavailabilities.

DATA BASE

The data base needed to implement the statistical correlation model (with SCORE) consists principally of estimates of three parameters: (1) mean unavailabilities, (2) variances of unavailabilities, and (3) correlations among unavailabilities. Mean unavailabilities can be calculated directly from data on single component failures; however, variances and correlation coefficients must be inferred from multiple component failures. The approach used to do this consists of performing parametric analyses with the statistical correlation model until a set of parameters is found that matches operating experience data. Since this same approach was used in the search for a joint distribution to implement the statistical correlation model, a by-product of that search was the beginnings of a data base.

To establish a data base for the statistical correlation model ideally requires that the totality of operating experience be examined to develop a set of parameter estimates that is both self-consistent and consistent with the data base. For some systems and components, such as diesel generators and emergency core cooling systems, there is data about how they perform during normal operation (as replicates), as well as how they perform in accidents (both as replicates and in conjunction with diverse items). For these items, the statistical correlation model parameters are uniquely determined and, in fact, the search for a set of parameter estimates that matches all operating experience constitutes somewhat of a test of the modeling approach.

However, for other items, the data base is sparse or nonexistent. Ideally, one would like to be able to estimate parameters for these items based not only on the sparse data available for them, but also on extrapolation from other items for which more data is available. To do this, it is desirable to define (and gather data on) new parameters whose values exhibit the smallest practical variability from one type of component to another. The parameters that have been defined for this purpose are as follows:

$$\text{Normalized variance} = \frac{\sigma^2}{\sigma^2_{max}}$$

$$\text{Normalized correlation} = \frac{r}{r_{max}}$$

where σ^2_{max} is the maximum variance of a double-delta function whose mean value is fixed, and r_{max} is the maximum correlation coefficient of a bivariate double-delta distribution whose marginal distributions are fixed. These parameters are therefore the ones that we are attempting to estimate in our data base development effort. The formulae for these parameters are as follows:

$$\sigma^2_{max} = \overline{Q}(1-\overline{Q})$$

$$r_{max} = \frac{\sigma_A(1-\overline{Q}_B)}{\sigma_B(1-\overline{Q}_A)} \text{ or its inverse, whichever is less}$$

where the subscripts A and B refer to components A and B, respectively, and $\overline{Q}$ = mean unavailability, σ^2 = variance of unavailability, and r = correlation between unavailabilities.

One last problem in developing a data base for the statistical correlation model needs to be addressed. Since the statistical correlation model accounts implicitly for CCFs not explicitly modeled, one must be careful not to double count potential common cause failure mechanisms. Therefore, considerable judgment is required, both in the development of the data base and in its use. We feel that tangible common cause failure mechanisms (e.g., hardware interactions) should be modeled explicitly in risk assessments and reliability analyses, whereas implicit methods such as the statistical correlation model may often be more appropriate for intangible failure mechanisms (e.g., design and operational errors and manufacturing and construction flaws). The data base that we are attempting to develop therefore reflects only the more intangible (and difficult to explicitly quantify) common cause failure mechanisms.

Table II summarizes our current generic estimates of the desired parameters of the statistical correlation model. This data base is very early in its formative stage and therefore subject to change as more operating experience data is evaluated. It nevertheless represents our best current estimates of the "residual" common cause failure potential of nuclear power plant systems and components, beyond that due to the more tangible failure mechanisms that are generally modeled explicitly in risk assessments and reliability analyses. It should be emphasized, however, that the specific parameter values depend on the amount of detail in the modeling approach used as well as on the number and type of common cause failure mechanisms explicitly modeled. For now, we are attempting to gather data on the same

TABLE II

GENERIC ESTIMATES OF STATISTICAL CORRELATION MODEL PARAMETERS

	DEGREE OF CORRELATION OR VARIABILITY	NORMALIZED CORRELATION		NORMALIZED VARIANCE
		DIVERSE	REPLICATE	
	VERY SMALL	0.01	0.2	0.05
BEST ESTIMATE RANGE ↕	SMALL	0.03	0.3	0.1
	MODERATE	0.1	0.5	0.2
	LARGE	0.3	0.7	0.4

level of modeling detail that we use in risk assessments and reliability analyses. A major future goal is to develop analytical procedures for selecting specific parameter values for a given level of modeling detail.

We are attempting to use this data base in two ways. First, by examining accidents that have involved the largest correlations among system and component unavailabilities, we are attempting to identify improved defenses against CCFs. The major causes of large correlations appear to have been design and operational errors (including maintenance and testing errors) rather than other causes such as manufacturing or construction deficiencies or external events. It is, therefore, in these areas that the most effective defenses against CCFs are being found.

Secondly, we are using the data base to quantify CCFs in risk assessments and reliability analyses. As can be seen from Table II, the range of normalized correlations for replicate systems and components is quite small (a factor of 3.5). Also, there is frequently good data available (from replicate items) with which to estimate these parameters. On the other hand, a much wider range of values are observed for normalized correlations among diverse items (a factor of 30), and the data available to estimate these parameters is much sparser. Therefore, the uncertainty associated with quantifying CCFs amongst diverse items is generally much larger than for replicates. This difficulty of quantifying CCFs among diverse items is not thought to be unique to the statistical correlation model but rather merely elucidated by it.

When little or no data is available, we use the generic data base in Table II, along with numerous sensitivity analyses, to help form a basis for decisionmaking. If a best estimate is needed for system reliability and if there is no data to the contrary, we obtain the best estimate by propagating the small to moderate variances and correlations through the model since most of the available data is in that range. In light of uncertainties, however, best-estimate reliabilities are viewed with caution and given only a modest role in the decisionmaking process.

CONCLUSIONS

We have discussed the theoretical basis for the statistical correlation model and summarized the status of current work to develop improved methods and data to implement it in risk assessments and reliability analyses. Although the methods and data discussed here are still early in the formative stage, a number of insights have been gained in the process of their development and application. The combined effect of these insights is to provide a fuller realization of the complexity and pervasiveness of CCFs and a better understanding of the resultant uncertainties in nuclear plant risk assessments and reliability analyses.

REFERENCES

1. E. P. Epler, "Common Mode Failure Considerations in the Design of Systems for Protection and Control," Nuclear Safety, 10(1):38-45, January-February 1969
2. E. P. Epler, "Diversity and Periodic Testing in Defense Against Common Mode Failure," Proceedings of the International Conference on Nuclear Systems Reliability Engineering and Risk Assessment, Gatlinburg, Tennessee, June 1977
3. E. P. Epler, "Common Mode Failure of Light Water Reactor Systems: What Has Been Learned," Oak Ridge Associated Universitites/Institute for Energy Analysis Report ORAU/IEA-80-7(M), May 1970
4. K. C. Hayden, "Common Mode Failure Mechanisms in Nuclear Power Plant Protection Systems," Oak Ridge National Laboratory Report ORNL-TM-4985, December 1975
5. M. E. Jolly and J. Wreathall, "Common Mode Failures in Reactor Protective Systems," Nuclear Safety, 18(1):624-632, September-October 1977
6. G. T. Edwards and I. A. Watson, "A Study of Common Mode Failures," UKAEA Safety and Reliability Directorate Report SRD-R-146, July 1979
7. A. J. Bourne, et al., "Defenses Against Common Mode Failures in Redundancy Systems," UKAEA Safety and Reliability Directorate Report SRD-R-196, January 1981
8. W. C. Gangloff, "Common Mode Failure Analysis," IEEE Transactions on Power Apparatus Systems, PAS-94(1), 1975
9. J. R. Taylor, "Common Mode and Coupled Failure," Danish Atomic Energy Commission Report RISO-M-1826, October 1975
10. T. Mankamo, "Common Mode Failures," Technical Research Centre of Finland, Electrical Engineering Laboratory Report 18, May 1976
11. R. G. Easterling, "Probabilistic Analysis of 'Common Mode' Failures," Proceedings of the ANS Topical Meeting on Probabilistic Analysis of Nuclear Reactor Safety, Newport Beach, California, May 1978

12. E. W. Hagan, "Common-Mode/Common Cause Failure: A Review and Bibliography," Oak Ridge National Laboratory Report ORNL/NUREG/NSIC-148 (NUREG/CR-0566), May 1979
13. A. M. Smith and I. A. Watson, "Common Cause Failures--A Dilemma in Perspective," Proceedings of the 1980 Annual Reliability and Maintainability Symposium, San Francisco, California, January 1980
14. "Reactor Safety Study: An Assessment of Accident Risks in U.S. Commercial Nuclear Power Plants," WASH-1400 (NUREG-75/014), October 1975
15. K. N. Fleming, "A Reliability Model for Common Cause Failures in Redundant Safety Systems," General Atomic Report GA-A13284, April 1975
16. "HTGR Accident Initiation and Progression Analysis Status Report," General Atomic Report GA-A13617, January 1967
17. W. E. Vesely, "Estimating Common Cause Failure Probabilities in Risk and Reliability Analyses: Marshall-Olkin Specializations," Proceedings of the International Conference on Nuclear Systems Reliability Engineering and Risk Assessment, Gatlinburg, Tennessee, June 1977
18. G. E. Apostolakis, "The Effect of a Certain Class of Potential Common Cause Failures on Reliability of Redundant Systems," Nuclear Engineering and Design, 36(1), 1976
19. C. L. Atwood and J. A. Steverson, "Common Cause Fault Rates for Diesel Generators: Estimates Based on Licensee Event Reports in U.S. Commercial Nuclear Power Plants 1976-1978," U.S. NRC Report NUREG/CR-2099 (EGG-EA-5359, Revision 1), June 1982
20. C. L. Atwood, "Common Cause Fault Rates for Pumps," U.S. NRC Report NUREG/CR-2098 (EGG-EA-5289), February 1983
21. J. A. Steverson and C. L. Atwood, "Common Cause Fault Rates for Valves," U.S. NRC Report NUREG/CR-2770 (EGG-EA-5458), February 1983
22. C. L. Atwood, "Common Cause Fault Rates for Instrumentation and Control Assemblies," U.S. NRC Report NUREG/CR-2771 (EGG-EA-5623), February 1983
23. J. A. Hartung, "A Statistical Correlation Model and Proposed General Statement of Theory for Common Cause Failures," Proceedings of the International Conference on Probabilistic Risk Assessment, Port Chester, New York, September 1981
24. A. McClymont and G. McLagan, "Diesel Generator Reliability at Nuclear Power Plants: Data and Preliminary Analysis," Electric Power Research Institute Report EPRI/NP-2433, June 1982
25. J. W. Minarick and C. A. Kukielka, "Precursors to Potential Severe Core Damage Accidents: 1969-1979, A Status Report," U.S. NRC Report NUREG/CR-2497 (ORNL/NSIC-182), June 1982

TURBINE ROTOR RELIABILITY:

A PROBABILITY MODEL FOR BRITTLE FRACTURE

Edward M. Caulfield*
Michael T. Cronin*
William B. Fairley**
Nancy E. Rallis***

*Packer Engineering Associates, Inc.
Naperville, IL

**Analysis and Inference, Inc.
Boston, MA

***Boston College
Chestnut Hill, MA

ABSTRACT

Rotors of electric turbines can fail catastrophically due to brittle fracture developing over many cycles of operation, resulting in injury and death and multi-million dollar economic losses. Probability distributions for measures of strength and stress of rotors are defined. A closed form solution is obtained for stress at any cycle from a dynamic model for the growth of crack size. Reliability at the nth cycle is defined within a Bayesian estimation framework as the posterior probability that true strength exceeds stress given estimates for both as of the (n-1)st cycle. Absolute and conditional reliabilities at any cycle are then obtained. Ways of reporting uncertainties in the reliability estimates are discussed, in particular, possible sources of bias from model specification and from parameter estimation procedures. Finally, suggestions for research are made.

KEY WORDS: Reliability; Turbine Rotors; Probability Model; Brittle Fracture; Bayesian Estimation.

1. INTRODUCTION

During the manufacture of a turbine rotor, in particular at the time of forging, flaws collect in the core material of the rotor. This part of the rotor is bored out not only to remove the discontinuities, but also to gain access for inspection purposes. Although not all flaws are removed, the rotor is then ready for operation. Its lifetime involves a series of starts and stops which are called cycles. At each cycle the rotor is subjected to two types of stresses--mechanical stresses due to the forces of rotation of the machine and thermal stresses. These stresses, specifically those in the vicinity of the bore's surface, affect the growth of cracks and flawed regions. As cracks grow and fields of discontinuity expand, the structure of the rotor is weakened. This in turn can lead to fracturing, projection of turbine missiles in the vicinity of the turbine, and subsequent catastrophic failure of the rotor, with loss of life and large economic losses due to turbine unavailability for service.

A review of failed rotors over the past thirty years points to brittle fracturing as the most likely cause of failure.[1] Recent developments in the nondestructive evaluation of rotors and in the theory of fracture mechanics have opened the way to studies in the reliability of turbine rotors.[2] Well's study[3] prepared at the Southwest Research Institute provides a computerized rotor lifetime prediction system that is based on stress and fracture models developed from physical assumptions and data.

The object of this paper is twofold: to develop a model for brittle fracture reliability of turbine rotors by combining a deterministic model of stress and crack size growth with probabilistic considerations; and to discuss the role of the model in practical risk estimation.

In the present analysis, reliability is defined in terms of the relationship of two quantities--fracture toughness of the rotor's material and stress intensity along a crack within the rotor (see Sections 2.1 and 2.2). We estimate these quantities and, in turn, reliability is estimated by applying the theory of interference.[4] Specifically, for each quantity--fracture toughness and stress intensity--a sampling distribution is based on uncertainties in the measurement and judgmental selections of parameters upon which the quantity depends. Reliability is defined in terms of the overlap (interference) of these two distributions.

In Section 2.3.2, we determine a useful closed form expression for the value of the true stress intensity parameter at any cycle in a sequence of cycles, based upon the assumed dynamics of crack growth and stress change. Use of this expression leads to a closed form distribution for the reliability criterion of fracture toughness minus stress intensity.

In Section 2.3.4, we show how a Bayesian point of view provides a useful framework for estimating reliability.[5] Here, the parameters of fracture toughness and stress intensity at given cycles are viewed as variables whose distributions reflect only uncertain beliefs about them, not real variation in the quantities themselves. Distributions are assigned to each quantity. A prior distribution expresses what is known about each quantity before measurement data is available. A posterior distribution describes our knowledge of the quantity once data has been obtained. We state the criterion for reliability in terms of the probability under the posterior distribution that fracture toughness exceeds stress intensity.

Next, in Section 2.4, we derive an expression for conditional reliability at a given cycle, that is, the probability that the rotor will operate at that given cycle if it has operated successfully up to that cycle. The conditional reliabilities are estimated as ratios of absolute reliabilities for successive cycles.

We discuss the reporting uncertainties in both absolute and conditional reliability estimates in Section 3, listing several ways of reporting these estimates.

In Section 4, we identify potential bias sources associated with model specification and statistical estimation and discuss the threat from each source.

To conclude in Section 5, we make recommendations for future research.

2. A PROBABILITY MODEL FOR RELIABILITY

2.1 The Definition of Reliability

In this analysis, the plane strain fracture toughness, an intrinsic property of the rotor's material, is used in determining the critical points in the growth of cracks and flawed regions. Specifically, the rotor is said to fail if the stress environment at a crack exceeds the plane strain fracture toughness. One objective of this analysis is to determine at each cycle the probability of failure due to a crack. The reliability is defined as 1 minus the probability of failure. In what follows, plane strength fracture toughness shall be referred to as fracture toughness.

2.2 Measurement of Fracture Toughness and Measurements of Stress Intensity

Given a rotor, its fracture toughness can be estimated. For rotors manufactured after 1950, the fracture toughness is correlated to certain fracture data (e.g., the charpy value) measured at the time of manufacturing and, in particular, during forging. This fracture data, however, was not accessible prior to 1950. For those rotors manufactured before 1950, the fracture toughness is estimated using several sources of data concerning the material of the rotor. In the present analysis, the estimate of fracture toughness is assumed to remain unchanged from cycle to cycle during the normal start-ups of the rotor. Here a normal start-up is one in which an appropriate pre-warming period has taken place in starting the rotor.

The stress analysis is obtained by utilizing finite element methods.[6] For a given crack, the stress intensity is a function of the size, shape, and location of the crack as well as mechanical stresses arising from the operation of the rotor. For any rotor the crack data involved in this estimate can be obtained during ultrasonic testing. It should be noted that by the nature of the testing procedure, some cracks may go undetected, in particular, those not near the bore region. The hoop stress (a mechanical stress), is determined from a function of quantities, which include, for example, the geometry and speed of the rotor. Thus, at the first cycle following an ultrasonic inspection, stress intensity at a detected crack is estimated by a function of measurable quantities. At the next cycle and succeeding cycles the crack parameters of size and shape are obtained in terms of a crack growth algorithm. Stress intensity at a detected crack is then determined at each cycle succeeding the first by an iterative process.

2.3 Mathematical Formulation of Reliability

We shall denote the fracture toughness parameter by $\mu_{K_{I_c}}$.

The stress intensity associated with a given crack at the nth cycle will be denoted by μ_{K_I}.

2.3.1 The iterative evolution of μ_{K_I}.

In this analysis, the collection of $\mu_{K_I}(n)$'s from cycle to cycle for a circular crack is described by a pair of simultaneous nonlinear equations:

$$(1) \quad \mu_{K_I}(n) = G\sigma(\pi a)^{1/2}$$

$$(2) \quad \frac{\Delta a}{\Delta n} = C(\mu_{K_I}(n))^r$$

---

where G = crack geometry size factor, σ = hoop stress, a = characteristic dimension of the crack, C = crack growth constant and r = crack growth rate exponent.

The origin of the first equation and appropriate modification in terms of elliptical cracks, are found in the works of (7), (8), and (9). Each work involves the combination of mathematical considerations with laboratory observations. The second relationship is rooted in the experimental work on crack growth in the aerospace industry.

The iteration of $\mu_{K_I}(n)$ is treated here as deterministic. Undoubtedly this is only an approximate representation of a stochastic process. Work utilizing stochastic differential equations has attempted to treat the situation stochastically.[10]

2.3.2 Closed formula for $\mu_{K_I}(n)$. By the form of the two-equation model a closed formula can be obtained for $\mu_{K_I}(n)$ for certain cases of the crack growth rate constant, r. To illustrate this point we obtained the following closed formula for $\mu_{K_I}(n)$ in the case r = 2.

$$(3)\ \mu_{K_I}(n) = G\sigma\sqrt{\pi a}\left(1+\sum_{i=1}^{n}[1+C\pi(G\sigma)^2]^{i/2}\right)\left(1-[1+C\pi(G\sigma)^2]^{-\frac{1}{2}}\right)$$

Appendix A gives a detailed derivation of this formula.

2.3.3 Examples of cyclic growth of μ_{K_I}. We observed the change $\mu_{K_I}(n+1)-\mu_{K_I}(n)$ for cracks of different radii over different cycles in Table I, where $C = .66 \times 10^{-8}$, $G = 2$ and $\sigma = 40$ ksi. We see that for cracks of radius a, 5 in. $\leq a \leq 20$ in., the $\mu_{K_I}(n+1)-\mu_{K_I}(n)$'s are bound by $0.05\sqrt{\text{ksi}}$ in. up to 100,00 cycles.

2.3.4 Bayesian interpretation of reliability estimation. In the present analysis the probability of rotor failure at the nth cycle, F(n), is defined by:

$$(4)\ F(n) = P(\mu_{K_{I_c}} - \mu_{K_I}(n) < 0)$$

Reliability, R(n), is then given by

$$(5)\ R(n) = 1-F(n)$$

TABLE I

CYCLE GROWTH IN μ_{K_I}

(BASED ON EQUATION 3 AND PARAMETERS DESIGNATED IN SECTION 2.3.3)

Crack Length Initial Value a_o (inches)	Cycle Numbers n (cycles)	Change in Stress $\mu_{K_I}(n+1) - \mu_{K_I}(n)$
5	100	.00007014
	500	.0000707
	1,000	.0000714
	5,000	.000077
	10,000	.0000855
	100,000	.00052
10	100	.00492
	500	.00497
	1,000	.00502
	5,000	.00544
	10,000	.00602
	100,000	.036682
20	100	.0069
	500	.0070
	1,000	.0071
	5,000	.0077
	10,000	.0078
	100,000	.05

TABLE I (Continued)

CONSTITUENTS OF DOWNTIME

1. Active repair time
   a. Detection time
      1. Estimated time failure started
      2. Fault detected
      3. Shutdown initiated
      4. Shutdown complete
   b. Diagnosis time
   c. Correction time
      1. Component isolation
      2. Access to faulty part achieved
      3. Decontamination
      4. Component removed
      5. Repair sequences completed
      6. Component replaced
   d. Verification time
      1. Checkout (including calibration, alignment, desolation, etc.)
      2. Normal operations resumed
2. Administrative time
   a. Processing
   b. Administrated verification (QA, safety, regulatory, etc.)
   c. No activity on this item, available work force concentrated on higher priority jobs.
   d. Administrated delay
3. Logistics delay time
   a. Supervisor/crew travel
   b. Replacement part re-order
   c. Preparation time
      1. Gather materials and technical data
      2. Gather equipment, special tools and instruments
      3. Study correction procedures
      4. Practice on prototype

If we assume that the unknown parameter $\mu_{K_{I_c}} - \mu_{K_I}(n)$ is regarded as fixed, then from a classical statistical point of view F(n) is either 0 or 1. A more appropriate interpretation of (4) is provided in terms of a Bayesian analysis. From a Bayesian standpoint, $\mu = \mu_{K_{I_c}} - \mu_{K_I}(n)$ is considered as a random variable.

Let us elaborate on this point. We may think of $\mu_{K_{I_c}} - \mu_{K_I}(n)$ as true but unknown deterministic values. Yet, from a Bayesian perspective, viewing $\mu_{K_{I_c}} - \mu_{K_I}(n)$ as a random variable reflects the uncertainty in our knowledge or beliefs about its value, either before obtaining some new data relevant to it (in the "prior" distribution) or after obtaining data (in the "posterior" distribution). Before any data at all on strength and stress have been collected, we have a priori relatively little information concerning $\mu_{K_{I_c}} - \mu_{K_I}(n)$. Consequently, we assign a noninformative prior distribution to $\mu_{K_{I_c}} - \mu_{K_I}(n)$. In particular, we select as our prior a uniform distribution. However, once data on both fracture toughness and stress intensity is available, we obtain new distribution for $\mu_{K_{I_c}} - \mu_{K_I}(n)$ reflecting information contained in the data. This new distribution, called the posterior distribution, is obtained by use of Bayes theorem.

We proceed now to explicitly describe the posterior distribution of $\mu = \mu_{K_{I_c}} - \mu_{K_I}(n)$. We denote the sampling estimators of $\mu_{K_{I_c}}$ & $\mu_{K_I}(n)$, respectively, by K_{I_c} and $K_I(n)$. Moreover, we denote the sample values of K_{I_c} and $K_I(n)$ respectively, by k_{I_c} and $k_I(n)$. The value of $k_{I_c}(n)$ is determined from correlation of fracture data, as discussed above. The value for $k_I(n)$ is determined by substitution of measured or judgmentally selected values of the parameters in a closed form expression for $\mu_{K_I}(n)$ of the type that appears in expression (3) above.

We assume, then, that K_{I_c} and $K_I(n)$ vary according to measurements errors and each is distributed normally as follows:

$$K_{I_c} \sim N(\mu_{I_c}, \sigma^2_{K_{I_c}}) \text{ and } K_I(n) \sim N(_{K_1}(n), \sigma^2_{K_I}(n))$$

where $\sigma_{K_{I_c}}$ and $\sigma_{K_I}(n)$ are the respective variances associated with measurement error. Specifically, $\sigma^2_{K_{I_c}}$ is estimated by a rule of thumb from correlation to fracture data $\sigma_{K_I}(n)$ and is estimated by use of an error propagation formula, assuming errors of component estimated parameters. (See Figure 1).

By the nature of the measurements made for $_{K_{I_c}}$ and $\mu_{K_I}(n)$ (see Section 2.2 above), we can assume that K_{I_c} and $K_I(n)$ are independent. So the sampling distribution of $S = K_{I_c} - K_I(n)$ is normal with density given for any n by:

$$(6) \quad h(s \mid \mu) = \frac{1}{\sqrt{2\pi\sigma}} e^{\frac{-(s-\mu)^2}{2\sigma^2}}, \quad s \in (-\infty, \infty)$$

where $= \mu_{K_{I_c}} - \mu_{K_I(n)}$ and $\sigma = \sigma^2_{K_{I_c}} + \sigma^2_{K_I(n)}$. For notational simplicity, we suppress the dependence of s, μ, and σ on n.

A noninformative prior on $\mu = \mu_{K_I(n)}$ can be described by:

$$(7) \quad g(\mu) = \begin{cases} \frac{1}{2m}, & \mu \in [-m, m] \\ 0 & \text{otherwise} \end{cases}$$

By Bayes rule the posterior density of $\mu = \mu_{K_{I_c}} - \mu_{K_I(n)}$ can be expressed by:

$$(8) \quad h(\mu|s) = \frac{h(s|\mu) \ g(\mu)}{\int_{-m}^{m} h(s|\mu) \ g(\mu) \ d\mu}$$

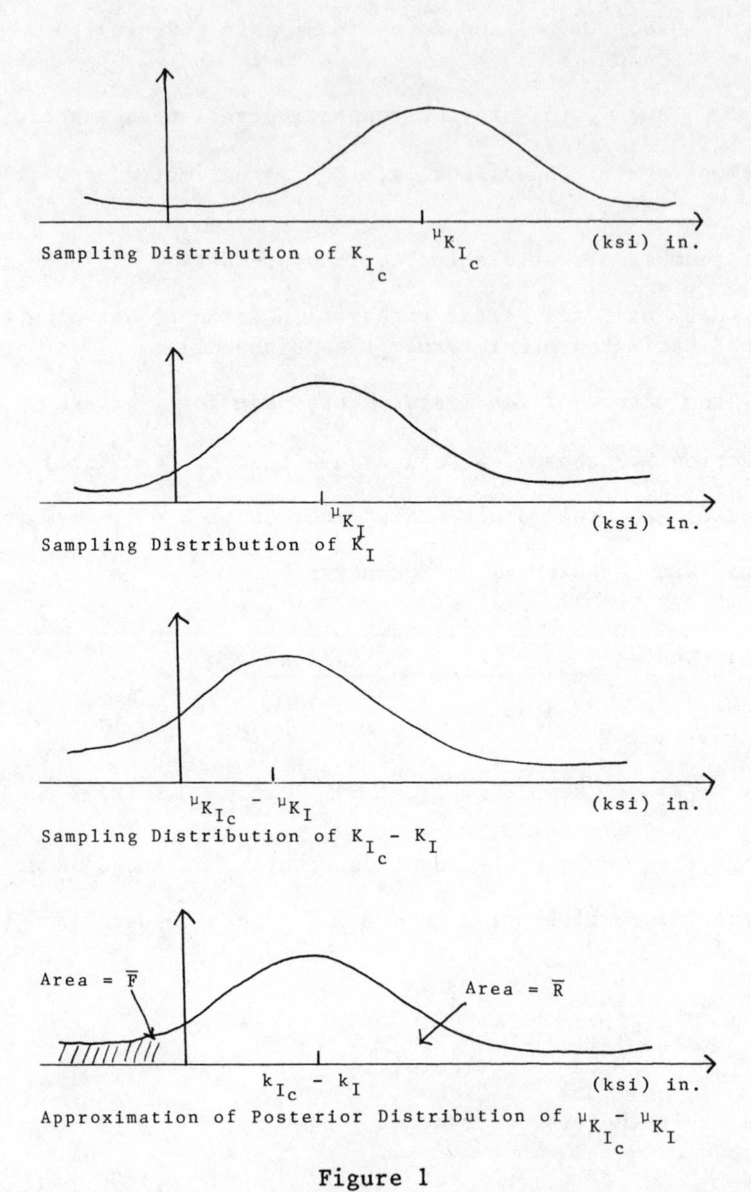

Figure 1

Sampling Distributions for Reliability

From (6), (7), and (8), expression (4) becomes:

$$(9)\quad F(n) = P(\mu_{K_{I_c}} - \mu_{K_I} < 0 \mid s) = P(\mu < 0 \mid s)$$

$$= \int_{-m}^{0} h(\mu|s)\, d\mu$$

$$= \frac{\frac{1}{2m}\int_{-m}^{0} h(s|\mu)\, d\mu}{\frac{1}{2m}\int_{-m}^{m} h(s|\mu)\, d\mu}$$

$$= \frac{\frac{1}{\sqrt{2\pi\sigma}}\int_{-m}^{0} e^{-(s-\mu)^2/2\sigma^2}\, d\mu}{\frac{1}{\sqrt{2\pi\sigma}}\int_{-m}^{m} e^{-(s-\mu)^2/2\sigma^2}\, d\mu}$$

Now, as $m \rightarrow \infty$, $\frac{1}{\sqrt{2\pi\sigma}}\int_{-m}^{m} e^{-(s-\mu)^2/2\sigma^2}\, d\mu \rightarrow 1.$

$$(10)\quad P(\mu_{K_{I_c}} - \mu_{K_I} < 0|s) \approx \frac{1}{\sqrt{2\pi\sigma}}\int_{-m}^{0} e^{-(s-\mu)^2/2\sigma^2}\, d\mu$$

But the right-hand side is approximately equal to

$$(11)\quad P(K_{I_c} - K_I < 0|\mu_{K_{I_c}} = k_{I_c},\ \mu_{K_I} = k_I),$$

Therefore, the probability of rotor failure at the nth cycle, $\overline{F(n)}$, can be approximated as follows:

$$(12)\quad F(n) \approx P(K_{I_c} - K_I(n) < 0|\mu_{K_{I_c}} = k_{I_c} \text{ and } \mu_{K_I}(n) = k_I(n))$$

For the sake of brevity, the expression on the right is denoted by $\overline{F(n)}$. An estimate of the reliability, R(n), is then given by:

$$(13)\quad \overline{R(n)} = 1 - \overline{F(n)}.$$

So R(n) is approximated by considering normal distributions centered respectively at the sample values of fracture toughness and stress intensity. This approximation is illustrated in Figure (1).

A Bayesian analysis then provides an interpretation of (4) that may be very useful when a turbine rotor has been analyzed once or more and subsequent testing provides an opportunity to revise the assessment of risk by failure. The posterior distribution for failure derived from the initial tests becomes the new prior distribution for the subsequent tests.

2.4 Reliability and Conditional Reliability Measures

Let the random variable X represent the rotor's time-to-failure in terms of cycles. The definition of reliability in (5) can also be given by:

$$(14)\quad P(X>n) = R(n)$$

Suppose that a rotor has operated successfully up to a given cycle. Under these conditions, we ask what the chances are that the rotor will operate successfully in the next cycle. This probability can be given in terms of absolute reliability measures R(n) as follows:

$$(15)\quad R(n) = P(X>n-1 \text{ and } X>n)$$

$$= P(X>n-1)\ P(X>n \mid X>n-1)$$

$$= R(n-1)\ P(X>n \mid X>n-1)$$

where $P(X>n \mid X>n-1)$ gives the probability that the rotor will operate up through the nth cycle given that it has operated successfully up through the (n-1)st cycle. So the conditional reliability can be expressed as:

$$(16)\quad P(X>n \mid X>n-1) = R(n)\ /R(n-1).$$

In the present analysis this conditional probability is estimated by $\overline{R(n)}/\overline{R(n-1)}$, where the estimates $\overline{R(n)}$ and $\overline{R(n-1)}$ are given by (13).

Although R(n) and R(n)/R(n-1) are measures of reliability, each measures reliability in a different sense. To expand this point, we use the sample in Table II. We start with Run 1. Suppose we are given 10,000 rotors all satisfying the conditions specifies under Run 1. The value $\overline{R(1)}$ = 0.61258 means that out of 10,000 rotors, the estimated number of rotors surviving through the first cycle is 6125.8. During the second cycle the estimated number of rotors surviving out of 10,000 is 6125.5 since $\overline{R(2)}$ = 0.61255. So the estimated number of failures between the 1st and 2nd cycles is 0.3.

The expected change in the number of rotors surviving from cycle to cycle can also be measured by the conditional reliability, $\overline{R(n)/R(n-1)}$. For example, the measure $\overline{R(1)}$ = .99995 means that out of the expected 6125.8 rotors surviving the first cycle, 99.994 percent or 6125.5 will survive the second cyle. Continuing in this way at the nth cycle, where n = 2,3,...,100, $\overline{R(n)/R(n-1)}$ gives the percentage of those rotors at the n-1 cycle which will survive the nth cycle. In this example this percentage remains high (specifically, 99.995 percent) over each of the 100 cycles since the expected number of failures from cycle to cycle is small.

The estimates $\overline{R(n)}$ and $\overline{R(n)/R(n-1)}$ in Run 2 can be interpreted similarly. What is interesting about this example is that R(n) is close to zero while $\overline{R(n)/R(n-1)}$ is close to one over a sufficiently large number of cycles. These results should not be viewed as contradictory. To see why this is the case again let us take 10,000 rotors satisfying conditions in Run 2. Then $\overline{R(100)}$ = 0.01605 means that out of the original 10,000 rotors the expected number of rotors surviving the 100th cycle is 16.05. The value $\overline{R(100)/R(99)}$ = 0.99949 means that out of the number of rotors that have survived the 99th cycle, 99.949 percent will survive through the 100th cycle. So as in Run 1, we once again observe a small expected number of failures from cycle to cycle.

3. REPORTING UNCERTAINTIES IN RELIABILITY ESTIMATES

Graphs and tables should report separately on absolute reliability by cycle number, conditional reliability by cycle number, and absolute reliability in the first cycle by temperature (and, therefore, fracture toughness) at start.

To highlight the importance of a normal start with pre-warming of the rotor a separate graph can display the estimated absolute reliability on the first start (cycle n = 1) at different temperatures and different assumed values of K_{I_c} at these temperatures.

A hybrid graph of cumulative conditional probabilities should be considered in place of a graph of absolute reliabilities for 20 cycles at a time. Thus, the cumulative amount of decrease in reliability for each group of 20 cycles would be plotted. That is, equivalently, the conditional probability of no failure in 20 cycles given success at the immediately preceding cycle would be plotted, the same for the next 20 cycles, etc.

For each graph upper, middle, and lower values can be reported. The middle values should be estimates from the model wherein input parameters are "best guesses" and not conservatively chosen. The upper and lower values can be determined either by sensitivity analyses or as 1 or 2 standard deviation limits.

4. POTENTIAL SOURCES OF BIAS IN SPECIFICATION AND ESTIMATION OF THE RELIABILITY MODEL

Estimated absolute reliabilities (see Table II) for an initial start (n=1) are often well below 1 and sometimes near 0 for parameter choices that do not seem on other grounds to imply a high probability of failure, suggesting that the model is substantially underestimating these. Hence, the direction of bias in estimation is downwards for absolute reliabilities. The source of such bias could be: (i) overly conservative estimates of standard deviations for fracture toughness and stress intensity creating excessive overlap of the two distributions; (ii) downward bias in the estimates of the difference in the mean fracture toughness and stress intensity creating excessive area below zero in the sampling distribution of the difference; and (iii) other sources listed in Table III below.

For conditional reliabilities the direction of bias is not clear. For the examples we have looked at, the conditional reliability from one cycle to the next decreased by about 3 per 10,000 (see Table II), or, for a turbine with 20 cycles a year, a decrease in reliability of about 1 in 1700 per year. That is, 1/1700 is the model-predicted failure probability over one year for a turbine with the assumed parameter values. With data on the distribution of parameter values for all turbines over a period of years, we could use the model to estimate the expected number of brittle fracture failures from normal starts and compare it to the observed number as a check on the model.

Table III summarizes the potential bias sources from both the model specification and statistical estimation procedure.

To begin, the model, as applied to a sequence of starts, holds only for normal starts, not for a sequence of abnormal starts (that is, starts with an inadequate pre-warming period). Abnormal or irregular starts, as noted in (1), appear, however, to be a leading or even sole source of actual brittle fracture failure. Either absolute or conditional reliabilities calculated for normal starts would be overestimates, possibly severe, of the absolute or conditional reliabilities under abnormal starts.

TABLE II

RELIABILITY AND CONDITIONAL RELIABILITY
(TWO RUNS OF THE PACKER COMPUTER PROGRAM
BASED ON EQUATIONS 10 AND 11)

RUN 1: For k_{I_c} = 56 ksi in., a_o = 2 in., b = 1 in., and σ = 36 ksi.

n(cycles)	a	b	k_{I_n}	$\overline{R(n)}$	$\frac{\overline{R(n)}}{\overline{R(n-1)}}$
1	2.00002	1.00005	53.00515	0.61258	———
2	2.00005	1.00010	53.00598	0.61255	0.99995
. . .	. . .	. . .	. . .	. . .	. . .
99	2.00230	1.00497	53.8706	0.60964	0.99995
100	2.00232	1.00502	53.08790	0.60961	0.99995

RUN 2: For k_{I_c} = 65 ksi in., a_o = 4 in., b = 1 in., and σ = 38 ksi

n(cycles)	a	b	k_{I_n}	$\overline{R(n)}$	$\frac{\overline{R(n)}}{\overline{R(n-1)}}$
100	4.00237	1.21603	101.41510	0.01605	0.99949
. . .			. . .	. . .	. . .
200	4.00481	1.23215	101.77183	0.01525	0.99949
. . .	. . .		. . .	. . .	. . .
400	4.00989	1.26463	101.79869	0.01379	0.99950

(where in both runs, C = $.66\ 10^{-8}$ and r = 2.25)

The reliability model could be simply modified to include abnormal starts by considering a total probability of rotor failure from both normal and abnormal starts:

(17) P(Rotor Failure) = P(Rotor Failure Normal Start) P(Normal Start)

+ P(Rotor Failure Abnormal Start)P(Abnormal Start)

The probability of rotor failure given a normal start has already been discussed in Section 2. The other probabilities in expression

TABLE III

POTENTIAL SOURCES OF BIAS IN SPECIFICATION AND ESTIMATION OF RELIABILITY MODEL

I. Scope of reliability model

  A. Normal starts vs. abnormal starts

  B. One crack vs. many cracks

  C. Merging of cracks

II. Assumptions in reliability model

  A. Circular crack

  B. Deterministic evolution of crack and stress

  C. Constancy of strength

  D. Stress everywhere equal to bore value

III. Estimation of parameters

  A. Judgmental or rule of thumb estimation

  B. Validation data

(17) would be obtained from frequency data (either from historical records or laboratory measurements) or by judgment.

Another potential source of bias in terms of model specification is the determination of reliability from a single crack fracturing (in particular, the crack which gives the worst possible case). Reliability estimated, however, for one crack will always be equal to or greater than that estimated from all cracks.

The model can readily be modified to treat this source of bias. We illustrate this modification as follows. Say we are given two crack sites in different regions of the rotors. Suppose the probability of failure from the first is given by P(F) = .63 and from the second by P(F) = .68. Then the probability that the first or second of both fail if failures are independent is given by:

$$(18)\quad P(F \cup F'') = P(F\ ) + P(F'') - P(F \cap F'')$$

$$= P(F\ ) + P(F'') - P(F\ )P(F'')$$

$$= .63 + .68 - .4284$$

$$= .8816$$

This discussion can be extended to any number of cracks. If independence between crack sites does not hold as in the case of merging cracks then only the first equality above holds. Conceivably two smaller cracks could merge into one large crack, and reliability is overestimated. This is not believed to be an important source of bias.

We next consider biases arising from assumptions in the reliability model. One assumption is that the evolution of crack size and stress intensity is given in terms of a deterministic 2-equation model. This assumption is not known to introduce a direction of bias. Moreover, the equation model holds only for circular or elliptical cracks. This assumption is not believed to be an important source of bias (9).

Changes in fracture toughness from cycle to cycle for normal starts are believed to be small. Therefore, the assumption of constancy of fracture toughness is also considered not to be an important source of bias (1).

The assumption, however, that stress intensity is everywhere equal to the core value (see Section 2) is conservative, tending to produce underestimates of the absolute reliability. The size of this effect requires consideration.

Another possible source of bias arises from the estimates of parameter of the reliability model. Table IV provides a summary of these estimates.

We note that a number of the estimates of standard deviations are made by rule of thumb. Further, in most cases these judgmental estimates are believed to be too large. Such estimates would help to produce the apparent bias towards unrealistically low absolute reliability estimates derived from the mode. (see Table II).

A comprehensive statistical study of error is needed in examining the contribution of error from each parameter to error in the final absolute reliability estimates.

We conclude by noting that the lack of observed frequency data (either from historical records or laboratory measurements) for validating the various parameter estimates makes the determination of bias a matter of guesswork.

5. RESEARCH RECOMMENDATIONS

Since the estimates for both absolute and conditional reliability are subject to potentially large but unknown bias, consideration should be given to using the model primarily for determining comparative risks between turbines, that is, relative levels, rather than

TABLE IV

ESTIMATES OF PARAMETERS OF RELIABILITY MODEL

	Parameters	Estimate	Method of Estimation	Confidence in Estimation Procedure and Direction of Bias
Mean Strength	$\mu_{K_{I_c}}$	k_{I_c}	Correlation of fracture data	Wide range of reported values
Error in Strength Standard deviation	$\sigma_{K_{I_c}}$	$0.10k_{I_c}$	Rule of Thumb	
Mean Stress				
Initial value	$\mu_{K_I}(1)$	$k_I(1)$	For Circular Cracks: $\hat{G}\,\hat{\sigma}(\Pi\,\hat{a}_o)^{\frac{1}{2}}$	Errors in ultrasonic testings
nth cycle value	$\mu_{K_I}(n)$	$k_I(n)$	Solution of non-linear equations for r=2 in Sec. 3.1. Closed formula given in Sec. 3.2.	Same
Crack length	a_o	$\hat{a}_o$	Length estimated by interpretation of ultrasonic information	Same
Crack width	b_o	$\hat{b}_o$	Width estimated by interpretation of ultrasonic information	Same
Crack depth	h_o	$\hat{h}_o$	Depth estimated by interpretation of ultrasonic information	Same

	Parameters	Estimate	Method of Estimation	Confidence in Estimation Procedure and Direction of Bias
Crack geometry	G	$\hat{G}$	Function of $\hat{a}_n$, $\hat{b}_n$, and $\hat{h}_n$	Mathematical function
Hoop stress	σ	$\hat{\sigma}$	Function of estimated parameters of rotor geometry, speed, etc.	Mathematical function
Crack growth rate	r	$\hat{r}$	Judgmentally selected	Not well known
Crack growth constant	C	$\hat{C}$	Judgmentally selected	Not well known
4. Error in Stress Standard deviation	$\sigma_{K_I(n)}$	$\hat{\sigma}_{K_I(n)}$	Determined from judgmentally determined standard deviations of $\hat{G}, \hat{\sigma}, \hat{a}_o, \hat{b}_o, \hat{h}_o$ and error propagation formula.	Standard deviations not well known. Use of error propagation formula to determine uncertainty in sampling distribution for $K_I(n)$ as a function of uncertainties in measurements and judgmental selections of parameters in the expression for $k_I(n)$ requires analysis.

absolute levels of probabilities. Accurate discussion of the comparative risk of specific turbines vis a vis entire populations of turbines in the U.S. requires a statistical compilation and analysis of data on U.S. turbines.

Another direction of research ensuing from the problems addressed is a thorough review of the methods used in the measurement of parameters and the determination of standard deviations. A thorough review would also be tied in with consideration of the economic value for run or retire decisions of gathering additional data or selecting large sample sizes during turbine testing. An important improvement would be made if data could be gathered to validate predictions of the dynamic model and the parameter estimates and their standard deviations.

A potentially useful source of risk estimates for turbine failure is a careful examination of the historical failure record. Table V shows a purely illustrative calculation, where no claim is made for the accuracy of the numbers used. The record of failures from abnormal and from normal starts needs to be considered.

TABLE V

HISTORICAL FREQUENCY OF TURBINE ROTOR FAILURE IN THE U.S.
ILLUSTRATIVE CALCULATIONS

(1)	Turbines in U.S.	400
(2)	Cycles per year per turbine	20
(3)	Years in operation	20
(4)	Turbine - cycles of operation in 20 years = (1) x (2) x (3)	160,000
(5)	Failures	5
(6)	Frequency of failures per turbine - cycle = (5) / (4)	1/30,000
(7)	Frequency of failure per turbine per year = (3) x (5) / (4)	1/1,600
(8)	Frequency of failure per year = (5) / (3)	1/4

Statistical methods for the analysis of survival by age of the equipment should be examined for their relevance to analyzing the historical record and for designing possible prospective collection of data on turbine failure.

Finally, statistical decision theory could be employed to help close the gap between the present numerical reliability reports and the decisions of utilities to run without testing, to test, or to retire after testing. Here testing refers to ultrasonic testing for cracks and flawed regions.

APPENDIX A

Derivation of a closed formula for $\mu_{k_I(n)}$

We set

$$f_1(a) = G\sigma(\pi a)^{1/2}$$

and

$$f_n(a) = \mu_{k_I}$$

where G = crack geometry size factor, a = initial crack radius, π = mechanical hoop stress

Then, from the above two-equation model in Section 2.3.1., assuming r = 2 and G constant,

$$f_2(a) = \mu_{K_I(2)} = f_1(a + C[f_1(a)]^2)$$

$$f_3(a) = \mu_{K_I(3)} = f_1(a_1 + C[f_2(a)]^2)$$

$$\vdots$$

$$f_n(a) = \mu_{K_I(n)} = f_1(a_{n-2} + C[f_{n-1}(a)]^2)$$

where, $a_n = a_{n-1} + C[f_n(a)]^2$ for n = 1,2, ...

From this it follows that:

$$f_n(a) = \mu_{K_I(n)} = f_1(a_{n-1})$$

Therefore, for any n, a_n can be expressed by:

$$\begin{aligned} a_n &= a_{n-1} + C[f_1(a_{n-1})]^2 \\ &= a_{n-1} + CG^2\sigma^2\pi^2 a_{n-1} \\ &= a_{n-1}(1+ CG^2\sigma^2\pi) \\ &= a_{n-1}(1+ C\pi(G\sigma)^2) \end{aligned}$$

So we obtain the recursive formula, for n = 1, 2, . . .

$$a_n = a[1+C\pi(G\sigma)^2]^n.$$

Hence, we have that,

$$\begin{aligned} \mu_{K_I(n+1)} - \mu_{K_I(n)} &= f_{n+1}(a) - f_n(a) \\ &= f_1(a_n) - f_1(a_{n-1}) \\ &= f_1(a[1+C\pi(G\sigma)^2]^n) - f_1(a[1+C\pi(G\sigma)^2]^{n-1}) \\ &= G\sigma\sqrt{\pi a[1+C\pi(G\sigma)^2]^n} - G\sigma\sqrt{\pi a[1+C\pi(G\sigma)^2]^{n-1}} \\ &= G\sigma\ \pi a([1+C\pi(G\sigma)^2]^{n/2} - [1+C\pi(G\sigma)^2]^{(n-1)/2} \\ &= G\sigma\sqrt{\pi a}\ [1+C\pi(G\sigma)^2]^{n/2}\ (1-[+C\pi(G\sigma)^2]^{-1/2}) \end{aligned}$$

Hence, the change in μ_{K_I} from the nth to (n+1)st cycle is given by the closed formula:

$$\mu_{K_I(n+1)} - \mu_{K_I(n)} = [G\sigma\sqrt{\pi a}[1+C\pi(G\sigma)^2]^{n/2}]\ [(1-[1+C\pi(G\sigma)^2]^{-1/2})]$$

It follows from the above expression that:

$$\begin{aligned} \mu_{K_I(n+1)} - \mu_{K_I(1)} = \ & \mu_{K_I(n+1)} - \mu_{K_I(n)} \\ & + \mu_{K_I(n)} - \mu_{K_I(n-1)} \\ & + \ \ ,\ ,\ , \\ & + \mu_{K_I(2)} - \mu_{K_I(1)} \end{aligned}$$

$$= \sum_{i=1}^{n} \sigma\sqrt{\pi a}\ [1+C\pi(G\sigma)^2]^{i/2}\ (1-[1+C\pi(G\sigma)^2]^{-1/2})$$

Therefore, we have the following closed formula for $K_I(n)$

$$\mu_{K_I(n)} = G\sigma\sqrt{\pi a}\ (1+\sum_{i=1}^{n}[1+C\pi(G\sigma)^2]^{1/2}\ (1-[C\pi(G\sigma)^2]^{-1/2})$$

REFERENCES

1. Caulfield, E. M. and Cronin, M. T., "Rotor Reliability Assessment in the Electric Power Industry." Proceedings of the Ninth Annual Engineering Conference on Reliability for the Electric Power Industry, 1982.
2. Golis, M. J. and Brown, S. D., "Nondestructive Evaluation of Steam Turbine Rotors - An Analysis of the Systems and Techniques Utilized for In-Service Inspection," Electric Power Research Institute (EPRI NP-744 Project 502-2), April, 1978.
3. Wells, C. H., "Reliability of Steam Turbine Rotors," Electric Power Research Institute (EPRI NP-923-SY Project 502), October, 1978.
4. Juvinal, R. C., Stress, Strain and Strength, McGraw-Hill, New York, 1967.
5. Box, G. and Tiao, G., Bayesian Inference in Statistical Analysis, Addison-Wesley Publishing Co., 1973.
6. Packer Engineering Associates, Inc., Packer Code - Main Route for Calculating Failure Probability and Crack Growth, 1981.
7. Kobayashi, S. S. and Shah, R. C., "Stress Intensity Factors for an Elliptical Crack Approaching the Surface of a Semi-Infinite Solid," International Journal of Fracture, Vol. 9, No. 2, June, 1973.
8. Sneddon, I. N., "The Distribution of Stress in the Neighborhood of a Crack in an Elastic Solid," Proceedings of the Physical Society of London, 187:229, 1946.
9. Irwin, G. R., "Crack-Extension Force for a Part-Through Crack in a Plate," Journal of Applied Mathematics, December, 1962, pg. 651.
10. Eckvall, J. C. et al., "Engineering Criteria and Analysis Methodology for the Appraisal of Potential Fracture Resistant Primary Aircraft Structure," AFDL-TR-72-80, Air Force Flight Dynamics Laboratory Wright-Patterson Air Force Base, Ohio, September, 1972.

A GUIDE FOR RELIABILITY AND MAINTAINABILITY DATA ACQUISITION

M. J. Haire*
E. H. Gift**

*Fuel Recycle Division
Oak Ridge National Laboratory[1]
Oak Ridge, Tennessee 37830

**Operations, Analysis, and Planning Division
Oak Ridge Gaseous Diffusion Plant
Oak Ridge, Tennessee 37830

ABSTRACT

The need to establish absolute values for availability risk assessments has added a new urgency to the collection of high quality reliability and maintainability data. This paper describes a logical format for the collection of the required data. The data to be collected can be acquired in three related recording sessions. Inventory and technical data are filed at the time the design is finished or the equipment received. An event report is completed at each shutdown. Finally, a maintainability report is filed on resumption of normal operations. Parameter values required for a complete and comprehensive equipment system availability analysis are defined. Priorities are established for the collection of data for these parameters--data are classified as basic, intermediate, or high quality. These data forms are designed such that a complete history of equipment and system uptime and downtime can be recorded systematically.

KEY WORDS: Data, Reliability, Maintainability, Data Forms, Collection, Quality.

INTRODUCTION

Probalistic risk assessment (PRA) techniques have been used successfully in studies when the objective is to make a choice between the safer (or more reliable) of two designs. Numerous examples exist which tell of PRA providing guidance in designing a system (e.g., helping to select the component or system assemblages of lowest risk). The worth of the technique is generally acknowledged when a choice must be made from among two or more systems which perform the same functions (i.e., when relative values are the product of the evaluations). Technique logic models such as event trees and fault trees can be, and are, resolutely defended.

Now pressures are being exerted to apply the PRA techniques to arrive at absolute numerical values which indicate that a design, safety, or perhaps even a licensing goal will be satisfied. This modest extrapolation of past successes seemingly requires only the application of failure and maintenance data input to established methods. However, this extrapolation is viewed with some apprehension because of the low quality of the basic data. For example, the U.S. Nuclear Regulatory Commission is engaged in an extensive study to establish mathematical techniques to handle the propagation of data uncertainties. Basic failure and maintenance data are the weak links of the PRA or reliability analysis. A number of data bases exist, yet there is often insufficient detail to determine failure rates for specific items. The alternative is to resort to generic data of "like" items since test programs are too expensive for all but the most important equipment. A search through data bases indicates, however, that failure data varies over wide ranges of values from source to source for apparently identical equipment. This variability results, in large part, from incomplete recording of the pertinent information (e.g., operating environment) required for a complete characterization of the failure or maintenance event.

Sources of much of these data are the operating records of a facility or test programs. Trying to glean detailed information from operation reports long after completion of operation or testing is futile. At best such data are of low quality and much information is lost because of oversight or from a lack of knowledge of what is important.

During the course of developing a failure and maintenance testing program for the Consolidated Fuel Reprocessing Program (CFRP) at the Oak Ridge National Laboratory, we concluded that to be most effective the data acquisition program must be defined before the test program actually begins. This conclusion required that the equipment, the system, the operating environment, and other system characteristics of importance be identified in advance; that the proposed method of data analysis be defined; and that data acceptance criteria be established.

This paper outlines the reliability and maintainability data that are considered necessary to ensure that the proposed comprehensive, quantitative evaluations of equipment and system availability will be of sufficiently high quality that design and flowsheet decisions can be made with confidence. System availability will be calculated during and following the test program using the accumulated reliability and maintainability data. These necessary data are presented in this paper as a set of data collection forms. These forms are designed to serve as a guide and can be tailored to any particular data acquisition or reporting program. The data to be collected by the reliability engineer can be acquired in three related recording sessions.

ITEM INVENTORY AND TECHNICAL DATA

The first of these recording sessions should preferably be done when design specifications for the equipment are set or during checkout of the items against the receipt. The Item Inventory and Technical Data form is completed for each equipment item to the level where repair or replacement with spare parts is made. It is important at the outset to uniquely identify all equipment items on which data will be collected and recorded. In addition to each inventory item being identified by number, the item should be characterized by manufacturer, model, size, type, and date of entry into service. The form contains a complete description of the environmental conditions under which the equipment item is expected to operate. The form lists the fabrication quality level, from off-the-shelf general industry to high reliability, military specifications. Known reliability information (e.g., failure rate data) is listed--usually, the best reliability data are that provided by the manufacturer. A base-line technical data set of operational data and design specifications is recorded for fault diagnosis should failure later occur.

EVENT REPORT

The second data collection period follows each significant interruption in operation. The principal reliability data form, entitled the Event Report, contains not only failure events from which failure rate data are acquired, but also provides a history of related events such as preventive maintenance shutdowns and equipment calibrations. However, if for example, an equipment item is rebuilt, and this fact is not recorded, reliability data collected on the item is worthless. The cause of the shutdown (the event type) is an important element of the form. The other most important question on the form is the operating time or number of cycles.

There are degrees of completeness to what is loosely known as reliability data. It is important to recognize the quality of information for what it is. The "basic grade" of data requires only that the failed item be identified and that its time of operation, or number of cycles, and the number of failures or breakdowns in the period be known. This information, obtained from operation records and available in most data bases, enables a failure rate of the item to be calculated. Note that this grade of data takes no account of qualifying conditions (e.g., environment) that affect the item behavior. The minimum information for the basic grade is (1) inventory number, (2) item name and description, (3) number of similar items being tested, (4) operating time (cycles), and (5) the number of item failures during the mission. Additional information needed for "intermediate grade" data is (1) the manufacturer, (2) equipment quality, (3) the component material, if relevant, (4) operating environment, and (5) the date of manufacture and entry into service. Note that some qualifying information has been introduced to differentiate between the performance of superficially similar items which may differ in age, environment, or manufacturer quality. In "high grade" reliability data collection, event information is important. An event can be in situ maintenance, workshop maintenance, testing and calibration, or equipment modification. In addition, it is useful to know certain facts associated with these events (1) the routine maintenance interval; (2) the effect on the system of item failure; (3) the effect of item failure on plant or installation; (4) the operating conditions at the time of failure (as opposed to design conditions); (5) replacement parts; (6) the number of failures, the mode of failure, and the cause of failure; and (7) the outage time for each event. It is clear that the Event Report is incomplete without the earlier Item Inventory and Technical Data form. Furthermore, the Event Report is missing some key information, for example, the spare parts, outage time, and fault classification which is requested in the maintainability data collection forms.

MAINTAINABILITY DATA

The third data collection period occurs at the end of each recovery from failure, or scheduled shutdown, with the completion of the necessary maintenance and the return to normal operating conditions. This is recorded as Maintainability Data. Just as reliability is a discipline unto itself, so is maintainability, with its own, more diversified elements. The difficulty of reliability analysis as a subject area is undeniable: reliability deals with the quality of equipment parts, component lifetimes, fail-safe design, redundancy, and the dependability of stand-by safety systems. These problems are, for the most part, quantifiable and measurable during equipment laboratory testing and field operation. Maintainability, however, is unique in every application; for example, the access space for repair will vary from plant to plant. There is a paucity of maintainability data bases. Yet, maintainability--mean time to repair--is one of the two parameters necessary to calculate system availability.

The human element is always present during fault correction processes. The interaction of humans with equipment systems is not as quantifiable as the physical aspects of equipment design. Variables which compose the maintainability discipline must be identified and measured to systematize the equipment design and facility operation process.

The primary variable of interest in maintainability analysis is the time interval between incipient failure and return to normal operating conditions. Table I lists the constituents of downtime. The general objective of the maintenance engineer is to reduce the downtime to as short a time as possible within the constraints of other considerations (i.e., safety, costs, etc.) The contributing elements to downtime must be known to identify the contributors which have the best possibility of being reduced.

The time interval between the start of failure and the equipment state of complete shutdown, ready for fault correction, is defined as the detection time. The time when a fault is noted always follows the occurrence of the fault. Frequently, the increment of time is very small, especially when a performance monitor is at or near the location of the fault. On the other hand, if the methods of detection of a fault is by product testing and the fault occurred near the input to the system, the time interval between incipient failure and its detection may be significant. Similarly, the failure of stand-by systems is detected at the worst possible moment--on system demand.

Detecting a fault does not always identify the location and cause of the failure. Consequently, diagnosis of the failure and deciding on the appropriate corrective action can be time consuming. Thus, the time interval between the complete system shutdown and the onset of preparing for the actual physical correction of the fault is defined as the diagnosis.

Correction time, the time interval required to physically correct the failure, includes (1) the preparation time to perform the repair, (2) the time to study the repair procedures and the technical manuals and data, (3) the time to collect the special tools, replacement parts, and (4) the time to acquire special skills through practice. The time to physically correct the fault is a component of correction time. Frequently, this time interval is incorrectly defined as the system downtime.

Once repair is complete, the corrected fault must be verified. Although verification is often a simple matter and can be done by the repairman rapidly, it may be time consuming. Additionally, if the diagnosis was incorrect, the whole process of fault correction may have to be repeated. Often substantial time is spent on administrative functions related to fault corrections. For example, management may require notification of a failure event and the appropriate report forms completed. Once the fault is corrected, approval for restart may be needed from quality assurance, quality control, and licensing groups. The number of shifts each day the facility is manned falls within the administrative downtime category. Clearly, the total clocktime (time interval between incipient failure and return to normal operations) should not be charged if the facility is manned for only 8 hours of a 24-hour day.

TABLE I. CONSTITUENTS OF DOWNTIME

1. Active repair time

   a. Detection time

      1. Estimated time failure started
      2. Fault detected
      3. Shutdown initiated
      4. Shutdown complete

   b. Diagnosis time

   c. Correction time

      1. Component isolation
      2. Access to faulty part achieved
      3. Decontamination
      4. Component removed
      5. Repair sequences completed
      6. Component replaced

   d. Verification time

      1. Checkout (including calibration, alignment, desolation, etc.)
      2. Normal operations resumed

2. Administrative time

   a. Processing

   b. Administrated verification (QA, safety, regulatory, etc.)

   c. No activity on this item, available work force concentrated on higher priority jobs.

   d. Administrated delay

3. Logistics delay time

   a. Supervisor/crew travel

   b. Replacement part re-order

   c. Preparation time

      1. Gather materials and technical data
      2. Gather equipment, special tools and instruments
      3. Study correction procedures
      4. Practice on prototype

It is useful to visualize the constituents of downtime as a critical path time-line. Some elements of downtime, shown in Table I, may proceed parallel to or overlap with other elements so that the sum of constituent-time intervals may be greater than the critical path time-line. Furthermore, a complete schematic of downtime constituents may contain feedback loops; for example, if it is shown that the fault was not corrected during the fault verification step. However, the total downtime for an event is the duration of the critical path timeline.

The "basic grade" of maintainability data consists of the critical path time-line, fault classification (break-in, design fault, random failure, or end of design life), and a description of the failure (the defective item, how fault was detected, what methods of diagnosis, and a brief description of the diagnosis).

The "intermediate grade" of maintenance data deals with questions concerning maintenance policy. The impact of policy is shown, for example, by the availability of redundant trains of equipment--the availability of the system is dramatically affected depending on whether the trains undergo preventive maintenance simultaneously, in sequence, or in sequence at different times by different maintenance personnel. Some other maintenance policy decisions are (1) whether the failed item is repaired or replaced, (2) the repair location, (3) the logistics of repair (special tools, spare parts inventory, etc.), (4) the completeness and usefulness of repair manuals and procedures, and (5) the precautionary procedures planned in advance of the event.

"High quality" maintenance data include additional information on the feature of the equipment design and layout that facilitate maintenance considering human factors and personnel requirements. Regarding equipment design and layout, special considerations are (1) accessibility (distances to nearest adjacent equipment) of the failed item, (2) complexity of the maintenance task (e.g., were components and parts modularized), (3) the need and frequency of preventive maintenance requirements, and (4) provisions for rapid resolution of the constituents of downtime (e.g., detection and diagnosis). Examples of human factors data are (1) confusion in labeling, (2) the possibility of incorrect connection or assembly, (3) excessive physical demands, (4) the job elements considered dirty, awkward, or tedious, and (5) ease of handling and mobility. Estimates should be given concerning the number of personnel required for the task and their respective skill level.

These data forms can be found in the work cited in Reference 2.

SUMMARY

The need to establish absolute values for availability risk assessments has added a new urgency to the collection of high quality reliability and maintainability data. This paper describes a logical format for the collection of the required data and suitable data acquisition forms. An inventory and technical data form is filed at the time the design is finished or the equipment received. An event report is completed at each shutdown. Finally, a maintainability report is filed on resumption of normal operations. Parameter values required for a complete and comprehensive equipment system availability analysis are defined. Priorities are established for the collection of data for these parameters--data are classified as basic, intermediate, or high quality. These data forms are designed such that a complete history of equipment and system uptime and downtime can be recorded systematically.

REFERENCES

1. Research sponsored by the Office of Spent Fuel Management and Reprocessing Systems, U.S. Department of Energy, under Contract No. W-7405-eng-26 with Union Carbide Corporation.
2. M. J. Haire and E. H. Gift, Reliability and Maintainability Data Acquisition in Equipment Development Tests, ORNL/TM-8525, in progress.

THE APPLICATION OF RISK-BASED COST-BENEFIT ANALYSIS IN THE ASSESSMENT OF ACCEPTABLE PUBLIC SAFETY FOR NUCLEAR POWER PLANTS

Thomas A. Morgan*
Alfred J. Unione*
George Sauter**

*Impell Corporation
350 Lennon Lane
Walnut Creek, California 94598

**Nuclear Safety Analysis Center
Electric Power Research Institute
3412 Hillview Avenue
Palo Alto, California 94304

ABSTRACT

In 1982, the U.S. Nuclear Regulatory Commission issued, for public comment, proposed safety goals for commercial nuclear power plants. In an effort to quantitatively evaluate these proposed goals, a methodology was developed for the Electric Power Research Institute's (EPRI) Nuclear Safety Analysis Center to assess the impacts of implementation of these goals and to identify inherent sensitivities in their interpretation. Utilizing existing risk assessment data for an operating nuclear plant, potential design and procedural modifications to improve plant safety were identified and evaluated for cost-effectiveness. Although the proposed safety goals which were used as the basis for this study have since been revised, the methodology developed is believed to be useful for providing an assessment of the impacts of regulatory compliance and may also be useful in determining the cost-effectiveness of any proposed plant modification. This methodology, however, requires that risk assessment data (e.g., modularized fault trees) be available for use.

For the example power plant used in this study, data from an existing probabilistic risk assessment was used as a basis for identifying representative procedural or hardware modifications that potentially could have been required, assuming that the proposed safety goals had been implemented. Estimates of the potential costs and benefits associated with each modification were developed in order to assess each modification's cost-effectiveness. Benefits were estimated based upon the expected reduction in the probability of occurrence of a major plant accident (i.e., core melt frequency) that would be expected to result if each modification was implemented. Modification costs were estimated for generic classes of equipment and procedural changes using currently available information and engineering judgment. These costs estimates considered design and installation costs, plant outage costs, and yearly operational maintenance costs. Variations in the assumptions and methods used in conducting the cost-benefit analyses were also evaluated to determine the sensitivity of possible modification decisions to these variations.

The risk assessment data indicated that several accident sequences existed with occurrence probabilities in excess of the proposed safety goal limits. However, the cost-benefit analyses performed using this methodology indicated that most major plant modifications that would reduce these accident sequence frequencies were not cost-effective. When incremental cost-benefit analysis techniques were used, it appeared that only simple procedural changes and minor design backfits were justified in almost all cases. It was also noted, however, that many of these cost-benefit decisions were sensitive to the calculational assumptions used, particularly the benefits to be expected from a reduction in core melt frequency, and the estimation of outage costs (i.e., replacement power costs) associated with the installation of a proposed backfit.

OVERVIEW

For several years, industry groups and regulatory bodies have sought to define quantitative goals for acceptable nuclear power plant safety. Sparked by the Three Mile Island accident, the U.S. Nuclear Regulatory Commission (NRC) officially committed to the development of such "safety goals." Public workshops were held in a number of key cities and suggestions and comments were sought from a number of industry and non-industry sources.

In 1982, a proposed Policy Statement was issued by the NRC with the stated objective of developing "goals for limiting to an acceptable level the additional potential risk which might be imposed on the public as a result of accidents at nuclear power plants." This policy statement included two qualitative goals supported by four provisional numerical guidelines. These qualitative goals were:

- "Individual members of the public should be provided a level of protection from the consequences of nuclear power plant accidents such that no individual bears a significant additional risk to life and health."

- "Societal risks to life and health from nuclear power plant accidents should be as low as reasonably achievable and should be comparable to or less than the risks of generating electricity by viable competing technologies."

The four supporting numerical guidelines were:

- "The risk to an individual or to the population in the vicinity of a nuclear power plant site of prompt fatalities that might result from reactor accidents should not exceed one-tenth of one percent (0.1%) of the sum of prompt fatality risks resulting from other accidents to which members of the U.S. population are generally exposed."

- "The risk to an individual or to the population in the area near a nuclear power plant site of cancer fatalities that might result from reactor accidents should not exceed one-tenth of one percent (0.1%) of the sum of cancer fatality risks resulting from all other causes."

- "The benefit of an incremental reduction of risk below the guidelines for societal mortality risks should be compared with the associated costs on the basis of $1,000 per man-rem averted."

- "The likelihood of a nuclear reactor accident that results in a large-scale core melt should normally be less than one in 10,000 per year of reactor operation."

In order to implement these goals, the Nuclear Regulatory Commission staff had developed a draft implementation plan, which proposed that all plants in operation or under construction have a total frequency of core melt of less than 1×10^{-3} per Reactor-year. In addition, it proposed that all plants meet the following "design objectives" subject to ALARA (As Low As Reasonably Achievable) risk reduction guidelines:

- The total core melt frequency (CMF) should not exceed 1×10^{-4}/Reactor-year

- The CMF of any individual accident sequence should not exceed 1×10^{-5}/Reactor-year

In an effort to quantitatively evaluate the proposed goals and NRC Action Plan, a methodology was developed by Impell Corporation for

EPRI's Nuclear Safety Analysis Center to assess the potential impacts to electric utilities and to identify inherent sensitivities in interpretation of these goals. This paper provides an overview of the methodology and study results which are more fully documented in the final report, NSAC-56, "Potential Impacts of NRC Safety Goals on the Nuclear Industry".[1] It should be noted that the Policy Statement and implementation plan were revised and formally issued subsequent to the completion of this study. However, the methodology developed is believed to be useful for providing assessments of the impacts of regulatory compliance and may also be utilized to determine the cost-effectiveness of any proposed plant modification that may affect the overall safety and operability of a plant.

The approach used for quantifying potential impacts of NRC Policy Statement and implementation plan compliance used probabilistic risk data to estimate the impact of plant modifications as benefits along with estimates of the costs of these modifications. Example calculations were performed using data from an existing plant-specific probabilistic risk assessment as a basis for identifying and evaluating representative procedural or hardware backfits that potentially could be required under the proposed implementation scheme. The Crystal River 3 IREP (Interim Reliability Evaluation Program) NRC-sponsored study[2] was used for these calculations. Estimates of the potential costs and benefits associated with backfits were developed for use in evaluations of the conditions under which possible modifications would be considered cost-effective. Variations in the assumptions and methods used for conducting the proposed cost-benefit analyses were also evaluated to determine the sensitivity of possible backfitting decisions to these variations.

The resulting analysis was intended more to show the elements of the methodology and to provide insights into the issues. Since the probabilistic benefits were not arrived at through a detailed probabilistic sensitivity analysis and the costs were estimated roughly, the quantification showed what the range of results of an analysis could be rather than what the results are for a particular plant.

METHODOLOGY

The data from the Crystal River 3 IREP study was reviewed to identify those accident sequences that might be subject to ALARA reduction evaluation under the guidelines of the NRC implementation plan. For this plant, seven core melt sequences were identified that exceeded the 10^{-5}/R-yr guideline, and the total CMF for the plant was estimated at 3.4×10^{-4}/R-yr (see Table I).

Table II details the failure frequencies for system-level failure events which contributed to the major functional sequence frequencies in the CR-3 IREP study. Table III indicates the contribution of system-level failure sequences identified from the PRA documentation.

Using this information, the reliability improvements which might result from the implementation of various system modifications were estimated as follows:

- Each contributing system-level failure identified in Table II was broken down by examination of the available PRA data to a component level of detail.

- At the component level of detail, a series of modifications were described which may consist of the addition of major hardware items such as an additional diesel generator), addition of minor hardware items (such as valving and control modifications to improve turbine pump starting), or procedural changes (such as implementing a revised recirculation mode initiation procedure with a commensurate level of training). The modifications identified were not intended to represent an in-depth development of system modification packages, but rather were intended only as illustrative examples for the purposes of the NSAC study.

- For each identified modification, the impact on system-level reliability was assessed using modularized fault trees from the IREP documentation. In some cases, the effect could be approximated by an increase in redundance (assuming some contribution from common-cause failures). In other cases, the impact was estimated by removing the affected system fault. In a few cases (such as the human error effects), the impact could only be estimated by assuming that an adequate program or design is established which will meet a reliability improvement objective.

The impact of each identified modification on the accident sequence CMF's and the total CMF were then evaluated. To account for uncertainties in the makeup of system failure sequences which were not identifiable from the IREP documentation (i.e., those sequences in Table III in which the identified system failures do not account fully for the frequnecy of the functional event), a weighted averaging procedure was used which pre-supposes that it is not known whether the modification has no impact on the magnitude of the residual unavailability or has the same level of impact as on the identified system failure sequences.

Once estimates were obtained for the reduction in core melt sequence frequency for each modification, monetary benefit values

TABLE I

DOMINANT ACCIDENT EVENT SEQUENCES FROM CRYSTAL RIVER 3 PRA (IREP)

Nomenclature	Description	Sequence Frequency*(/R-yr)	Major Sources of Failure
$T_{2A}T_{10}$	Loss of Offsite Power (T_{2A}) plus failure of secondary cooling & containment heat removal (T_{10})	1.0×10^{-4}	Hardware
B_4S_2	Small LOCA (B_4)plus failure of coolant recirculation (S_2)	1.7×10^{-5}	Human
B_4S_{23}	Small LOCA (B_4)plus failure of coolant injection (S_{23})	1.3×10^{-5}	Human
$T_{2A}T_8$	LOSP (T_{2A})plus failure of secondary cooling (T_8)	7.4×10^{-5}	Hardware, Human
$(T_2\text{-}T_{2A})T_8$	Transient other than LOSP (T_2-T_{2A}) plus failure of secondary cooling (T_8)	6.1×10^{-5}	Hardware, Human
$(T_1\text{-}T_{1A})T_8$	Transient where access to condensor is not initially lost (T_1-T_{1A}), plus failure of secondary cooling (T_8)	2.9×10^{-5}	Hardware, Human
$T_{2A}T_9$	LOSP (T_{2A}) plus failure of secondary heat removal & radioactivity removal (T_9)	2.2×10^{-5}	Hardware
TOTAL		3.2×10^{-4}	
TOTAL (all sequences $>10^{-6}$/R-yr)		3.4×10^{-4}	

*Contribution to point estimate of core melt frequency

TABLE II

SYSTEM LEVEL UNAVAILABILITIES CONTRIBUTING TO DOMINANT ACCIDENT SEQUENCE FREQUENCIES IN THE CRYSTAL RIVER 3 PRA

Nomenclature	Description of Unavailability Event	Unavailability (per demand)
ACA	Unavailability of onsite AC power train A	5.1×10^{-2}
ACB	Unavailability of onsite AC power train B	5.2×10^{-2}
DCB	Unavailability of DC power train B (given LOSP*)	3.2×10^{-3}
E1	Emergency Feedwater System (EFS) turbine pump train failure (excluding power loss)	2.2×10^{-2}
EF1	EFS full system failure (not initiated by LOSP)	3.4×10^{-4}
DB	Decay Heat Close Cycle Cooling System train B failure (excluding power loss)	4.3×10^{-2}
HEFB	Human error, failure to establish feed & bleed given a loss of secondary side cooling to steam generators	1.0×10^{-1}
HELPI	Human error, failure to initiate coolant recirculation after a LOCA	1.0×10^{-2}
HERE	Human error, inadvertant early initiation of coolant recirculation after a LOCA	1.0×10^{-2}

TABLE III

SUMMARY OF DOMINANT SYSTEM LEVEL FAILURE SEQUENCES FOR THE CRYSTAL RIVER 3 PRA

Sequence[a]	Sequence Pt. Est.	System Level Sequences[b]	System Level Frequency Evaluation	System Pt. Est.	% Total
$T_{2A}T_{10}$	1.0×10^{-4}	$T_{2A}\cdot ACA\cdot DCB$	$.32\times5.1\times10^{-2}\times3.2\times10^{-3}$	5.2×10^{-5}	52
		$T_{2A}\cdot ACA\cdot ACB\cdot E1$	$.32\times4.9\times10^{-3(d)}\times2.2\times10^{-2}$	3.4×10^{-5}	34
					86
$T_{2A}T_8$	7.4×10^{-5}	$T_{2A}\cdot ACA\cdot E1\cdot HEFB$	$.32\times5.1\times10^{-2}\times2.2\times10^{-2}\times.1$	3.6×10^{-5}	49
$(T_2-T_{2A})T_8$	6.1×10^{-5}	$(T_2-T_{2A})\cdot EF1\cdot HEFB$	$1.78\times3.4\times10^{-4}\times.1$	6.1×10^{-5}	100
$(T_1-T_{1A})T_8$	2.9×10^{-5}	$(T_1-T_{1A})\cdot PCS\cdot EF1\cdot HEFB$	$8.48\times.1\times3.4\times10^{-4}\times.1$	2.9×10^{-5}	100
$T_{2A}T_9$	2.2×10^{-5}	$T_{2A}\cdot ACA\cdot DB^{(c)}\cdot E1$	$.32\times5.1\times10^{-2}\times4.3\times10^{-2}\times2.2\times10^{-2}$	1.5×10^{-5}	68
B_4S_2	1.7×10^{-5}	$B_4\cdot HELPI$	$1.3\times10^{-3}\times1.0\times10^{-2}$	1.3×10^{-5}	76
B_4S_{23}	1.3×10^{-5}	$B_4\cdot HERE$	$1.3\times10^{-3}\times1.0\times10^{-2}$	1.3×10^{-5}	100

(a) Accident Event Sequence from the PRA
(b) See Table I and Table II
(c) Assumes that AC power is available.
(d) Common cause effect produces less reduction in emergency AC unavailability than assumption of independence for diesel generator sources.

(assessed as yearly benefits realized over the remaining life of the plant) were calculated using various assumptions based upon interpretations of the NRC Policy Statement and Implementation Plan. Benefits were assigned using various combinations of the following assumptions:

- \$20,000 per 10^{-5} reduction in CMF was assumed to be the yearly benefit of a reduction in early and latent public fatalities. (This value was proposed in the implementation plan.)

- An additional \$8,000 per 10^{-5} reduction in CMF was assumed to be the yearly benefit due to the aversion of off-site property damage. (This value was derived from estimates in Starr and Whipple[3] and WASH-1400[4].)

- An additional \$20,000 per 10^{-5} reduction in CMF was assumed to be the yearly benefit due to the aversion of on-site (i.e., power plant) property losses. (This value was implied in the implementation plan.)

Estimated costs (both yearly and one-time installation costs) for each of the modifications were then developed, taking into account such factors as:

- engineering design and hardware costs
- construction labor costs
- replacement power costs incurred while installing the modifications
- yearly maintenance, operational, and training costs associated with the modification

The order-of-magnitude cost estimates were based upon available plant cost data and engineering judgment. For each cost category, a range of values was used to account for uncertainties in the cost data and estimated outage durations for installation. Outage costs were calculated using two assumptions. In the first case, it was assumed that the modification would be scheduled into an existing refueling maintenance outage. Hence, outage costs would be incurred only for that portion of the installation that would result in an extension of the outage. In the second case, it was assumed that a special, unscheduled outage was required to install the modification.

USE OF COST-BENEFIT ANALYSIS TECHNIQUES

For each of the potential modifications that were identified, a cost-benefit comparison was used as a means of selecting viable modification options based upon a comparison of expected benefits versus estimated costs. In a simple decision situation, it can be assumed that those actions with a greater benefit value than cost impact will be worthwhile.

Graphically, the cost-benefit data were illustrated in a series of simple charts (such as that shown in Figure 1), which presents a proposed modification and its associated costs and benefits. Yearly monetary values for these costs and benefits were derived in accordance with the various guidelines presented in the NRC Policy Statement and implementation plan (e.g., regarding treatment of inflation effects, remaining plant life, discounting of plant costs, and the monetary benefits resulting from a reduction in core melt frequency).

Clearly, the use of different calculational assumptions resulted in significant variations in the perceived costs and/or benefits expected for a given modification. In addition, because a range of costs was estimated for each modification, the cost-benefit plots frequently extended across both sides of the breakeven (i.e., "Do-Don't") line, indicating that a more detailed cost-benefit analysis would be required to evaluate the viability of the modification.

An incremental cost-benefit methodology was also applied to several cases to investigate the effects of the diminishing returns that might occur if several modifications (which might produce similar effects in terms of reducing the frequency of one or more system failure sequences) were considered simultaneously. This concept, in which modifications are prioritized by their effectiveness and each modification is considered only as an addition to previously accepted modifications, is demonstrated in Figure 2. In this case, the incremental assessment of the viability of a second modification is negative, although it may have appeared cost-effective if evaluated on an individual basis.

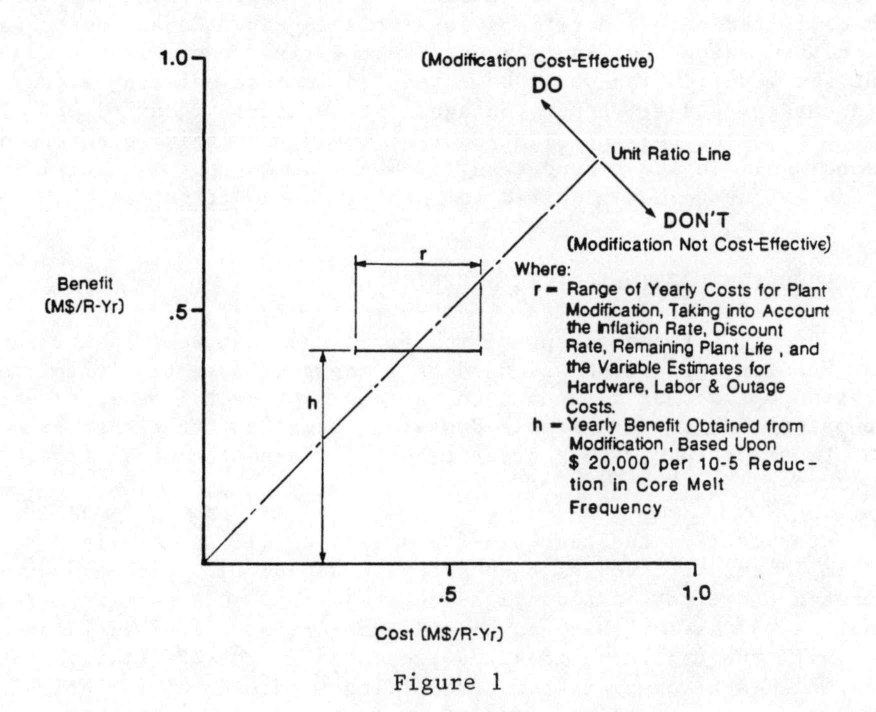

Figure 1

Cost-Benefit Calculation: Standard Cost-Benefit Comparison Technique

RESULTS

The cost-benefit analyses performed for each potential modification used the following "base case" assumptions:

o Yearly benefit values were indexed to inflation

o Plant costs were annualized over the plant lifetime (using a 10 percent real discount factor)

o A 40 year plant life was assumed (i.e., a new plant)

It should be noted that the above assumptions were selected because they represented the most likely interpretations of the implementation plan guidance. Use of assumptions different from those above (e.g., non-indexing of yearly benefits, no discounting of costs, etc.) yielded substantially different results, but will not be further discussed in this paper.

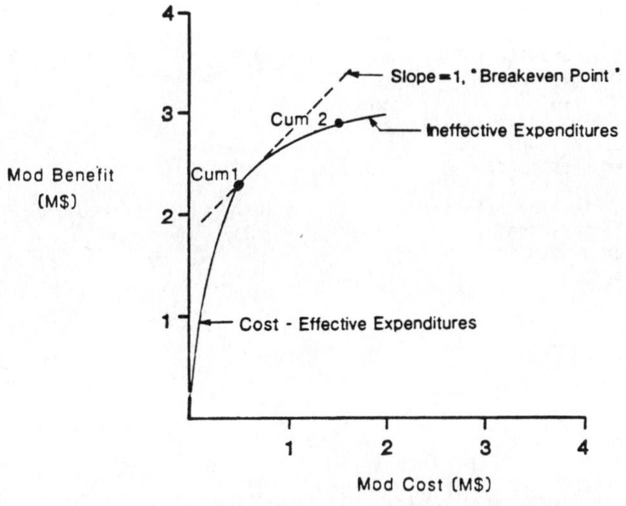

Figure 2

Cost-Benefit Analysis: Incremental Assessment

TABLE IV

REVISED CORE MELT FREQUENCIES AFTER IMPLEMENTATION OF VARIOUS IDENTIFIED MODIFICATIONS

Modification	Total Mean Core-Melt Frequency (/R-Yr) (Design Objective = 1×10^{-4}/R-Yr)	Largest Individual Core Melt Sequence Mean Frequency (/R-Yr) (Design Objective = 1×10^{-5}/R-Yr)
Base case (no modifications)	3.16×10^{-4}	1.0×10^{-4}
Install additional diesel generator	1.83×10^{-4}	6.1×10^{-5}
Install additional auxiliary feedwater system train	1.46×10^{-4}	5.9×10^{-5}
Install additional DC battery train	2.57×10^{-4}	7.4×10^{-5}
Procedural changes for feed & bleed and high-pressure injection	1.6×10^{-4}	1.0×10^{-4}
Procedural changes plus installation of additional diesel generator	6.9×10^{-5}	2.6×10^{-5}
Procedural changes plus installation of additional DC battery train and auxiliary feedwater system train	8.8×10^{-5}	4.1×10^{-5}
Procedural changes plus installation of additional DC battery train, auxiliary feedwater system train, and raw water pump	8.1×10^{-5}	4.1×10^{-5}

The modifications evaluated in this study included AC and DC electrical system modifications, auxiliary feedwater system modifications, service water system modifications, procedural improvements, and increased operator training. The impacts of various single and multiple modifications are presented in Table IV. As can be seen, no single modification reduced the core melt frequencies (total or individual sequence) to below the NRC design objectives, and most modification groups could not satisfy both (i.e., total and individual sequence CMF) design objectives.

Figure 3 presents an example cost-benefit plot using only the base case assumptions. The plot is typical of the many developed in this study in that most major modifications (e.g., installation of a new diesel generator) do not appear to be cost-effective. In addition, many other modifications (other than inexpensive procedural and training upgrades) are only marginally cost-effective and are strongly dependent upon the magnitude of the assumed cost and benefit values.

Figure 4 illustrates the sensitivity of two of the modifications shown in Figure 3 to variations in benefits assumptions. Here, the inclusion of off-site and on-site property damage benefits makes one modification (improved diesel starting reliability) much more cost-effective. The major modification (new diesel), however, still remains not cost-effective.

In Figure 5, a number of potential groups of modifications are evaluated. While the procedural changes are clearly cost-effective the other modification groups' viabilities are dependent upon the cost values assumed (e.g., a high outage cost assumption would make these groups seem non-cost-effective). In Figure 6, the same modification groups are evaluated using an incremental cost-benefit analysis. The

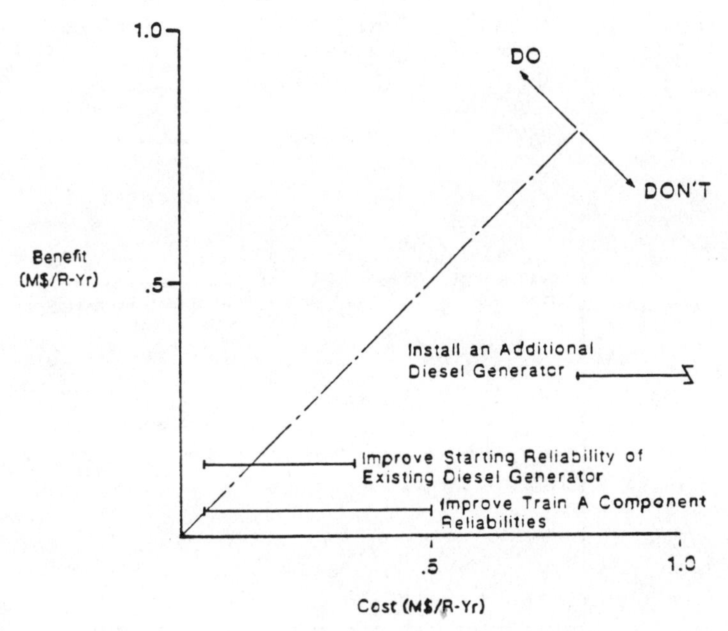

Figure 3

Cost-Benefit Calculation: Cost-Benefit of Increasing the Reliability of AC Power Train A Using Various Modification Packages

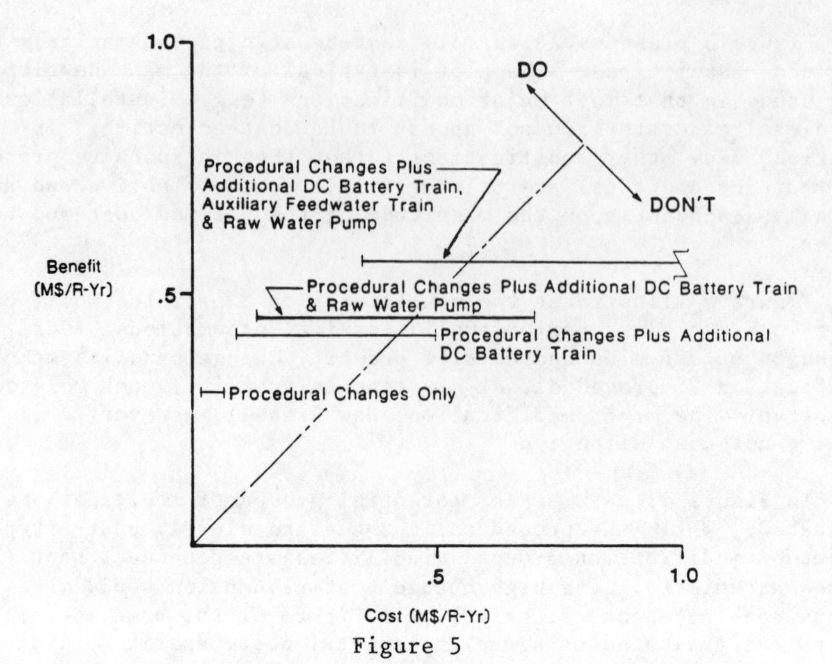

Figure 5

Cost-Benefit Calculation: Groups of Procedural and Hardware Modifications

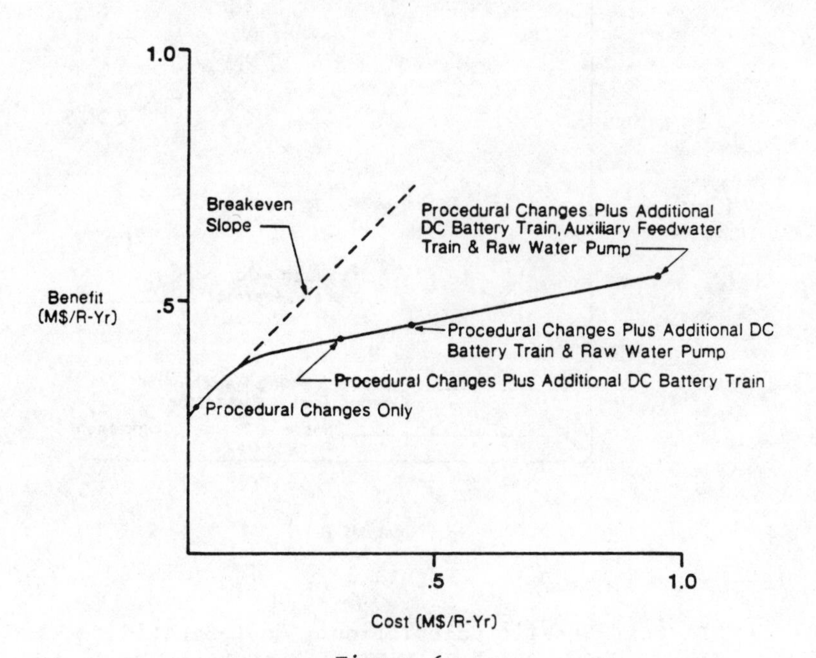

Figure 6

Cost-Benefit Calculation: Incremental Cost-Benefit Analysis for Various Groups of Modifications

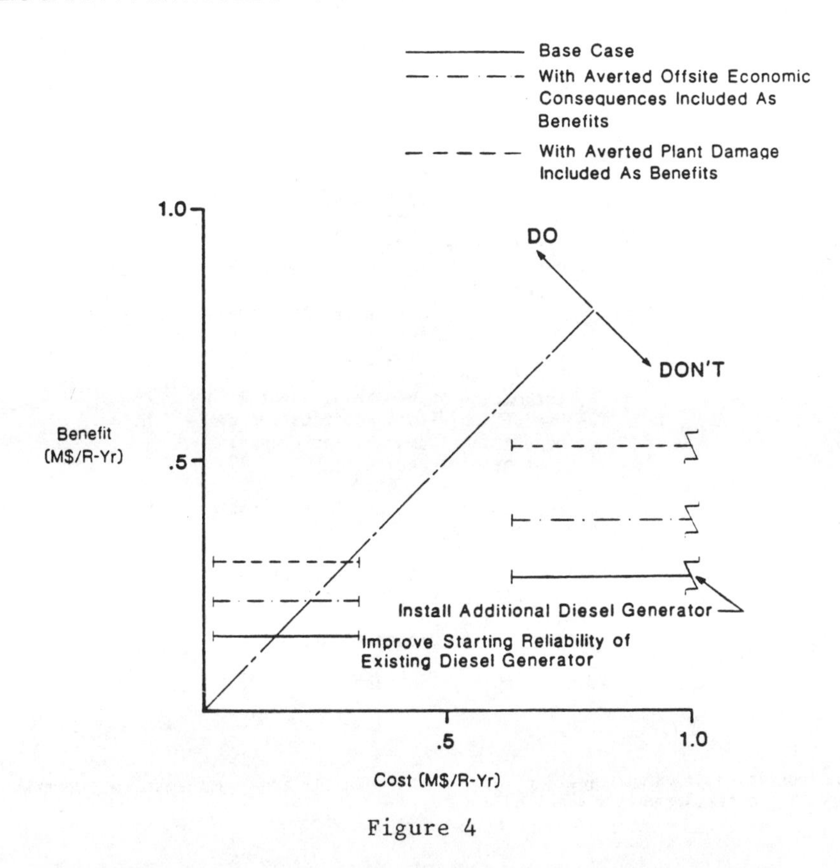

Figure 4

Cost-Benefit Calculation: Comparison of Cost-Benefits for Several Emergency AC Power Train A Modifications

point values used for each modification cost are the mean values of the cost ranges shown in Figure 5. In this case, it is quite apparent that only the procedural changes appear justifiable. The incremental benefits gained from additional modifications are far outweighed by the incremental modification costs.

In summary, the results of the example problem indicated that, in many cases, major plant modifications were not justified, largely due to their high implementation costs. The study also revealed, however, that a number of sensitivities exist, particularly concerning the treatment of outage costs, on-site and off-site property damage, and various cost-benefit "accounting" assumptions that may profoundly influence the results of such cost-benefit evaluations. Table V provides a qualitative summary of the magnitudes of each sensitivity examined in this study.

TABLE V

SUMMARY OF IDENTIFIED QUANTITATIVE SENSITIVITIES

Issue Quantitatively Assessed	Identified Key Parameters	Sensitivity[a] of Cost-Benefit Decisions to Key Parameters
Manner of performing financial calculations	Accounting Assumptions:	
	- Indexing of benefits to inflation, discounting of plant costs	High
	- Treatment of onsite averted damages (i.e., "negative" costs vs. "positive" benefits), if included in the cost-benefit analysis	Low
	- Estimation of remaining plant life	Low
	- Use of individual modification cost-benefit analysis vs. incremental cost-benefit analysis	High
	Assumed outage costs for implementation of modifications	High
	Assumed hardware and labor costs for implementation of modifications	Moderate
Inclusion of benefits other than averted off-site casualties	Averted offsite property damage	Moderate
	Averted onsite plant damage	High

[a]The sensitivities shown are derived from the example and are themselves intended only as a general range of variability.

REFERENCES

1. J. C. Guibert, E. A. Hughes, T. A. Morgan, A. J. Unione, "Potential Impact of NRC Safety Goals on the Nuclear Industry", NSAC-56, Electric Power Research Institute (May 1983).
2. "Crystal River-3 Safety Study", Science Applications, Inc. Report SAI-002-80-SE, Working Draft (April 1980).
3. C. Starr and C. Whipple, "Coping with Nuclear Power Risks: The Electric Utility Incentives", Nuclear Safety, Vol. 23, No. 1 (January-February, 1982).
4. "Reactor Safety Study: An Assessment of Accident Risks in U.S. Commercial Nuclear Power Plants", WASH-1400, U.S. Nuclear Regulatory Commission (October 1975).

INSURANCE MARKET ASSESSMENT OF TECHNOLOGICAL RISKS

William B. Fairley
Michael B. Meyer
Paul L. Chernick

Analysis and Inference, Inc.
10 Post Office Square, Suite 970
Boston, MA 02109

ABSTRACT

Market risk assessments (MRAs) are defined as probability and consequence products, estimated by parties with substantial financial interests in the accuracy of the estimates, such as insurance companies. Several potential problems with traditional quantitative risk assessment procedures are discussed, such as the potential problems of adversary bias, of cost-of-error and value-related biases, and of hard-to-quantify risks. Potential advantages of MRAs are discussed, such as their potential lack of bias, their potential ability to combine quantitative and qualitative risk assessments, and their potential availability. Several requirements for future research are outlined, and the current state of knowledge and current research activities are described.

I. INTRODUCTION

Public concern about scientific and technological risks is great and probably increasing. As a result, there is often a demand that risks be managed or assessed by public agencies. This demand has often been met in the past by regulatory agencies either producing or requiring risk assessments by traditional analytical methods which have focused on probability and consequence estimation. Insurance markets may provide an independent source of risk assessments, which may be extracted from coverage and premium information, and which may provide useful frequency (probability) and severity (consequence) estimates. This paper will discuss several aspects of MRAs stemming from insurance arrangements. This paper will focus on risks which might be termed "non-actuarial" in that actual accident experience data for the risk are sparse or non-existent. It is clear that MRAs work satisfactorily where data are plentiful and of high quality; life, health, auto, and homeowner's insurance provide four examples. Similarly, the state-of-the-art is probably adequate for natural disasters of large consequence for which much data on events exist (such as hurricanes); or for any risks, artificial or natural in origin, which have been dealt with in a standard way by the traditional actuarial methods of life or casualty insurance ratemaking. It is by no means obvious, however, that MRAs would be useful or helpful in areas where data are sparse, non-existent, or of poor quality. An example would be technological risks which cause rare events with serious consequences, which are a subset of events of ten, loosely referred to as "catastrophes" or "low-probability/high-consequence" (LP/HC) events. The remainder of this paper will, therefore, refer to "catastrophic" or "LP/HC" risks for which experience data on events are sparse or non-existent.

Three basic types of work remain to be done to validate the use of MRAs. First, some further analytical work will be required to define the conditions required for the existence of MRAs, and to explore the expected characteristics of MRAs. Second, a model for insurance premium prices must be developed to predict premiums as a function of expected losses, expected expenses, required return on capital, and any compensation for risk aversion. Third, case studies must be conducted in each relevant application area to explore existing insurance arrangements (in terms of coverage design and extent, insurable event definition, premium design and amount, etc.) in order to discover what information is available for the extraction of risk assessments from existing arrangements. These last two tasks (constructing the premium model and conducting the case studies) will together form the basis for exploring three issues: (1) which methodologies are appropriate for extracting MRAs from existing insurance programs; (2) how future insurance programs could be designed to enhance their use as sources of MRAs; and (3) exactly how, if at all, MRAs can be appropriately used for regulatory and public policy purposes.

II. IDENTIFICATION AND SIGNIFICANCE OF THE ISSUE

RISK RESOLUTION IN COURTS AND REGULATORY AGENCIES

Introduction

Courts and regulatory agencies face an increasingly heavy burden of resolving factual uncertainties and legal issues in the determination and evaluation of risks to health and safety from a variety of environmental, technological, or commercial sources. Green (1981) and Meyer (1982) cite several examples. Much of this burden constitutes the ordinary business of factfinding in legal and regulatory proceedings. For example, the treatment of uncertainties in the determination of events leading to an automobile injury or a product liability dispute is the main concern of these proceedings.

Areas of risk determination in legal and regulatory proceedings that have met with less satisfactory treatment are risks in which large numbers of persons are put at risk, but either (1) the risk of some future possible catastrophic event is in doubt, as for commercial nuclear power plants, or (2) where the identity of persons harmed by the specific cause is difficult to determine, but some harm is generally conceded (as in the health effects of air pollution) and, in either case, (3) where experience with the specific event or with the causation of harm is sparse or non-existent. See, for example, Fairley and Mosteller (1977) and Crandall and Lave (1981).

Major Problems of Factual Uncertainties

Three major problems confront courts and regulatory agencies which attempt to resolve factual uncertainties about risks of low probability and high consequence events with sparse or non-existent data. These problems are: (1) adversary bias; (2) cost-of-error or other value-related bias; and (3) difficulties surrounding the quantification of inherently hard-to-quantify risks. These three problems will be discussed in turn.

Adversary Bias. The parties to adversary proceedings put forward widely divergent risk estimates. The uncertainties in these estimates are rarely frankly described, because the parties advocating the risk estimates fear attacks on the soundness of their estimates and wish to convince the decisionmaker of the correctness of facts and analyses favorable to their side. Still worse, biased risk estimates are often explicitly produced solely for use in a particular regulatory proceeding. This problem is not often discussed openly in the professional literature, but it occasionally surfaces in print (U.S. Nuclear Regulatory Commission, 1980, NUREG-0739, p. 44) and is often discussed orally within the risk analysis profession (numerous discussions by conference participants on LP/HC risk analysis). As a result, decisionmakers often have a difficult time in reconciling widely divergent risk estimates, and in discovering (even in a range) where the most reasonable value of various risk measures lie.

Cost-of-Error and Other Value-Related Biases. It is common for parties to cite "conservative" or "prudent" estimates of risk on the grounds that the cost of the error of under-estimating (or, alternatively, of over-estimating) the risk requires an estimate deliberately biased in a known direction. These deliberate biases are often not quantified, as in the possible bias resulting from linear extrapolations of cancer risk from high dose results in animals to low dose results in humans; and may cascade if they appear at several steps of a multi-step process. The decisionmaker is left to imagine how much the estimated risk is over-stated and how to estimate the "true" risk. The grounds on which those biases are introduced are admittedly value judgments; for example, where human life is at stake, it is often stated that an estimate of a risk should be higher than what one might believe was the expected value. Raiffa (1982) and the National Research Council (1980) discuss "conservative" estimates of carcinogenesis.

Hard-to-Quantify Risks. Catastrophic risks of complex new technologies are inherently difficult to estimate. Historical data on the frequency of interest is generally quite limited. For example, Fairley (1982) shows how risk estimates from nuclear accident records yield only very imprecise upper bounds on risk that are far from the safety levels required by accepted standards. As an alternative to such direct data, risks can be estimated from strong detailed theories of event causations. The frequencies of various limitations preclude such strong models, either because the basic causation is not well understood (as in many biomedical issues), because it cannot be determined that all event paths have been traced (as in nuclear safety issues), or because the input data necessary to drive the model does not exist (as in failure rates for new system components). In the absence of either direct data or strong models, a large measure of expert judgment must be used to determine and estimate risks, building on whatever weak data and models are available or on intuition and experience. Unfortunately, experts sometimes treat the best available data and models as if they provided precise risk estimates. For example, the results of complex quantitative models, such as Probabilistic Risk Assessments (PRAs) for nuclear power plants, are generally reported without any reference to such uncertainties as the quality of the data, the analyst's degree of confidence in the model, or the analyst's prior beliefs about the relevant probabilities.

INSURANCE MARKET RISK ASSESSMENTS

Definition

As part of their ordinary course of business, insurance companies make countless risk estimates every day. These estimates have not figured prominently in legal and regulatory risk determinations to date even though they appear to have several attractive features. For example, these MRAs are made independently of considerations of use in adversary proceedings, and independently of cost-of-error value considerations (other than considerations concerning the compensation necessary to induce insurance companies to sell the coverage and assume the risk). They quantify risk in a particular way by putting a price on the assumption of risk. Fairley (1982) defines and discusses MRAs.

Insurance prices (premiums) must include as components estimates for several distinct types of costs and profits. Specifically, estimates or assumptions must be made about the frequency of accident events; about the severity (in dollar damages) of the accident events that do occur; and about the size of necessary loss adjustment expenses, general overhead expenses, and compensation for profit and risk assumption. Information on risk contained in insurance prices has probably not been prominently used to date because such information is contained in coverage prices that mix together different elements along at least three dimensions. First, insurance coverage prices mix together information on estimated frequency and severity, two quantities that are usually considered and estimated separately in more traditional risk analyses. Second, insurance coverage prices mix together loss estimates (from combined frequency and severity estimates) with expense, profit, and risk aversion estimates or loadings. Third, insurance coverage prices often mix together estimates for a wide variety and number of distinct types of accident events (in terms of their initiating chain of events) into single estimates for more aggregated types of outcome events useful for particular types of accident (insurable event) definitions. For example, insurance coverages tend to be for outcome events as generalized as fire, theft, collision, or property damage events, while more traditional risk analyses tend to focus more narrowly on particular types of events, usually defined as causation-outcome pairs. Finally, insurance coverage prices are the result of rather complicated actuarial methodologies, which may not be particularly accessible to non-actuaries, and are also at least partially the result of non-quantified business judgment.

The implicit frequency and severity estimates represent insurers' best available outcome estimates. Because insurers stand to lose money if their frequency and severity estimates turn out (after the fact) to be too low, these estimates might, if anything, be expected to be biased on the high side. On the other hand, to the extent that workable competition exists in the insurance industry and between the insurance industry and other providers of financial assurance (self-insurance pools, reinsurers, true self-insurers, etc.), these estimates could not become too biased on the high side as competing insurers should assume the risks at lower prices. Estimates of commercial nuclear power plant accident risks provide perhaps the best example of "insurance market risk assessments" or "market risk assessments," which merely mean that insurance premiums may be interpreted (via a model and necessary collateral information) as risk estimates.

Although using insurance coverage prices as the sources of risk estimates involves some difficulties, discussed above, these difficulties can perhaps be overcome. To estimate certain types of nuclear power plant frequencies (probabilities), Denenberg (1973), Wood (1981), and Chernick et al., used collateral information on the estimated costs of payouts in the event of accidents, on estimated expenses, on return on equity, and on risk aversion. This was in conjunction with a standard analytical model for insurance pricing to estimate the expected frequency of accidents that underlies the premiums charged to the plant's owner/operators.

Potential Advantages

Given that market risk assessments are made routinely by insurance companies, and that at least one example exists (commercial nuclear power plant accident risks) in which these market risk assessments provide quite different estimates from PRAs, [see Denenberg (1973) and Wood (1981)], are there compelling reasons for further study of their feasibility, the methodology for extracting them from premiums and coverage information, and their use in regulatory and policy decisionmaking? We believe, in brief, that the potential advantages of MRAs, when used in conjunction with other types of risk assessments, are sufficiently attractive to warrant a closer look. We briefly discuss four apparent or potential advantages of MRAs.

Potential Lack of Bias. Estimates of risk measures (probabilities and consequences) produced by special purpose studies to be used in adversary proceedings, and even those produced in agency decisionmaking, are highly vulnerable to partisan influence. This partisan influence may come into play because of the high stakes involved in the outcome of the decision or because of the values which the various participants to the decisionmaking process are attempting to pursue. For these reasons, risk estimates produced for or by regulatory proceedings are frequently either admittedly biased or subject to a strong suspicion of bias in known directions. MRAs, by contrast, should tend to offer independent, arm's-length estimates that should not be subject to these kinds of influences.

Combination of Quantitative and Judgmental Estimates. MRAs combine both statistical or actuarial and other kinds of information on risk that are inputs to coverage and pricing decisions. This may be a particularly valuable quality for LP/HC risk for which inadequate data preclude complete reliance on actuarial methods. This problem (of actuarial methods being inadequate due to lack of relevant data) does not exist for some traditional life and casualty insurance lines, but appears in such casualty insurance coverages as hazardous waste disposal site insurance and commercial nuclear power insurance. MRAs thus provide a solution, albeit pragmatic, to the difficult problems of combining information from several diverse sources and diverse methodologies, including purely intuitive judgments. By contrast, more traditional analytical disciplines have recognized the formidable obstacles in the way of formal combinations of such diverse evidence, including pure intuition or judgment, to reach conclusions about final risk estimates. Further, traditional analytic methodologies are difficult to review and evaluate as to whether they form an adequate basis for decisionmaking.

Interest Represented by MRAs. In one dimension, MRAs represent a polar case: parties who make MRAs are making serious bets, in that large financial gains or losses ride on the accuracy of the MRAs. Financial incentives thus appear to operate to favor the more unbiased and more precise estimates. Thus, there is an incentive mechanism, strongly rooted in general experience, which should operate to promote the accuracy of MRAs. Where the events insured are expected to occur with sufficient frequency, there will be actual periodic feedback that should operate to correct MRAs by experience.

By contrast, traditional quantitative risk assessments, such as PRAs, ostensibly represent the other polar case (along the interest dimension) in that they are produced by expert researchers who are assumed to be disinterested in the outcome of the regulatory decisions which will be made in partial reliance upon the PRA. In practice, of course, it may be difficult to separate advocates from truly disinterested researchers. In any case, neither truly disinterested experts nor experts who are acting as advocates have the strong financial incentives provided by the risk market.

MRAs as an Untapped Resource. Various regulatory agencies, and regulated industries or parties, have committed very substantial resources toward producing and improving quantitative, analytic risk assessments. This commitment of resources is laudable and almost certainly represents a wise investment of research resources. The source of this activity is perhaps well represented by the NRC which is currently sponsoring or supervising major research efforts on quantitative safety goals (U.S. Nuclear Regulatory Commission, NUREG-0739, NUREG-0880, and Strip, 1982b), on PRAs (U.S. Nuclear Regulatory Commission, NUREG/CR-2300), and on quantitative consequence estimates (Strip, 1982a), all relating to commercial nuclear power risks. Market risk assessments, by comparison, are an untapped resource based upon information already collected and analyzed, which should be subject to extraction from existing premium and coverage data with relatively modest effort and expense, compared to PRAs and quantitative consequence estimates. As is true of any source of information about risk, if risk assessments derived from other methodologies differed significantly from MRAs, decisionmakers should want to note these differences, understand the reasons for these differences, and weigh the reliability of the alternative sources before forming opinions about risks. In short, MRAs may provide a relatively inexpensive and relatively available alternative to traditional, analytic risk assessments, which can be used to supplement, but not necessarily to replace, the analytic risk assessments.

III. REQUIREMENTS FOR FUTURE RESEARCH

Four areas of interest concerning MRAs need to be explored. These four areas of interest are: (1) what are the characteristics of

MRAs, and what conditions will allow the determination of useful MRAs; (2) what methodologies are appropriate for extracting MRAs from existing insurance programs and premiums; (3) how can insurance programs, coverage designs, and insurable event definitions be best structured so as to make the resulting insurance premiums interpretable as MRAs; and (4) how, if at all, are MRAs appropriately used for public policy purposes. These four research areas will be discussed in turn in this section.

WHAT ARE THE CHARACTERISTICS OF MRAs AND WHAT CONDITIONS ARE REQUIRED FOR THE EXISTENCE OF MRAs?

Five characteristics of MRAs need to be investigated: (1) whether, and under what conditions, MRAs are unbiased; (2) what precision MRAs embody; (3) whether MRAs can be used for risk classification (as well as total risk assessment) purposes; (4) how, if at all, MRAs combine quantitative and qualitative risk assessments; and (5) whether, and to what degree, MRAs separate factual judgments from value judgments.

Each of these five characteristics might be investigated in light of formal models for insurance premiums as these are developed; of the theory of risk and insurance developed in the literature; of portions of modern financial theory, such as portfolio theory and the capital asset pricing model; and of insurance industry practice.

With respect to bias, an obvious attraction of MRAs is their potential for supplying statistically unbiased estimators of risk parameters. A stochastic specification for the insurance premium model appears appropriate for determining when and under what conditions data on premiums and collateral information can be used to define unbiased, or nearly unbiased, estimates of risk parameters. For example, the simplest theoretical condition for obtaining unbiased estimators for risk parameters will probably result from conditions of pure competition in the insurance industry. Under pure competition conditions, premiums should be bid to a level that has a simple structure. In the real world, of course, things may not be so simple. More rough-hewn operational approaches may be necessary for studying whether conditions in existing insurance markets provide a basis for some notion of workable competition and for reasonably unbiased estimators. For example, there may be attributes of the markets in which arm's-length, bargained-for transactions for premiums are believed to occur; under some conditions such transactions may define unbiased or nearly unbiased estimators. Alternatively, examination of the attributes of insurance markets may allow analysts to adjust for or bound the size of the bias which is present in the risk parameter estimates in such premiums. The theory of contestable markets may provide a helpful theoretical framework for considering these questions (see Bailey and Friedlander, 1982). The most extensive recent treatments of the economic organization of the insurance industry are Hill (1978), Joskow (1973), and National Association of Insurance Commissioners (1974).

With respect to precision, the quality, character, and size of the database upon which insurance premiums are based will obviously affect the precision of the premium's implied risk parameter estimates. The insurance premium model(s), as they are developed, will form the basis for exploring ways of measuring the precision of risk parameter estimates in MRAs primarily as a function of the size of the database and of other features of the data. Precision is, of course, a central concern when making risk estimates where actual accident experience is sparse or non-existent.

With respect to risk classification purposes, there are several relevant questions. Existing insurance markets supply risk estimates only for those classifications of risks that are separately priced. That is, insurance premiums reflect the structure of insurance companies' adopted risk classification schemes. The insurance premium model(s) should indicate how the cells of the existing risk classification schemes are priced currently. In addition, it is necessary to determine how fine the risk classification process should be, how the fineness relates to other qualities, such as bias and precision, and how the fineness of the classification scheme in terms of number of rating variables employed or classification cells may be balanced off against rating methods which combine information from other cells of the classification scheme. An example is the empirical Bayes techniques for rating, for which one should see Morris and Van Slyke (1978) and Tomberlin (1982).

With respect to combining statistical with other information, there is a major unexamined question of how insurance premiums combine statistical (or other narrowly actuarial) data with "pure intuition" or other types of qualitative information. It appears that insurance ratemakers in fact use heuristic analogs to formal Bayesian techniques to perform this combining function. Credibility theory is explicity used by casualty actuaries now for estimating individual risk classification cell parameters where adequate data exists for the population as a whole, but not for the risk classification cell alone. Whether some analog of credibility theory is used to mix quantitative with qualitative risk estimates remains an open question. This is largely an empirical issue, to be answered by determining what major types of information are used in insurance price setting, and how these disparate types of information are combined. This question should probably be addressed in connection with case studies on MRA applicability.

With respect to separating factual judgments from value judgments, it must be determined whether risk parameter estimates used in MRAs are, or can be, constructed to be independent of value considerations. Both insurance premium models and case studies will be helpful in examining whether elements of value such as the costs of errors of estimation or the degree of risk aversion of the insurer have been separated from other risk parameter estimates (empirically, in the real world) and whether one should expect them to be so separated under our model (as a matter of theory).

Beyond these five characteristics of MRAs which need to be investigated at some level, five possible pre-conditions, or requirements for practically feasible determination of MRAs need to be examined: (1) the existence of some notion of workable competition, contestable markets, etc., among providers of financial assurance (insurance companies, reinsurers, self-insurance pools, etc.); (2) the discount rate and the degree of risk aversion for providers of financial assurance should be roughly equivalent or in a known relation to society's discount rate and degree of risk aversion; (3) the party providing the financial assurance should have a true financial interest in the event's outcome; (4) information on the probabilities and the consequences of the event should exceed some minimal threshold level to permit intelligent risk assessment by any party, whether in an MRA or in a PRA; and (5) the existence of sufficient total insurance capacity (including re-insurance and self-insurance) to permit sufficient "laying off" of the assumed risk to keep any risk aversion term in the insurance model tolerably small. Two pertinent discussions of insurance capacity in relation to large risks are Solomon, Whipple and Okrent (1978) and Stone (1973). For each of these five possible conditions, two questions must be answered: (1) to what extent, if any, are they necessary prerequisites for MRAs; and (2) are they largely met in the real world?

WHAT METHODOLOGIES ARE APPROPRIATE FOR EXTRACTING MRAs FROM EXISTING INSURANCE PROGRAMS AND PREMIUMS?

Insurance premiums clearly depend upon estimates of the expected frequency of insured-against events, the estimated cost of these events, the expenses of writing insurance and adjusting losses, and the profit required on the insurance business for the particular line of insurance, including compensation (if any) for risk aversion resulting from risks of losses over and above expected losses. A number of models of insurance premiums have been proposed in insurance ratemaking regulatory hearings, insurance rate calculations in insurance companies, or in academic journal literature.

Examples are shown in Chang and Fairley (1979), Fairley (1979), Hill (1979), Joskow (1973), Seal (1969), Tomberlin (1982), and Weisberg and Tomberlin (1982). These listed models have been applied to date, however, only in "actuarial" situations where adequate data existed. They differ in the precise theories by which the elements are determined and in how they combine various elements to reach an overall premium.

The various models differ in whether or not, and how well, they specify parameters for the measures of risk that were previously selected as being of interest for purposes of decisionmaking, that is, probability and consequence estimates. Models for premium estimation may or may not correlate well with impliit models used for underwriting decisions made by insurers on how risks should be classified. In short, insurers may not necessarily act in the real world exactly in the way the model might show that they "should" act.

We believe that models for insurance premiums can be specified which meet two criteria simultaneously. The preferred model(s) must: (1) be consistent with currently accepted economic, financial, and risk theory; and (2) be capable of specifying estimable parameters for the chosen measures of risk. This includes, for example, components representing returns on equity and risk aversion compensation amounts, if any. The specification of this general model for insurance premiums (which will then be used to extract probability/consequence estimates from premiums) has been stated in our previous work for the Nuclear Regulatory Commission. [See Chernick et al. (1981, pp. 63-84), but much work remains to be done.]

HOW CAN INSURANCE PROGRAMS, COVERAGE DESIGNS, AND INSURABLE EVENT DEFINITIONS BE STRUCTURED SO MRAs CAN BE BEST EXTRACTED FROM PREMIUMS?

As previously discussed, one of the problems that makes it difficult to extract consequence and probability estimates from insurance premiums stems from the fact that coverages often include relatively broad groups of particular accident sequences within the defined insurable event. One does not wish to design insurance coverages solely for the ease of the analyst who wishes to use the resulting premiums as sources of MRAs, to the detriment of the appropriateness of the coverages and the coverage prices. This would truly be an example of the tail wagging the dog. However, it is possible that insurance coverage definitions (which include dollar amounts of coverage, and, as applicable, deductible amounts, co-insurance percentages, coordination of benefits provisions, and the like) and the insurable event definitions can be modified to improve the accuracy with which the resulting premiums can be interpreted as MRAs, without causing any corresponding detriments.

We expect that this issue can be studied most effectively through two or three case studies of potential applications of MRAs, such as commercial nuclear power, hazardous waste storage, liquefied energy gas transport, and biomedical engineering. Especially as the issues are clarified by the work described above of defining the characteris tics of and prerequisites for MRAs, and of modeling insurance premiums, we believe that it will be fruitful to examine the existing insurable event definitions, coverage definitions, and premium structures available in each application area. This will be done to determine whether or not improvements are possible to permit more accurate extractions of MRAs from the premiums. The insurance premium models may indicate that certain modifications to the coverage definitions and insurable event definitions would result in more accurate MRAs without otherwise degrading the efficiency of, or the financial assurance provided by, the insurance program.

HOW, IF AT ALL, ARE MRAs APPROPRIATELY USED IN PUBLIC POLICY AND REGULATORY DISPUTE RESOLUTION?

The ultimate purpose of studying MRAs is to draw appropriate public policy conclusions which may improve the public risk assessment process. This section will discuss a few areas in which public policy conclusions may be appropriate.

First, decisionmakers will need an outline of potential application areas, either in terms of types of activities or types of existing economic arrangements, in which MRAs can be used to estimate the sizes of loss risk externalities--which can be internalized if an insurance program is in fact put into operation, and charged to the activity generating the loss risk. The other areas in which MRAs can be used as sources of probability/consequence estimates of loss risks--whether or not coverage ends up being put into place or internalized--must also be considered. On the other hand, there may be very substantial areas in which MRAs cannot be put to either one of these two possible uses.

Second, the accuracy of MRAs must be compared to that of traditional analytic probability and consequence estimation methods, to determine whether MRAs are credible sources of information in various application areas. It must be remembered, of course, that experience data in these areas may be quite sparse, and thus may preclude all but the weakest inferences from the data. Despite this serious handicap, this analysis can determine whether MRAs and analytic risk estimates in the various study areas are "within each others' confidence intervals" or are more discrepant, and, similarly, whether either the MRAs or the analytic risk estimates are inconsistent with experience data. Some very general conclusions about the credibility or usefulness of MRAs in risk assessments may be possible.

Third, some highly desirable conclusions may be possible concerning the desirability or possibility of combining MRA results with other data, including analytic risk assessment results, in order to capture the information contained in both methods. One of the subsidiary questions to be resolved in this regard is whether or not MRAs somehow already contain within them the results of prior, analytic risk assessments.

IV. THE CURRENT STATE OF MRA RESEARCH

There is a limited level of activity in the areas discussed in this paper, or in closely related areas. This section will briefly describe the related work being done currently, and also recently completed related work.

Investigators at Analysis and Inference are in the process of completing a research project entitled, "The Application of Modern Statistical Principles to Automobile Insurance Risk Classifications," supported by the Division of Applied Research of the National Science Foundation. The purpose of this project is to study traditional actuarial approaches to the classification and premium rating of risks in insurance, and to propose modern statistical models and estimation methods to improve actuarial practice. This project has determined that the currently-used additive and multiplicative models (which are employed to combine relativities derived from univariate analyses of data cross-classified in several dimensions) can be improved upon by mixed additive and multiplicative models, which, in turn, can be improved by empirical Bayes estimators. (For details, see Tomberlin (1982) and Weisberg and Tomberlin (1982).) This work forms an important part of the theoretical under-pinning for public policy applications for MRAs. The performance of insurance classifications and premium rating is critically important to MRAs; if insurers do not assess risks well, their assessments are not very useful to the public. Improvements in technical understanding of the strengths and weaknesses of the risk classification and rating process will allow for better estimation of MRA accuracy. One of the goals, in fact, of the project on statistical theory in actuarial practice is to develop methods that are broadly applicable in risk assessment. The approaches used by casualty actuaries are similar in many respects across different lines of insurance. There are, for example, commonalities in risk classification and rating approaches between the automobile insurance line and the commercial general liability (CGL) line which is the insurance line most relevant to most MRAs (see Williams, Head, and Glendenning, 1979).

The present authors and colleagues at Analysis and Inference completed a project for the U.S. Nuclear Regulatory Commission which analyzed self-insurance pools as financial assurance mechanisms for assuring the adequacy of funds for decommissioning expense. This project developed the most detailed MRA methodology in existence to date, building upon earlier work by Denenberg (1973) and Wood (1981), for extracting catastrophic probability and consequence estimates from insurance premiums. This work was published by the NRC as NUREG/CR-2370; see Chernick et al. (1981).

On the policy side of the present proposal, Michael Baram recently completed a project, "Alternatives to Government Regulation," and is currently working on a project, "An Assessment of the Use of Insurance for Managing Risks to Health, Safety, and Environment," both funded by NSF. The first study was a broad review from a public policy perspective of alternatives to direct governmental regulation for managing technological risks. It reviewed a wide variety of proposed alternatives, including common law actions for negligence and strict liability, industrial and professional self-regulation, insurance mechanisms, other compensation mechanisms, efficient charges, and tax systems. The second study is presently focusing on a general investigation of the qualities of insurance as a regulatory mechanism, through four case studies. Although this work may identify many of the problems MRA will face, it does not address the study of the properties and conditions for market risk assessment which is the focus of the work proposed here.

The project at Clark University and Decision Research, Inc., on "Improving the Social Management of Technological Hazards," has worked on exploration of: (1) the size and trends in technological hazards; (2) taxonomies of technological hazards; (3) causal models for risks as a basis for planning managerial strategies of intervention; (4) judgmental basis in risk perception; and (5) generic approaches to "acceptability of risk" decisionmaking. These topics are tangentially related to this proposed work.

Similarly, the International Institute for Applied Systems Analysis (IIASA) in Vienna has sponsored research for several years on the general subject of risk management. This work has recently focused on decisionmaking for LNG safety, and is again tangentially related to this proposed work.

The Division of Risk Analysis at the U.S. Nuclear Regulatory Commission is sponsoring or directing work currently on three related topics: (1) quantifying safety goals for nuclear power plant regulation [see NUREG-0880, NUREG-0739, and Strip (1982b)]; (2) improving probabilistic risk assessments, the traditional fault-free methodology (see NUREG-2300); and (3) improving consequence of accident estimates [see Strip (1982a), and Ritchie et al. (1982)]. The results of these projects may define better both the goal of future MRAs and the extent of competition from conventional engineering risk assessments, at least in the nuclear power field.

Several other U.S. governmental agencies are sponsoring work on risk assessment. For example, the Environmental Protection Agency has conducted a study of private insurance availability for "non-sudden" accidents resulting from hazardous waste disposal. The Department of Commerce has conducted a study on availability of insurance, and competition in the insurance markets, with respect to oil pollution insurance for oil spills from off-shore oil drill activity. The Department of Energy's Office of Pipeline Safety has conducted a study of LNG and LPG production, transmission, and storage risk assessment, and on the various methods of assuring financial responsibility of parties engaging in LNG and LPG operations.

CONCLUSIONS

Market risk assessments, extracted from information contained in insurance premiums, may provide a valuable source of risk assessments which may be used to supplement more traditional types of quantitative risk assessments. Although substantial work remains to be done on the qualities, the interpretation, and the use of MRAs, there is reason to be optimistic that these problems can be overcome to a great enough extent to permit MRAs to become a valuable supplement to more traditional quantitative risk assessments.

REFERENCES

1. Buhlmann, H. (1970), Mathematical Methods in Risk Theory, Springer-Verlag, 1970.
2. Bailey, Elizabeth E. and Friedlaender, Ann F. (1982), "Market Structure and Multiproduct Industries, Journal of Economic Literature, September 1982.
3. Chang, Lena and Fairley, William B. (1979), "Pricing Automobile Insurance Under a Multivariate Classification," Journal of Risk and Insurance, March 1979.
4. Chernick, Paul L., Fairley, William B., Meyer, Michael B., and Scharff, Linda C. (1981), "Designs, Costs, and Acceptability of an Electric Utility Self-Insurance Pool for Assuring the Adequacy of Funds for Nuclear Power Plant Decommissioning Expenses," NUREG/CR-2370, U.S. Nuclear Regulatory Commission, December 1981.
5. Crandall, Robert W. and Lave, Lester B., Editors, (1981), The Scientific Basis of Health and Safety Regulation, The Brookings Institution, 1981.
6. Denenberg, Herbert S. (1973), "Testimony Before the Atomic Safety and Licensing Board, Licensing Hearings for Three Mile Island Nuclear Power Plant," November 1973.
7. Fairley, William B. (1981), "Assessment for Catastrophic Risks," Risk Analysis, Vol. 1, No. 3, 1981.
8. Fairley, William B. (1982), "Market Risk Assessment of Catastrophic Risk," in Kunreuther, Howard and Ley, Eryl, Editors, The Risk Analysis Controversy, Springer-Verlag, 1982.
9. Fairley, William B. (1979), "Investment Income and Profit Margins in Property-Liability Insurance: Theory and Results," Bell Journal of Economics, Spring 1979.
10. Green, Harold P. (1981), "The Role of Law in Determining Acceptability of Risk," in Nicholson, William J., Editor, Management of Assessed Risk for Carcinogens, Annals of the New York Academy of Sciences, Volume 363, 1981.
11. Hill, Raymond (1978), Capital Market Equilibrium and The Regulation of Property-Liability Insurance, Ph.D. Dissertation, Massachusetts Institute of Technology, 1978.

12. Hill, Raymond (1979), "Profit Regulation in Property-Liability Insurance," Bell Journal of Economics, Spring 1979.
13. Joskow, Paul (1973), "Controls, Competition, and Regulation in the Property-Liability Insurance Industry," Bell Journal of Economics and Management Science, Autumn 1973.
14. Meyer, Michael B. (1982), "Catastrophic Loss Risk: An Economic and Legal Analysis, and a Model State Statute," Proceedings of a Workshop on Low-Probability/High Consequence Risk Analysis, sponsored by the Society for Risk Analysis, the U.S. Nuclear Regulatory Commission, the U.S. Department of Energy, and the U.S. Environmental Protection Agency, Arlington, Virginia, June, 1982; Plenum Press (1983).
15. Meyer, M. B. (1983), "Regulating Catastrophes Through Financial Responsibility Requirements: A Model State Statute," Harvard Journal of Legislation, Summer 1983.
16. Morris, Carl and Van Slyke, L. (1978), "Empirical Bayes Methods for Pricing Insurance Classes," Proceedings of the Business and Economics Section, American Statistical Association, Annual Meeting, San Diego, 1978.
17. National Association of Insurance Commissioners (1974), Monitoring Competition: A Means of Regulating the Property and Liability Insurance Business, Volumes 1 and 2, May 1974.
18. National Research Council (1980), Regulating Pesticides, A Report Prepared by the Committee on Natural Resources, National Research Council-National Academy of Sciences, 1980.
19. Raiffa, Howard (1982), "Science and Policy: Their Separation and Integration in Risk Analysis," in Proceedings of the Sixth Symposium on Statistics and the Environment, American Statistician, August 1982.
20. Ritchie, L., Johnson, J., and Blond, R. (1982), "Calculations of Reactor Accident Consequences, Version 2: User's Guide," NUREG/CR-2326, SAND 81-1994, Sandia National Laboratories, 1982.
21. Seal, Hilary (1969), Stochastic Theory of a Risk Business, Wiley, 1969.
22. Solomon, Kenneth A., Whipple, Chris, and Okrent, David (1978), "More on Insurance and Catastrophic Events: Can We Expect De Facto Limits on Liability Recoveries," Rand Corporation, P-5940, March 1978.

THE "THIRD MODE": TECHNOLOGY REGULATION BY INSURANCE

Christopher Mallagh*
Jerome R. Ravetz**

*Analysis and Inference, Inc.
Research and Consulting
10 Post Office Square, Suite 970
Boston, MA 02109

**Department of Philosophy
Leeds University

ABSTRACT

Insurers offer reduced premiums for improved risk taking behavior. The financial benefit to firms through internal risk management may be matched by an external benefit to the public. Economic pressures help to mitigate moral hazard, and the problems of thresholds and "best practical means" clauses. Underwriters perform many direct inspection functions. "Retrospective" inspection takes place when claims are made: insurers, supported by strong legal rights, may repudiate cover when conditions of policies on risk taking have not been complied with. Self-regulation is forced on the insured and low frequency risks can be controlled as a result. The obligation to disclosure by the insured when making policy applications creates a situation in contrast to that under statutory regulation. The insurance mechanism involved risk analysis by reference to actuarial data. Risks as reflected through premium levels are empirically validated in this way. Scientific evidence can supply the insurer with information on risks before actuarial data builds up. Laboratory results are validated under practical conditions by actuarial data. Public values influence the system through claims generated by third parties and by levels of damages set by juries. More detailed research on the role of insurance as a technology control mechanism is projected.

KEY WORDS: Technology-regulation; Insurance; Inspection; Risks; Public-benefit.

INTRODUCTION AND HISTORICAL BACKGROUND

Risks are typically treated as being concerned merely with events that it wishes to eliminate. Insurance is then only a device which helps to somewhat mitigate the effects of the failures that do inevitably occur.[1] Yet underwriting can act as a technology control mechanism. Frequently insurers are able to offer reduced premiums in return for reduced risk taking and better safety practices. In fire prevention the installation of sprinkler systems can achieve substantial reductions in premiums; in certain industries in Britain, up to 90 percent or more.(2)

This function of insurance in technology control is not new. During the nineteenth century the traditional regulatory mechanism depending on nuisance proceedings at common law (the "First Mode") was largely replaced by statutory legislation (the "Second Mode"). Legislative acts directed toward the mitigation of specific technological problems became the norm, with the assignment of the required enforcement powers to public or quasi-public bodies.[2] But the same period saw the growth of the "Third Mode" of technology regulation based on insurance.

The Manchester Association for the Prevention of Boiler Explosions ran a successful boiler inspection program using the incentive of insurance to achieve the full internal inspections necessary to the prevention of explosions.(3) Certain fire offices of the time were only willing to offer coverage on farm property on the condition that the workers did not carry the very unstable phosphorus matches of the time loose in their pockets. (4,5)

The use of insurance in risk management is a familiar theme. Analysis of the issues involved in economic terms is the accepted practice.(2,6) This is understandable. Financial benefit is the primary motivation in risk management. Improved behavior on risk taking and safety practices may lead to significant reductions in premium costs. Competition encourages underwriters to quote reduced premiums in return for better safety practices.[3] Moreover, in selected cases, such as the fire losses on insured property, the total cost to society over a period can be quantified economically.

However, internal risk management by firms and businesses under a drive for reduced costs may lead also to an external benefit to society of a general kind. In a highly technological society individuals are continually subjected to risks as they go about their daily living, whether in the form of a chemical complex in their neighborhood, through the use they make of facilities provided by industry, or in connection with their employment. Financial considerations cannot tell the whole story here. Any reduction in such risks as a result of pressure from the insurance mechanism may be considered worthwhile irrespective of the economic benefits that ensue. Non-smokers may now be rewarded by more favorable insurance conditions of life insurance coverage.(7) Any reduction of smoking that results is a desirable outcome in public health terms whatever the accompanying economic effects.

The "Third Mode" of technology regulation based on insurance and its influence on risks thus needs to be examined in a broader context.

THE FUNCTIONAL ROLE OF ECONOMIC PRESSURES

"Moral hazard" is often considered to be a problem with insurance; that is, there is a danger that the insured may change his behavior, by becoming careless, for example, when he is covered against financial loss.(2,8,9)

But the attraction of reduced premiums for improved behavior on risks acts in the opposite way. To gain such reductions the insured must act prudently and underwriters have the power to load premiums in the case of individuals with a bad claims record.[4] A drive for better cost performance by way of decreased premiums, buttressed by the interest of underwriters in ensuring that lower premiums are justified, generates a system of rational decisionmaking by risk managers in accord with the public interest.

Business and industry are familiar with the management of costs on a day to day basis; costs are the very units of their operations. Where reduced insurance premiums are available for improved risk taking behavior, different risk levels with different cost implications enter into general cost equations as variables with determined values. Risk levels become part of the whole complex of decision-making and play their part in determining the best allocation of resources in the planning and operation of projects.

From the point of view of the public exposed to risks from industry, this has significant advantages. The mechanism involved continually favors any improvement in risk behavior. Any reduction in risk taking accompanied by a lower frequency of unwanted events will enable the underwriter to quote reduced premiums because he will have fewer claims to meet. As long as the costs to business of reduced risk taking are outweighed by gains in terms of premium reductions, the pressure toward improved risk behavior will continue. Where this does not occur, the economic advantage switches to technological innovations, which by decreasing costs of risk reduction provide opportunities for further gains in another round of premium reduction.[5]

Today it is frequently accepted that the notion of a "safe" threshold is an extremely problematic one.(10) If a specific emission or dosage level leads to a particular risk of disease, a lower dosage or emission level may only lead to a reduction in the risk of the disease, not to its elimination. Under the statutory mode, however, threshold levels for all sorts of emissions have to be set, whether of chemical or biological pathogens, radiation, or other pollutants. Design and operation of projects is then governed merely by the need to achieve the statutory limit set. When this level has been reached, motivation to achieve further reductions disappears.

Similarly, "best practical means" clauses--so much a part of the statutory tradition--tend always to legitimize the existing levels of risks to which the public is exposed.[6] What is the best practical means at any particular time has to be defined very much by reference to the best existing practices of a trade or profession. Where those levels are achieved, there is little motivation to further improvements in practice.[7] In contrast, under the "Third Mode" in both these situations there is always commercial pressure to increase the effectiveness of risk containment, whether through technical innovation or otherwise. This often may lead to a benefit to the public.

Thus the insurance mechanism, acting under the "Third Mode," latches on to the day-to-day commercial practices of trade and industry. By making use of the drive of individual firms for cost reduction, it provides significant adavantages not otherwise available.

THE INSPECTION FUNCTION UNDER THE "THIRD MODE"

Statutory or quasi-statutory regulatory agencies are all too prone to overidentify or otherwise become involved with their regulatory "client."(13) Under the "Third Mode," many of the functions of such agencies are taken over by the underwriter. The basic relationship between insurer and insured remains commercial, however, and serves to prevent many such difficulties.

Insurers undertake direct inspection and allied functions in a whole range of situations. Traditionally, in the U.S.A. and Britain, fire risks and resultant premium levels have been set by reference to a detailed schedule, although often completed by relatively untrained personnel. In Britain, the Fire Offices' Committee, a rate setting body, is responsible for such activities as specifying approved sprinkler systems suitable to give reduced premiums.(2) The U.S. Standard for the Installation of Centrifugal Fire Pumps, produced by a committee with a strong representation from the insurance field, specifies a whole series of operations for the regular testing of such systems.(14) In Britain, "Most engineering insurers will undertake the examination of second-hand boilers and machinery for purchase.... as part of the service given under an engineering policy."(15) Lloyds certifies the competence of welders. Pollution Liability Coverage in the U.S.A. is today often only granted after an approved survey has been carried out.(16)

However, probably the most important function of insurers in this area is not direct inspection. It is rather after-the-fact, or retrospective inspection. Where the conditions of a fire policy specify that paint shall be kept in a proper paint store, and the insurance adjuster on scrutinizing a claim discovers a pile of burst paint cans "higgledy-piggledy" in a corner of the burnt-out building, he is all too likely to hold the claim void. The sweeping up of waste and sawdust in a "woodworking risk" at least once daily is an often quoted example of this type of situation.(9)

The effect here is to throw the responsibility for day-to-day inspection and regulation onto the insured, generating an important type of self-regulation. This is only reasonable. Consider the historical example of farm workers with matches in their pockets: the idea that the insurer should have regularly gone around peering into the pockets of the workers in a direct inspection role is ludicrous. Only the insured in such situations is in a reasonable position to see that such conditions are adhered to. The danger that a claim may be held void ensures that the policyholder does keep his risks at a low level as required.

Insurance law provides a battery of mechanisms to give the insurer substantial powers to void policies or claims when the insured does not behave on risks as called for. Under English law an applicant for coverage is under an absolute obligation to make complete disclosures of all material factors relevant to the insurance contract. "If he conceals anything that may influence the rate of premium which the underwriter may require, although he does not know it would have that effect, such concealment entirely vitiates the policy."(17) The insured must, therefore, be very open about all the risks he takes in negotiating with the underwriter. This situation must be contrasted with that under statutory regulation. There the inspector of a regulatory agency always tends very much to be seen as an adversary attempting to enforce existing regulations: any obligation to disclose rapidly degenerates into merely an obligation to allow access and inspection. Clearly in practice one does not go out of one's way to disclose one's risktaking to an inspector which might be an offense against a statutory regulation. Whereas in the insurance situation,[8] one must do this or risk defectiveness in coverage purchased.

But perhaps the strongest weapon in the insurer's armory is the use of "continuing warranties." That is the writing into policies of conditions which the policyholder agrees to abide by. By means of such clauses insurers are "able to grant policies without troubling to make inquiries about matters covered by warranties." Potential retrospective inspection again provides the pressure to ensure compliance.[9]

But further, "a breach of warranty entitles him (i.e., the insurer) to repudiate whether it is material or not."(20) That is, coverage may be void because a condition of the insurance contract has not been complied with even though such failure had nothing to do with the occurrence of the event for which a claim under the policy is made. In the case of a claim consequent on a fire caused by the ignition of paint, that claim may be void even though the paint was kept in a proper paint store as required, because one's electrical equipment, which had nothing to do with the fire, was not maintained at regular intervals in a workmanlike way as also was called for under the policy.

There is an important principle of technology control involved here. Poor behavior on a risk which leads to an unwanted event is not always easy to discover by retrospective inspection. But norms of risktaking, whether good or bad, are likely to be relatively constant throughout a company or project. Evidence on performance relative to conditions called for in a policy, even though those conditions are not material to the occurrence of the event in respect of which a claim is being made, can thus provide a good indicator of probable behavior in respect of a condition which is material. There is thus good reason why coverage should be void through non-performance on non-material conditions. In the field of very low frequency risks this principle has an important role to play. Such risks are all too likely to be subject to a tendency to believe that "it won't happen here," or, "we'll get away with it this time." But if general risk behavior is judged as a whole, then performance on low frequency risks has to be kept up in order to prevent breaches of coverage on other risks; of perhaps a somewhat higher probability frequency, of which at least one is likely to occur within the life of the project. Linked warranties effectively do this. Low frequency risks are then brought within the compass of[10] the motivating force created by potential retrospective inspection.

Such legal rights amount to a very powerful tool in the hands of the insurer which can do much to ensure self-regulation on the part of the policyholder. Moreover, the retrospective inspection process is extremely efficient in terms of man-power. The need for the underwriter to inspect on a random basis is obviated and much of the time spent inspecting situations where good risk behavior is practiced, is saved. Inspection takes place when claims are made and hence in the cases where bad risk behavior is the most likely to have been involved. The larger the claim, the worse the behavior is likely to have been, and the more closely the underwriter will want to scrutinize the situation.

The power of this process and the self-regulation it generates clearly can do much to reduce the danger of moral hazard. In fact, the strength of the insurers' legal rights can be seen to arise because of such dangers and to provide the necessary antidote.[11] It then turns out that under the "Third Mode," regulation is forced on the individual technically most competent and in the best position to undertake this function; in other words, the risk taker who is in the best position to control risk taking on a day to day basis. This person is also likely to be the best informed about the technical and economic implications arising from different risk levels in his own practice.[12]

Thus the "Third Mode," using the inspection and regulation functions set up by the commercial practice of insurance, is potentially a very powerful mechanism for the control of risks.

THE INTEGRATION OF RISK ANALYSIS, RISK MANAGEMENT, AND TECHNOLOGY CONTROL UNDER THE "THIRD MODE"

The writing of insurance and the setting of premium levels is part of a highly integrated and continuously monitored system, or at least should be. Premiums have to be set at levels which reflect the frequencies of unwanted events, so that the insurance fund is sufficient to meet the claims that arise. The collection and interpretation of actuarial data by which this is accomplished is a type of risk analysis. A type of analysis, moreover, empirically validated (or falsified) by the frequencies at which unwanted events do in fact occur and the claims that ensue. In this process perceived risks, influenced by a host of factors--the total surrounds of the situation, styles of life, and valuations of many sorts--have to match up to the reality of the number of claims that cross the insurance adjuster's desk.

This is in contrast to any abstract numerological approach to risk analysis wherein frequencies so low that they never can be validated empirically by actuarial data can become the subject of "scholastic" debate. Under the "Third Mode," risk analysis occurs as part of a dynamic system, results being continually constrained by feedback from claims. The writing of conditional policies and the frequencies which result, is a rich source of information about risks and the many variables that affect outcomes. As risk managers seek to minimize premium costs using different technological approaches to risk reduction, their decisions can be assessed by reference to the shifting responses of underwriters in setting premium levels. The process is not one where a unique figure quantifies the risk attached to a specific activity. Outcomes depend on the quantification of risks in certain ways, the use of particular technological devices to reduce risks as a result, and the use of these quantifications in situations governed by standards which are a part of the given social and technical milieu.[13] The total interaction of all these factors is then what is validated by actuarial data. The underwriter's role in this situation is to function as the required information conduit.[14]

At the same time at the premium setting level, evidence from scientific research and experiment plays an important role. Where new types of risk appear (new hazardous chemicals for example), or, when new methods of risk reduction become available, it takes time for actuarial data to build up so as to justify the insurer's selection of a particular premium level. Here scientific evidence from research and other allied activities is about the only dependable source of information available to the underwriter. But the scientific information supplied as to the probable risk associated with particular materials, mechanisms, or activities is only tentative to the underwriter. Actuarial data is necessary to justify dependence on figures so derived; this is of great importance. Under conditions of laboratory experiment and similar research, every attempt is made to monitor and control all the relevant variables to the greatest possible extent. In contrast, in practical application the situation is characterized more by the unpredictable and uncontrollable nature of all the mutlifarious conditions that have to be coped with. These conditions in turn are influenced by standards of behavior on the shop floor and at management level quite different from those the scientist in his laboratory is able to exercise.[15] As actuarial data becomes available and the underwriter adjusts his premium levels from those suggested by scientific research, he takes account of these factors.

Into this complex technical and commercial system another element enters which makes the whole system responsive to social values and gives it the force of a technology control system. In cases where insurance coverage relates not to the direct losses of the insured but rather to claims from third parties which arise as a result of his activities, public values have an important role to play. The public, as instigators of third party claims, makes known those aspects of technology it considers most damaging. When public concern appears about certain types of technological effects, the prudent underwriter must adjust his premium levels accordingly if he is to allow for the potential severity and frequency of the claims he is likely to have to meet as a result. Additionally, where claims get into court and damages are assessed by juries, public values again play a role. The jury, drawn from a cross-section of the public, signals a public evaluation of the unwanted effects involved by setting damages at high or low levels. These damages have to be paid out of the general insurance fund. By these means, public values can affect premium levels which, in turn, influence decisions by risk managers as to the comparative allocation of resources to different types of risk reduction. As a result, the responsiveness of the whole control system to social norms and values is enhanced.

The "Third Mode" of technology control thus involves an integrated system in which risk analysis, risk management, and public evaluations of technological effects all play their parts. At the same time commercial pressures serve to motivate participants in useful ways even as the whole system obtains empirical validation from the pattern of claims that arises. Clearly a potentially very powerful process is at work when all these aspects fuse together. It remains to be seen whether in application, the "Third Mode" can live up to its apparent advantages.

CONCLUDING REMARKS: PRESENT PROBLEMS AND FUTURE PROSPECTS

Under the "Third Mode," technology regulation is fueled by the motivations of individuals and companies as they seek their own personal profit. Yet public benefit can ensue from such motives if the requisite institutional mechanisms are present.[16]

One possible problem with the more extensive use of this mechanism is how these motivations may shift when compulsory insurance is introduced as a way of ensuring that the requisite mechanisms go to work. The signs are also that the insurance industry in the U.S.A. is not altogether happy with the regulatory role it is being asked to play with the introduction of the "Third Mode" in the field of hazardous wastes.(24,25) The problem of the amount of coverage any company should be required to hold is another that requires solution.

We hope to study these and other problems at Leeds in the future. We will watch, with interest, developments in the regulation of the chemical industry in the U.S.A. where a radical shift towards dependence on insurance is taking place. Technology control mechanisms are highly dependent on their social and technical milieu. We look forward to making more detailed comparisons between practices in Britain and the U.S.A., particularly with regard to these radical developments, in order to clarify more usefully the general working of the "Third Mode" and insurance under different social and technical circumstances, and within different administrative and legal systems.

REFERENCES

(1) H. Kunreuther & J.W. Lathrop "Siting Hazardous Facilities: Lessons from LNG" Risk Analysis 1, 289-302 (1981)
(2) Neil A. Doherty Insurance Pricing and Loss Prevention (Saxon House, Farnborough, Hants. 1976)
(3) Manchester Association for the Prevention of Steam Boiler Explosions A Sketch of the Foundation and of the Past Fifty Years' Activity (Manchester 1905)
(4) Andrew Wynter "Fires and Fire Insurance" Curiosities of Civilization (London 1860) 401-40
(5) Andrew Wynter "The Sufferings of the Lucifer Matchmakers" Curiosities of Civilization (London 1870) rpr. in pt. in Technology and Toil in Nineteenth Century Britain ed. Maxine Berg (London 1976)
(6) C. Authur Williams, Jr. & Richard N. Heims Risk Management and Insurance (N.Y. 1964)

(7) Advert. Post Magazine and Insurance Monitor 144, No. 27 (1983) "Good News for Non-Smokers" Economic Insurance Co., Life Office
(8) "Sweeping Changes Suggested for Britain's Fire Protection" Fire International 68, 53-58 (1980)
(9) J. Vann "Hazards in Insurance: The Physical and Moral Dangers" Post Magazine and Insurance Monitor 144, 1388-9 (1983)
(10) Giandomenico Majone The Logic of Standard-Setting: A Comparative Perspective (IIASA Laxenburg, Austria 1983)
(11) R. A. Buckley The Law of Nuisance (London 1981)
(12) Christopher Harvey The Law Relating to the Pollution of Rivers: A Functional Study of Prevention and Redress from Feudal to Modern Times (Unpub. Ph.D. Thesis, University College, London 1975)
(13) Bruce M. Owen & Ronald Braeutigam The Regulation Game: Strategic Use of the Administration Process (Camb. Mass. 1978)
(14) National Fire Protection Association Standard for the Installation of Centrifugal Fire Pumps NFPA No. 20 (1974)
(15) J. Vann "An Engineering Wonder" Post Magazine and Insurance Monitor 144, 1026-8 (1983)
(16) C. S. Gilpin "Gradual Pollution Liability: The Insurers and Their Policies" Rough Notes 124, 24 66-70 (1981)
(17) J. Wormell "The Changing Concept of 'Utmost Good Faith'" Post Magazine and Insurance Monitor 144, 1026-8 (1983)
(18) Newcastle Fire Insurance Co v MacMorran & Co (1815) 3 Dow 225
(19) McEwan v Guthridge (1860) 13 Moo P.C.C. 304
(20) Raoul Colinvaux The Law of Insurance (4th ed. London 1979)
(21) R. M. Hall "The Problem of Unending Liability for Hazardous Waste Management" Business Lawyer 38, 593-621 (1983)
(22) Merchant Shipping Oil Pollution Act (c. 59, 1971)
(23) Alfred D. Chandler & H. Daems (eds.) Managerial Hierarchies (Camb. Mass. 1980)
(24) J. C. Smith & D. Godfrey-Thomas "Insurers Re-Enter the Hazardous Waste Game" Best's Review, Property Casualty and Insurance Ed. 83, 20-26 (1982)
(25) J. W. Milligan "EPA to OK Coverage for Dumps' Closure Costs" Business Insurance 16, 3, 52 (1982)

FOOTNOTES

1. "One possible direction for future study is the role of insurance as a way of protecting potential victims against potential property losses and physical injury"(1)
2. In the fields of smoke control, river pollution, and public health particularly.
3. In the U.S.A., two large underwriting associations have come to specialize in the business in "highly protected risks" which can be generated in this way.(2)
4. "....premiums might be loaded until moral hazard improves."(9)

5. In the 19th century, many trades were considered so dangerous as to be uninsurable.(4) Conditional insurance as in the case of farm workers and matches is a development from this. But in the same period, safety matches were introduced by Bryant & May.(5) We thus get a developmental schema: Uninsurable--Conditional insurance--Technological innovation--Reduced need for insurance. This ensures that with developing technology and new risks, the insurance load on firms does not become too heavy.
6. Such clauses emerged in the 19th century particularly in smoke control. Under the 1974 Control of Pollution Act, "In respect of noise emanating from a trade or business, it is a defense to prove that the best practical means have been used for preventing, or for counteracting the effects of the noise."(11)
7. In the historical period, the older nuisance tradition had to be employed to force local authorities into technological innovation to get reduction in river pollution from the sewage they discharged.(12)
8. Allied to disclosure is description which must be accurate on the same grounds. In the case of a cotton mill described as "First class," i.e., "having the stoves not more than one foot from the wall with pipes or flues not more than two feet in length," the mill not being of this standard the insurer was held to be not liable.(18)
9. In one case, the Privy Council held that the insured could not recover when he kept more than 56 lbs of gunpowder on the insured premises contrary to a condition in the policy.(19)
10. Recently in Britain, there has been pressure for curtailment of insurer's rights on linked warranties as inequitable. (17,20) For the reasons given it would be unfortunate if this went too far. The account given is based on English law: emphasis differs in other legal systems on which more study is required. A tendency in recent legislation on hazardous wastes in the U.S.A. to curtail insurers' rights to repudiate,(21) should also be watched with concern.
11. One British anti oil-pollution act in requiring compulsory insurance recognizes that, under normal conditions, claims against insurers are voided when there is non-performance in a general way by the insured. It allows that in the case of claims against insurers by third parties, it is a defense "to prove that the discharge or escape was due to the wilful misconduct of the owner himself."(22)
12. Independent technical decisions of professionals are increasingly subject to litigation. Under the "Third Mode," technical decisions are highly constrained by a whole network of commercial practices. This should improve acceptability.
13. In air transport, working with fatalities per journey can give very different results as against working with fatalities per mile.

14. Large scale operations have been important historically in the development of managerial skills.(23) Large firms should, therefore, contribute disproportionately to developing risk management skills. The underwriter, in quoting premium levels with relevant technical backup, can provide a conduit also for transfer of the experience of large firms to smaller concerns.
15. In the 19th century in the field of smoke control, laboratory type experiments showed that it was all too easy to achieve smoke-free running. But again and again, manufacturers were unable to achieve such results in practice. The factors at work were very much those indicated here.
16. "Private Vice, Public Benefit," as Bernard de Mandeville subtitled his well known <u>Fable of the Bees</u> of the early 18th century in supporting ideas that were the background of the development of <u>laissez-faire</u> capitalism. But the "Third Mode" is not totally dependent on a free market economy. When a chemical plant explodes and demolishes X houses which have to be rebuilt, this occurs whether a market economy exists or not. Insurance is simply a mechanism reflecting the resource allocations involved. Where no market economy exists the same resource transfers can be handled as "book transactions." The same control mechanism can then function, although here motivations have to be plugged in rather differently.

ENVIRONMENTAL IMPAIRMENT LIABILITY (EIL)
INSURANCE RISK ASSESSMENT SURVEYS

Robert M. Wenger
Joseph V. Rodricks

Environ Corporation
Washington, DC 20006

EVOLUTION OF THE EIL INSURANCE MARKET

The market for Environmental Impairment Liability (EIL) insurance has developed over the past ten years. The first step in the evolution of the market took place in the early 1970s when underwriters excluded non-sudden, accidental environmental occurrences from coverage under Comprehensive General Liability (CGL) insurance policies. Initially, this action had only a limited impact as most companies did not focus on the extent of their financial risks as a result of this exclusion. In recent years, however, scientists have increasingly linked exposure to certain chemicals and chemical wastes to adverse environmental and human health effects; and the public's awareness of the potential dangers of hazardous substances to health and the environment has risen drastically, fueled by a series of front-page stories of innocent victims of alleged toxic wrongs. As a result, corporate risk managers have become more concerned about the significance of this exclusion.

In addition, the Federal Government and many states have responded to the public's outcry for increased protection of health and the environment by passing laws requiring industry to monitor its practices, to insure safer management of hazardous materials, and to make companies financially responsible to the public for the damages that their activities may cause. The Resource Conservation and Recovery Act (RCRA), in particular, imposes financial responsibility requirements on treaters, storers, and disposers of hazardous waste to protect against sudden and non-sudden occurrences which may result in environmental impairment. With regard to non-sudden, accidental occurrences, owners and operators of surface impoundments, landfills, and treatment facilities must evidence financial responsibility by either purchasing commercial insurance or by meeting financial tests for self-insurance with limits of $3 million per occurrence and $6 million in the aggregate. These requirements are to be phased in over a three-year period beginning in January 1983. In addition, the increasing numbers of toxic tort lawsuits have caused other companies whose activities involve hazardous substances, but who are not required by regulation to evidence financial responsibility, to become increasingly concerned about their potential exposures.

These changing market conditions have spurred the emergence of an active group of domestic and international EIL underwriters. These underwriters have filled the gap left by the CGL policy exclusion. EIL liability insurance has become a welcome way for companies to share some of the risks of financial exposure as a result of their environmental practices. In general, these policies cover claims and defense costs associated with off-site environmental impairment. They normally do not cover the costs of on-site cleanup, although the scope of coverage is a negotiable item in several markets.

Today, there are two separate groups of facilities seeking EIL coverage. The first group consists of those treaters, disposers, and storers of hazardous waste who must meet the financial responsibility requirements either through self-insurance or the purchase of an EIL policy. The second group consists of companies whose activities involve the use of hazardous materials that are not under any regulatory mandate to evidence financial responsibility but are concerned that they have significant areas of exposure which could result in future civil liabilities.

THE ROLE OF RISK ASSESSMENTS

The risk assessment survey is an essential element in the process of purchasing or deciding to purchase EIL insurance. Whereas time-tested procedures and data bases exist to assist underwriters of other forms of insurance (e.g., life, fire) in understanding the risks they are underwriting, no parallel established procedures and data base exist for predicting future gradual and non-sudden environmental impairment liability claims. In a sense, an EIL data base is evolving each time toxic tort litigation is begun. Models are being developed by EPA to predict the likelihood of releases from certain kinds of facilities, e.g., landfills. However, it will be some time before underwriters will feel confident that they understand the likely risks associated with the operations of a facility without the benefit of an on-site risk assessment by a qualified assessment firm.

The purpose of the EIL risk assessment is to highlight for the underwriter and insurance applicant those aspects of an applicant's operation that are most likely to give rise to environmental impairment, third party damages, and ultimately insurance claims and defense costs. By defining major corporate exposures, the EIL risk assessment also gives corporate officials a better sense of what deductibles and limits of coverage they should be seeking (the limits contained in the RCRA requirements fall well below the likely outcome of successful litigation of third-party claims). For those companies not required by law to obtain coverage, the EIL risk survey provides an analysis of existing exposures and answers the question of whether they have the types of exposure that merit insurance coverage. Once that question is answered, corporate management can then decide whether to insure through self-insurance, purchase of commercial insurance, or a mixture of both.

The market will bear survey costs of between $2,500 and $7,000 per facility. As a result, the assessment is necessarily of limited nature; no independent data gathering is involved. These surveys are by no means the same as exhaustive environmental audits, where every permit violation, no matter how minor, may be identified. Rather, the major risk areas of a company's operation are identified and recommendations for reducing the risks, if appropriate, are made. For example, occasional pH violations of an NPDES permit will likely have insignificant long-term impact on the receiving stream and would not be an area of concern in this kind of assessment. Because these surveys are typically a "quick and dirty" assessment of activities at a facility based on a one-day site inspection, interviews with a company's employees, and a review of often incomplete data, they are largely reflective of the experience, judgment, and instincts of the individuals involved in the assessment.

In some cases, the focus of the risk assessment is an "eye-opener" for corporate officials. In a recent site survey performed by Environ, the client was most concerned about the potential for groundwater contamination associated with an on-site lagoon. However, the risk assessment made corporate management aware for the first time of the potential adverse effects of air emissions from the site to the nearby population. While the chemical being emitted was known to be of concern for worker exposure and precautions in the plant had been implemented, the potential impact on the population living near the plant had been largely ignored. In another survey at an incinerator, the risk manager was most concerned, as would be expected, about air emissions at the site. However, the storage of leaky barrels directly on the ground without a pad, pallets, or protective diking, was identified as the major area of risk, not the air emissions. Further, site surveys often uncover areas of risk no longer at the forefront of management's "worry list." For example, an assessment of a chemical company identified prior benzene emissions as an area of concern, even though a feedstock change had recently eliminated such discharges.

Because there have been very few EIL claims to date, the significance or relevance of certain information is sometimes unclear to both the risk assessor and the underwriter. For example, in addition to providing a description of operations that take place at a site, the risk assessment report identifies the use of the land surrounding the site. If a manufacturer is located in a heavily industrialized area, this may cut two ways for an underwriter. On the one hand, if a pollution incident occurs involving commonly used chemicals, it may be difficult to locate the source of the pollution within the individual area. On the other hand, it is likely that all parties that could have conceivably been the source of the pollution will be named as defendants in a lawsuit.

In addition, because of the newness of the coverage and the small number of claims made to date, the distinction between coverage under CGL and EIL policies has not yet been completely defined. For instance, in the case of a major catastrophe (e.g., a ruptured pipeline, a flood), the initial cleanup would typically be covered in part by a CGL policy. However, it is unclear which policy would cover claims for environmental impairment made weeks, months, or years after the intial cleanup. Issues such as this will most likely be settled in court.

Components of a Risk Assessment

Environ employs a multi-disciplinary procedure for evaluating the potential for adverse impact on public health and the environment arising from the activities at a facility. The analysis considers the following:

- Inherent hazard or toxicity of raw materials, by-products, and products
- The means and media through which contaminants may move through the environment
- Population at risk and the route of their potential exposure
- Management practices employed at the site

As risk is a function of toxicity and exposure, the initial step in any assessment is to identify the toxicity or inherent hazard of the material produced, handled, stored, discharged, or disposed of at the site. A toxic substance is one that has an inherent capacity to cause injury to living organisms including human beings. There is, of course, enormous variation in both the types of injury chemicals can cause and the levels of exposure to the chemicals in which injury is caused.

Toxicologists classify toxic injury as either acute or chronic. Acute toxic injury is that which results from a single exposure to a chemcial which becomes apparent immediately or very soon after the exposure. With regard to gradual environmental impairment, chronic toxic injury can arise following repeated exposures to a chemical sometimes involving many years or from just a few exposures at levels well below those capable of producing acute toxicity. Factors evaluated with regard to chronic toxicity include the potential to induce teratogenesis, carcinogenesis, mutagenesis, and other chronic toxic effects in either humans or animals.

The nature of the damage potentially caused by chemical exposure depends on many factors, however, including the molecular structure of the chemical, the nature and susceptibility of the exposed person, and the route and duration of exposure. Thus, the means by which toxic substances on-site may be released to the air, soil, or water and their fate and transport in the environment need to be identified. Accordingly, major exposure pathways and likely dispersion or dilution rates are evaluated. Factors significant in evaluating the pathways that materials will travel in the event of gradual release include the topography of the site, the proximity to surface water and ground water, the hydrogeology of the site, the type and porosity of the soil, local drainage patterns, relevant climatological factors, and ambient air conditions.

As mentioned before, the level of analysis is largely limited to the data available for a site. For example, where estimates of annual air emissions can be made, the so-called Pasquill-Gifford equations or Gaussian equations may be used for dispersion estimates and the calculation of worst case conditions. However, without stack data and downwind measurements, the exact magnitude of the exposure cannot be pinpointed. Similarly, the level of analysis for groundwater contamination is limited by available hydrogeological data. In most cases, only well boring logs are available for review. Finally, the potential for surface water contamination is analyzed considering the proximity to surface water, the nature of direct discharges (e.g., treated wastewater effluent), and stormwater runoff controls. The flow of a receiving river or stream will be considered in estimating contaminant dilution and dispersion rates.

After completing the exposure pathways analysis, attention is focused on identifying the population at risk. No matter how toxic a substance is, it will not have an adverse effect unless a susceptible population is exposed to it. There are basically three likely routes of exposure--through the lungs (inhalation), through the mouth (ingestion), and through the skin. A gas, vapor, mist, or dust in the air may be inhaled and enter the body through the lungs just as oxygen from the air enters the body. If the substance in question gets into food or water it will be taken in through the mouth into the digestive system (ingested). This may happen, for example, if a chemical leaks out of a waste disposal site into a stream or river that is used as a source of drinking water downstream.

Thus, a major thrust of any risk assessment is the identification of those individuals who may be potentially exposed and therefore potentially injured by discharges from a facility. For example, if groundwater contamination is of concern, those persons on well water (or who could potentially be on well water) would be identified and the potential risk evaluated. While possible adverse effects to agricultural foods or injury to aquatic life will often be identified in an EIL risk assessment, the major emphasis of the report is on human exposure. From an underwriting perspective, however, it is still a matter of some debate as to the significance of the actual number of people in a potentially exposed population.

Finally, past and present management practices at a site are evaluated, as such practices can significantly reduce the probability of environmental impairment. Factors evaluated are the personnel in charge of environmental practices at the site, corporate oversite of environmental management, recordkeeping procedures, preventative maintenance, employee training, spill plans, closure plans, site security, compliance with federal and state environmental permits, lawsuits and citizen complaints, if any. In addition, above-ground storage tanks would be inspected to see if they are properly mounted and diked. PCB transformers on-site would be inspected to see if they are on concrete pads, and routinely inspected. Finally, consideration is also given to those aspects of past management practices that could impact potential environmental impairment (e.g., unauthorized dumping of liquid wastes on-site, improper landfill design).

CONCLUSION

There is an attempt to quantify the assessment by assigning a score to the facility. To assure consistency and quality control for the underwriting markets, this score reflects Environ's experience with similar operations and practices and state-of-the-art procedures. The overall evaluation also recommends modifications to current procedures to minimize exposure, wherever practical. In summary, an EIL risk assessment provides corporate management with an independent evaluation of their environmental liabilities and need for pollution coverage, and underwriters with a basis for writing coverage and determining which aspects of a company's operation should be excluded from coverage.

RISK COMPENSATION, BEHAVIOR CHOICE, AND THE ENFORCEMENT OF DRUNK DRIVING STATUTES

Lloyd Orr
Alan Lizotte
Frank Vilardo

Indiana University Institute for Social Research

ABSTRACT

Combining well-known research results from the alcohol and driving literature with the concept of risk compensation, this paper provides some insight on the relationship between public information and the effective deterrence of drunk driving. Results suggest that our public information programs, although changing in the right direction, may be seriously counterproductive.

Results from a variety of statistical studies show death and injury rates to be the same or even less at moderate levels of blood alcohol content than they are at zero levels. These results have usually been regarded as an anomaly to be explained away. But, the statistics are consistent with risk compensation theory--which implies a driver's ability to compensate behaviorally for moderate alcohol impairment. Most drivers intuitively understand this and drive within safe (and legal) limits. But, the weight of media messages is still that all drinking-driving is dangerous and probably illegal--with strong moral overtones. The result is cognitive dissonance. The resolution of this dissonance is likely to result in lenient and inconsistent treatment of drunk drivers by the community. Our media messages have little affect on alcohol abusers. They seem to affect other citizens in a counter-productive way--convincing them that their typically responsible behavior places them in legal jeopardy. Consistency with well-established research results and new survey data is demonstrated, and media alternatives are discussed.

KEY WORDS: Alcohol and Driving, Cognitive Dissonance, Drunk Driving Deterrence, Risk, Risk Compensation

INTRODUCTION

The basis for this paper is the observation that (1) risk compensation theory, (2) the numerous statistical studies on the relationship of accidents to blood alcohol content (BAC), and (3) individual experience and decisionmaking on alcohol consumption and driving, constitute a consistent and mutually reinforcing theoretical-empirical package. This package can be used as a foundation for a critique of the role of mass communication in the enforcement of drinking-driving statutes. Stated in simple terms, the data indicate that the overwhelming majority of our citizens handle drinking and driving responsibly and safely. However, the public message they receive is that all drinking-driving is dangerous, irresponsible, and probably illegal. The resulting cognitive dissonance could be an important explanation of our failure to sustain effective enforcement of Driving While Intoxicated statutes (DWI).

DRINKING, DRIVING, AND ACCIDENTS

Risk Compensation

Risk compensation is the interaction of people with their environment in the search for an optimal level of risk. Risk is optimized in many dimensions as people strive to reach complex human goals. In the presence of constraints it is necessary to balance the acceptance of risk against the attainment of other goals. To an important extent people choose their level of risk. This is a very ordinary problem in constrained utility maximization and comparative statistics.(1,2) The theorem states that, in the absence of a change in the preference for safety, a parametric improvement in the safety environment will be at least partially offset by behavior adaptations.

The usual policy implications of risk compensation lead to predictions of disappointing results from changes in the safety environment--for example, the various mandates applied to the workplace or the roadway through OSHA or NHTSA. The implicit assumption in models of such regulations, and associated benefit-cost studies, is that there are no compensating responses to the mandates. To the extent that risk compensation occurs, benefit-cost studies of safety mandates will have an upward bias that will be reflected by disappointing results in the field. The empirical importance of risk compensation is a controversial issue.(2,3,4,5) The issue is important because of the potentially large impact of the concept on our approach to the management of risk.(3,4,5)

However, the importance of risk compensation is broader than the prediction of behavior offsets to mandated changes in the safety environment. Individuals use experience in adapting to risk on a continuous basis. Their adaptation has components that range from the momentary to the very long run. Our current research deals with compensation for alcohol impairment. Risk compensation provides an important theoretical basis for well-known results of statistical studies on the relationship between BAC and accidents.

Alcohol and Physical Impairment

Most of the research investigating the amount of physical impairment related to alcohol concentrations has utilized laboratory situations. These tests have usually measured alcohol effects on such things as vision, standing ability, coordination, concentration, and comprehension.(6,7,8,9) There are also studies on actual driving performance under experimental conditions.(10,11,12) The questions raised with the first group of studies have to do with whether or not any simple or direct relationship exists between these measures and actual driving performance. In the second group of studies there is a tendency to set tasks that are difficult to do perfectly under ideal conditions or to push performance levels of specific tasks. This makes the appropriate variance in the experimental setting easier to generate, but again reduces valid inferences to the real world where individuals can influence the requirements of the driving environment.

Individual influence on the driving environment, of course, is the principle of risk compensation. Its probable effect on drinking and driving performance was suggested by Hurst (12) well before the current surge of interest in the topic. Hurst states that our information for dealing with the drinking-driving problem should be based on "epidemiological" findings--data on the relationship between BAC and highway crashes.

The pioneering work in this area is the Borkenstein, et al., 1964 "Grand Rapids Study".(14) The accident rate related to BAC does not rise rapidly until levels in excess of .06 percent are reached (Figure 1). There is, in fact, a slight decrease in the relative probability of a crash associated with low levels of BAC as compared with the standard of BAC = 0. It should be noted that when the data are divided into subgroups by self-reported drinking frequency there is a monotonic increase in the BAC-accident relationship for each subgroup. (12) However, the self reporting feature may introduce a bias that would explain the monotonic increase. Also, the general shape of the relationship does not change very much, and the low relative probability of accidents with BAC in the .00 - .06 percent range is a characteristic of other epidemiological studies.(Figure 2) This contrasts with some of the predictions and most of the folklore based on the physical impairment studies noted at the beginning of this section.

These statistical results are clearly consistent with risk compensation behavior, where the driver realizes that there is some impairment and adjusts his driving behavior accordingly. In fact, the very rapid increase in accident rates associated with increases in BAC

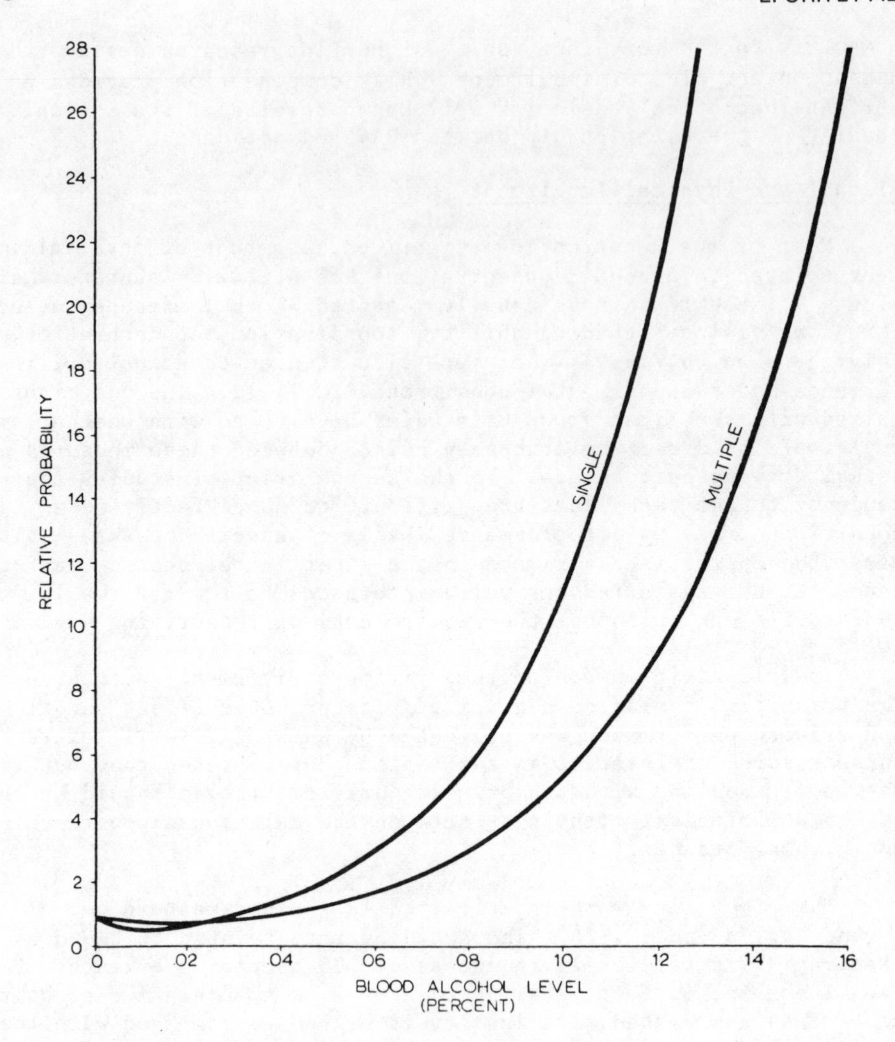

From Borkenstein et. al. (14) p. 175

Figure 1

Relative Probability of Involvement in
Single or Multiple Vehicle Accidents

above .08 may be characterized as a progressive loss in the ability to perceive and adapt to impairment.

Individual Experience and Decisionmaking

A large majority of the adult population that drives, does so with BAC's that are zero or well below the typical legal limit of .10

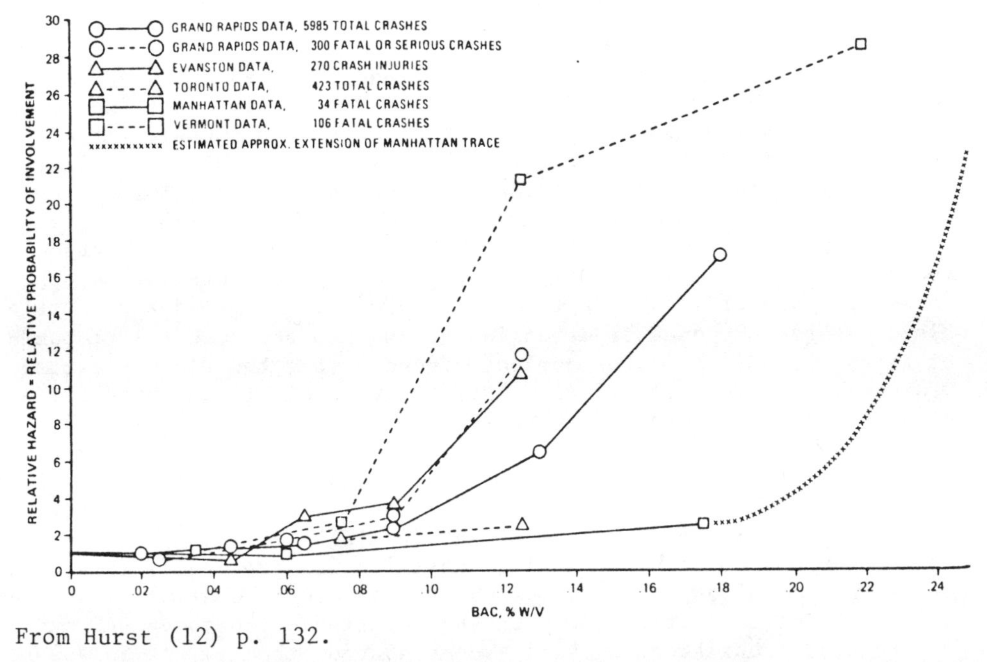

From Hurst (12) p. 132.

Figure 2

Relative Probability of Crash Involvement as a Function of BAC
Where 1.0=Relative Probability at Zero Alcohol

percent w/v. The most conservative evidence for this statement comes from studies on drinking patterns. Conclusions from these studies are fairly consistent. Of those in alcohol-related situations, 66 percent never consume more than four drinks according to one survey.(15) In another survey, 55 percent of the respondents never consumed as many as four drinks, and 78 percent drank four or more drinks two times a month or less.(16)[2]

In a study based on quantity, type, and frequency of drinking, the breakdown in drinking classifications was:(17)

Heavy	12%
Moderate	13%
Light	28%
Infrequent	15%
Abstainers	32%

Seventy-five percent were classified in the lowest three categories. Most surveys classify about 30 percent of the respondents as abstainers, although the percentage for one was as low as 16 percent.

Considering the drinking BAC relationship broadly illustrated in Table I--together with the data in the previous paragraph, and factors such as body weight and decisions not to drive after heavy drinking--a statement that 75 percent of the adult driving population rarely or never drive when legally intoxicated should be conservative.

The previously cited Grand Rapids study looks at the questions of decisions on drinking and driving from a different perspective. A control sample of 7,590 was taken in proportion to times and sites of accidents for purposes of establishing the BAC accident relationship of Figure 1. Of these, .6 percent of the sample had BACs in excess of .10 percent. The statistically "safe" BAC level of .05 percent was exceeded by 2.5 percent of the sample.

Our initial conclusions are the following: Although the problem of drunk driving is a serious one, the vast majority of drivers make responsible decisions on drinking and driving. Risk compensation in the context of constrained utility maximization provides a theoretical basis for decisions on the compatibility of moderate drinking and safe driving. The statistical work on the relationship between BAC and accidents is consistent with risk compensation theory. It is apparent that those who drive after drinking are able to effectively compensate for alcohol impairment at relatively low BAC levels.

TABLE I

RELATIONSHIP OF DRINKING AND BLOOD ALCOHOL CONTENT[1]

Body Weight in Pounds:	100			150			200		
Hours Spent Drinking:	1	2	4	1	2	4	1	2	4
Number of Drinks									
1	.03	0	0	.02	0	0	0	0	0
2	.06	.03	0	.04	.01	0	.03	0	0
4	.13	.10	.07	.09	.05	.04	.07	.03	0
6	.19	.16	.13	.13	.10	.07	.10	.06	.02
8	.26	.23	.19	.17	.14	.11	.13	.10	.07
10	.32	.29	.26	.21	.18	.15	.16	.13	.10

[1]BAC given as a percentage, rounded to two digits. A "drink" is a standard beer, 4 oz. of table wine, or 1 oz. of 86 proof liquor. (adapted from Carroll reference 18).

THE EFFECT OF PUBLIC MESSAGES ON DRINKING AND DRIVING

In recent years, both NHTSA and several state agencies have made efforts to provide some useful information on drinking and BAC, and to distinguish drinking and driving from drunk driving. However, casual observation as well as survey results demonstrate that the overwhelming content of public messages on alcohol and driving condemns all drinking and driving.

The data for our empirical work comes from a survey conducted by the Indiana University Institute for Social Research. Telephone interviews were conducted with 375 respondents selected by random digit dialing from the Indianapolis metropolitan area. Responses to a question on recall of media messages are reported in Table II.

Sixty-five percent of the respondents could not recall any message under the interview conditions. This, of course, does not imply that their behavior was free of the influence of messages. What is striking, but not surprising, is the dominance of messages recalled that rigidly proscribe drinking and driving. We would place messages 1, 3, and 4 of the most frequently recalled messages in this category. Of the 105 respondents in the first four categories, 93 percent selected 1, 3, or 4. This is consistent with some results from pretesting the question on a group of males concentrated in relatively high income and education levels. A higher proportion of the sample recalled messages--65 percent compared to 35 percent in Table II. But 38 of 41 respondents (93 percent) recalled proscriptive messages. The apparent purpose of these messages is to control behavior by warning or threat, rather than by meaningful information on safety or legality.

This issue was addressed by Harold Mendelsohn in an address to an International Symposium on Alcohol and Driving.[3] Professor Mendelsohn was pointed in his denunciation of typical media traffic safety messages. In his address, he touched on several of the elements of our critique. Audiences have the right to accurate information, and they

TABLE II

Media Message Recalled	n	%
1. If You Drink Don't Drive.	78	22
2. Drinking and Driving Don't Mix.	14	4
3. Friends Don't Let Friends Drive Drunk.	7	2
4. Gasoline and Alcohol Don't Mix.	6	2
5. Other	20	6
6. No Message Recalled.	231	65

n = 356

have the right not to be insulted, told lies, or subjected to a missionary attitude that demands unthinking response. "Most of your audience does not consume alcohol at all, or consumes it wisely. Stop turning them off!"

Our characterization of the potential harm of such messages is in the cognitive dissonance created in the audience. Dissonances always aroused by a discrepancy between two or more relevant cognitive elements, where one deals with behavior and the other with the internal or external environment. On the one hand, there is the decision to drive after drinking. For most people this decision is based on sound judgment and is within safe limits. On the other hand is the overwhelming media message that these decisions are dangerous, immoral, and probably illegal.

Figure 3 models some potential effects of these messages. It focuses on two ways through which the tension created by the cognitive discrepancy may be reduced. The first is the assumption of effective risk compensation--the judgment that moderate drinking and safe

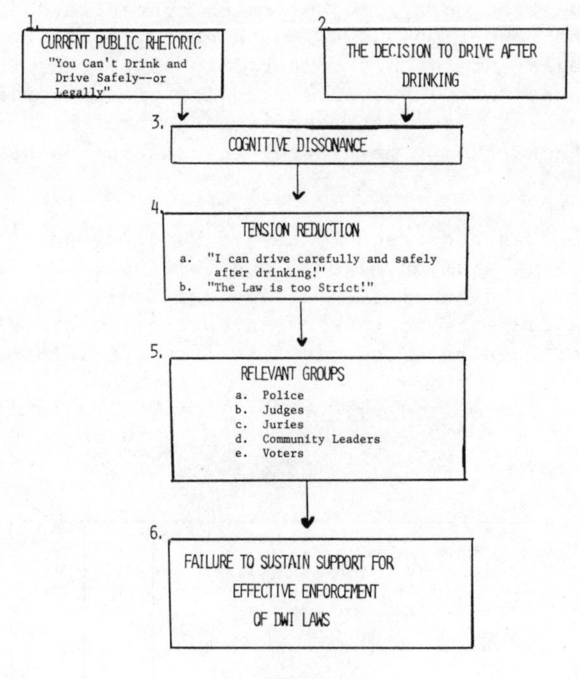

Figure 3

The Generation and Resolution of Cognitive Dissonance on Drinking and Driving

driving can be combined! We have argued that this view is substantially correct.

A second means of reducing tension is the socially destructive conclusion that the law, as it is understood, is too strict. This phenomenon has long been recognized in the alcohol research community, but with little insight into origins or solutions. It is characterized by the attitude toward the DWI arrestee that is captured in the phrase, "There, but for the grace of God, go I!"--the fear of the typical citizen of his own legal jeopardy, and his consequent tolerance of the drunk driver.[4]

Although the citizen is the backbone of law enforcement in a free society, it should be clear that judges, juries, and policemen are also a direct subject of media messages. Some of the Alcohol and Safety Action Programs funded by NHTSA over the past decade experimented with groups of judges throughout the country. More than half of the judges regarded themselves as legally intoxicated after being given two drinks in a one-hour time period, and thought that a legal limit of BAC = .10 percent w/v was too strict. Let us add for emphasis that these are the individuals sentencing people convicted of DWI violations.

For an additional illustration, a visit to Indiana University by a County Sheriff's Department research officer resulted in impromptu comments that were amazingly congruent with our hypothesis. Knowing only that we were engaged in a research project on alcohol and driving, the officer made the following comments: "If You Drink, Don't Drive" type of slogans are counterproductive. They indicate that anybody who drinks and drives should be arrested--an impossible task. They lessen citizen support for police, and, thus, police effectiveness. Both of these factors tend to lower police morale and create a "why bother" attitude.

We should note, further, that the average BAC of a DWI arrestee is in the range of .17 percent to .20 percent. Locating these values in Table I should persuade most people that there is indeed a substantial difference between the DWI arrestee and the typical adult social drinker. Mass education policies have, according to our model, created a destructive myth. The failure of DWI laws as a deterrent is certified by the high average BAC associated with arrest and by the high proportion of highway deaths associated with legal intoxication. The DWI problem is currently receiving a lot of media, policy, and legal attention. The attitude of most experienced researchers is "We've been through these cycles before."

EMPIRICAL EVIDENCE

The hypothesis that media messages lessen community support for effective law enforcement has testable implications. We have just begun the analysis of the previously cited survey data from the Indianapolis metropolitan area and can report some initial results.

One result of media messages that proscribe rather than inform should be that people overestimate their legal jeopardy when they drink and drive. Numerous surveys show this phenomenon, although most do not address the question precisely. Our own survey included questions on the estimated number of clearly defined "drinks" to reach the legal limit (BAC = .10 percent w/v in Indiana). We also asked for the respondent's body weight. We were thus able to calculate the implied BAC of the individual's response to the drinking limit question as a ratio to the legal limit. Table III reports the results for a one-hour drinking period.

Interpreting these results, we note the following:

- Forty percent of the respondents overestimated their BAC by 100 percent or more
- Sixty-six percent of the respondents overestimated their BAC by 50 percent or more
- Eleven percent estimated limits that were above the actual legal limits

These results are consistent with previous survey results. When asked to estimate three-hour limits, the results were markedly higher ($\bar{x}$ = .813) and not consistent with previous survey findings. Perusal of the data suggests a "learning problem" in the survey. That is, having given a one-hour estimate, many respondents simply tripled it

TABLE III

Ratio of Estimated to Actual One-Hour Drinking Limit	f	
0 - .30	28	
.31 - .40	27	
.41 - .50	24	
.51 - .60	38	
.61 - .70	22	
.71 - .80	24	n = 199
.81 - .90	8	$\bar{x}$ = .641
.91 - 1.00	6	
1.01 - 1.10	5	Med. = .546
1.11 - 1.20	4	
1.21 -	13	

for a three-hour period--resulting in a much higher three-hour estimate than would have been given in isolation. We will check this observation by splitting the sample for purposes of asking one-hour and three-hour questions on the next survey. The question of degree of underestimation of legal limits is important in itself for our hypothesis. However, we want to emphasize that use of the one-hour or three-hour limit makes little difference in the covariate analysis that follows.

We have built an elaborate multi-equation model testing various interactions of legal limit estimates, willingness to drive after drinking, and willingness to support enforcement of DWI statutes. All of these relationships are conditioned on respondents' socio-economic characteristics. Thus far, we have estimated only isolated parts of the model using simple regression techniques. The results are nevertheless worth reporting, and are presented in Table IV.

These results support implications of our hypothesis. The greater the respondents' underestimate of legal drinking limits, the more likely they are to drive after reaching those limits, and the smaller their perceived likelihood of being stopped and arrested. Most important, the greater the respondents' underestimate of legal drinking limits, the less they would support tougher DWI statutes. We emphasize the preliminary status of this work and the need for further analysis of our survey results and other data sources.

TABLE IV

Dependent Variable	Independent Variable	B Coeff.	F
Willingness to Drive after Reaching Estimated Legal Limit	Ratio of Estimated to Actual Legal Drinking Limit	-1.04	35.9
Perceived Likelihood of Being Stopped and Arrested if Driving after Reaching Estimated Legal Limit	Ratio of Estimated to Actual Legal Drinking Limit	.92	10.2
Willingness to Support Tougher DWI Laws	Ratio of Estimated to Actual Legal Drinking Limit	.32	10.96

ALTERNATE POLICY

Our present media messages seem to have little effect on problem drinkers. If further empirical work sustains the preliminary conclusion that messages affect other citizens in counterproductive ways, then a strong shift in our approach to educating the public on drinking and driving would be indicated. Risk compensation and the epidemiological studies suggest that we reenforce the behavior of the relatively responsible citizen as a means of building support for removing truly dangerous drivers from the highways.

Figure 4 is an illustration of this approach. This would not be a simple task. Box 4 attempts to address some of the issues that would arise in connection with legitimate concern over the implied permission to drink and drive inherent in such a policy.[5] There is research being done on effective and accurate means of conveying information on the complex relationship between drinking and BAC.

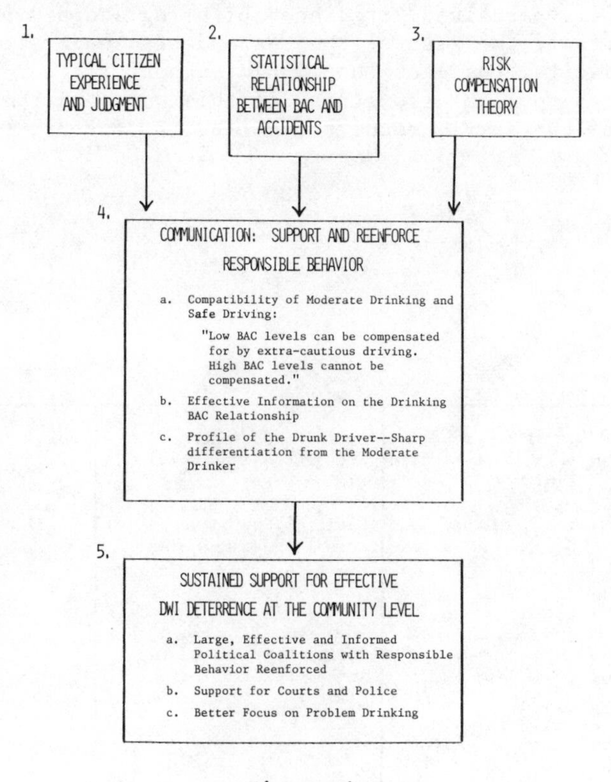

Figure 4

Reducing Dissonance: Information Consonant with Reality

Although it is not the purpose of this project to determine specific deterrents to drunk driving, the most persuasive current research suggests that we need to raise the probability of arrest. It currently takes 45 to 90 unpleasant minutes to make a DWI arrest. Per se laws, mobile breathalyzer units, hand-held breathalyzers, citing rather than arresting offenders, and mandatory sentences, are techniques designed to make things easier and more rewarding to the police officer. Stiff sentences are probably counterproductive, at least in the absence of a very strong consensus on the appropriateness of the law. Placing the BAC limit below .10 percent, as some states have done, may also be a mistake. We would do well, in terms of Figures 1 and 2, to enforce the higher limits most states now have. However, the specific solution is less important than sustained support at the community level to find programs that deter the relatively small group of citizens who are persistent and serious violators.

CONCLUSION

The first point to be made clear is that this research is at an early stage. Secondly, a "successful" completion of this project will simply lead to an elaborate, and perhaps necessary rationale for doing what we ought to be doing, given some very basic tenets from Theories of Law and Information. We are simply suggesting that law and public policy tend to work better through informed consent of the governed. Who would argue against accurate information as an important basis for compliance with a law? Should we favor moralistic propaganda that has little to do with the law or social reality? Should we expect such an approach to be an effective means of gaining compliance in a pluralistic society that is dependent on states and local communities for enforcement? Who is surprised when we try to control through misinformation--and manage to shoot ourselves in the collective foot? The work we are pursuing deals with a narrow area of public policy. However, it may have broad application to situations where safety-related information misrepresents the hazard in an attempt to control behavior. Compliance, morale, and even risk may be adversely affected.

REFERENCES

1. Barry O'Neill, "A Decision Theory Model of Danger Compensation," Accident Analysis and Prevention, Vol. 9 (1977), pp. 157-165.
2. Sam Peltzman, "The Effects of Automobile Safety Regulation," Journal of Political Economics, Vol. 83 (August 1975), pp. 677-725.
3. John Adams, "The Efficacy of Seat Belt Legislation," Department of Geography, University College, London (January 1981).
4. Lloyd Orr, "Incentives and Efficiency in Automobile Safety Regulation," Quarterly Review of Economics and Business, 22, (1982), pp. 43-65.

5. G. Wilde, "The Theory of Risk Homeostasis: Implications for Safety and Health," Risk Analysis, Vol. 2, No. 4 (December, 1982), pp. 209-225.
6. J.A. Carpenter, "Effects of Alcohol on Some Psychological Processes," Quarterly Journal of Studies on Alcohol, 23 (1962), pp. 274-314.
7. H. Moskowitz, and D. DePay, "The Effect of Alcohol Upon Auditory Vigilance and Divided Attention Tasks," Quarterly Journal of Studies on Alcohol, Vol. 29 (1968), pp. 54-63.
8. M.W. Perrine, "Alcohol Influences on Driving-Related Behavior," Journal of Safety Research, Vol. 5 (1973), pp. 165-184.
9. H. Wallgreen, and H. Berry, Actions of Alcohol, Vol. I: Biochemical, Psychological Aspects, Amsterdam Elsevier (1970).
10. L. Bjerver, and L. Goldberg, "Effect of Alcohol Ingestion on Driver Ability; Results of Road Tests," Quarterly Journal of Studies on Alcohol, 11 (1950), pp. 1-30.
11. M. Huntley, "Alcohol Influences Upon Closed-Course Driving Performance," Journal of Safety Research, 5 (1973), pp. 145-164.
12. P. Hurst, "Epidemiological Aspects of Alcohol in Driver Crashes and Citations," Journal of Safety Research, 5 (1973), pp. 130-148.
13. National Highway Traffic Safety Administration, 1978 Alcohol and Highway Safety (1968).
14. R.F. Borkenstein, R.F. Crowther, R.P. Shumate, W.B. Ziel, and R. Zylman, "The Role of the Drinking Driver in Traffic Accidents," Department of Police Administration, Indiana University (1964).
15. Grey Advertising Agency, Communication Strategies on Alcohol and Highway Safety, 2 Vols., NHTSA (1975).
16. A. Wolfe, and M. Chapman, An Analysis of Drinking and Driving Survey Data, Ann Arbor, University of Michigan Highway Safety Research Institute (1973).
17. D. Chalan, I. Cisin, and H. Crossley, American Drinking Practices, New Brunswick, NJ. Rutgers Center of Alcohol Studies (1969).
18. C.R. Carroll, Alcohol Use and Abuse, Wm. L. Brown, Co., Dubuque, IA (1970).

FOOTNOTES

1. For a comprehensive review of alcohol and driving, see reference 13.
2. These are averages of the survey results gathered in 1971 and repeated in 1973.
3. Washington, DC, November 17-18, 1982. Professor of Mass Communications, University of Denver.

4. Professors Joseph Palladino, Indiana State University--Evansville, and Bernardo Carducci, Indiana University Southeast, have conducted research that confirms this phenomenon empirically. We have had only telephone communication on this research and have not yet received a report.
5. Cognitive dissonance may be resolved in part by drinking less when driving is planned. The solution to this potential problem must be based on effective impairment information, improved law enforcement, and perhaps lower legal limits if the problem occurs and sufficient community support can be generated.

COGNITIVE MAPS OF RISK AND BENEFIT PERCEPTIONS

Adrian R. Tiemann*
Jerome J. Tiemann**

*Digital Interfaces & Systems
2 Wilburn Avenue
Atherton, CA 94025

**Electronics Research Laboratory
Stanford University
Stanford, CA 94035

ABSTRACT

The work reported herein is part of a continuing effort to delineate important dimensions for the measurement of public approval and/or disapproval of various activities in which society is engaged. A battery of questions designed to elicit opinions relating to various types of risks and benefits is posed to the respondent for an ensemble of activities and technologies. If these questions are approached in terms of a dominant cognitive map, answers to specific pairs of questions will be correlated over the ensemble of activities and technologies.

A procedure derived from factor analysis was used first to identify the principal components of the correlation matrix of each subject's responses, and then to analyze these components in terms of four major themes (or cognitive maps). These themes can be described as: (1) Polarized, (2) Benefit oriented, (3) Risk oriented, and (4) Trade-off oriented. A respondent characterized by a "Polarized" cognitive map sees activities having high benefit as having a low element of risk (and vice-versa), and therefore, appears to be totally in favor of the activity or technology or totally opposed to it. A person with a "Benefit" orientation evaluates an activity as having either many benefits at once or relatively few. Similarly, a "Risk" oriented respondent will see an activity or technology as having either many risks or very few risks. The "Trade-off" respondent sees many activities that are acknowledged as risky as having more than average benefits and many low risk activities as having lower than average benefits.

Survey items tapping perceptions of benefits and risks of a variety of activities and technologies derived from previous research were used in preliminary studies, and the results of these studies will be presented. Since observed factor loadings are used to weight the questions employed, this technique is more independent of specific assessment questions than previous approaches. Because of this, questions which do not require complex judgments on the part of the respondent can be used. The stability of the method was checked by varying the questions, and the activities and technologies with respect to which they were asked. Further stability checks were obtained by using different groups of respondents. Consistent results were obtained in over 75 percent of the cases, suggesting that the cognitive mappings uncovered in this research are both pervasive in the general public and important determinants of perceptions of risk and benefit.

INTRODUCTION

This paper is based on the assumption that people usually approach complicated tasks by first simplifying them as much as possible; then they deal with the simplified versions. We also assume that dealing with the myriad decisions concerning risk and benefit that people must constantly make is one of these complicated tasks, and that people simplify them greatly. One might argue that to do otherwise would require so much time and effort as to severely limit the amount of one's participation in the activities under consideration! We will attempt to show in this paper that one of the strategies people use to simplify the problem of perceiving and evaluating risks and benefits is the formation of a modest number of relatively simple cognitive maps, and that many perceptions of seemingly independent aspects of risk and benefit perceptions are thereby linked together at the individual level. A conceptual model of this mechanism is shown schematically in Figure 1, in which an individual's social/psychological characteristics and perceptions influence the formation of values and beliefs; these, in turn, determine the formation of cognitive maps, which then operate to influence that individual's perceptions of further experiences. The closed loop character of this mechanism is corroborated by the observation that incompatible perceptions of some activities exist among individuals, and that these perceptions conform to and reinforce the underlying incompatible beliefs. Since the model does not consider the effects of external events on the individual's sociodemographic or psychological characteristics, it is not complete. The model does show (via lightly dotted lines) some effect of an individual's beliefs on his/her social and personal characteristics, but this is not explored further. The aspect of the model which is the subject of the present research is the role of cognitive maps in

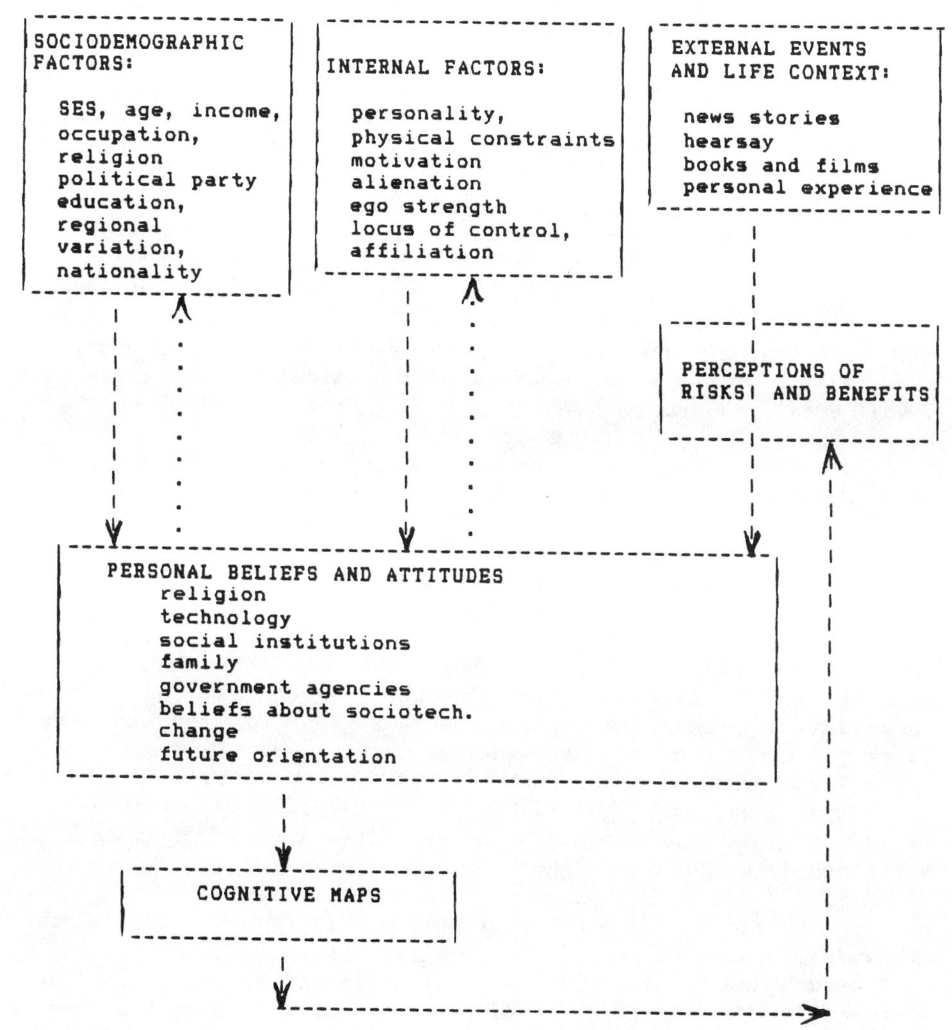

Figure 1. Schematic Diagram Depicting the Interactions Between Perceptions, Beliefs, and Attitudes, and Cognitive Maps.

simplifying the application of large numbers of values and beliefs to an even larger number of cognitions.

Previous research on the subject of the relationship between perceptions of risks and benefits has been aimed at determining how much risk is acceptable, and has sought to measure the correlation between various risk and benefit measures. Starr,[1] using aggregate capital investment as a measure of benefit and mortality figures as an indicator of risk, found a positive correlation.[1] This result was disputed by Otway, Cohen, et al.,[2] Slovic et al.,[3] and by Gardner et al.,[4] the latter two of whom found weakly negative relationships at the aggregate level between perceptions of overall risk and overall benefit of various activities by groups of respondents.

Other studies suggest that "societal pressures appear to be different for risks to which society has grown accustomed and for those risks that are new and potentially catastrophic,"[5] and that technologies and activities embodying these risks appear to acquire different cognitive characteristics to their perceivers. In other words, the relationship of risks and benefits of a particular activity is conditional on other perceived attributes of that activity.

Moreover, a variety of sociodemographic characteristics appear to be correlated with differing perceptions of certain technologies and activities. One of these characteristics is group membership[4] or what are called "reference groups"--those collections of people with whom significant characteristics are shared, and to which appeal is made in identifying oneself.

The current work addresses these same issues by attempting to identify and quantify the "cognitive maps" which underlie individual responses among selected populations by straightforward survey methods using generally accepted mathematical techniques. This latter restriction reminds one that risks and benefits need to be assessed for policy purposes,[6] and that risk assessment techniques "must be abstract and general...[and] must be presented so that it is rational for somebody to choose on this basis."[7]

We, therefore, examine two testable questions: first, whether an individual applies a consistent cognitive approach in evaluating the risks and benefits of a wide range of different activities, and to what extent such cognitive mappings will vary across groups of people occupying different status and role positions in society. Second, one questions whether groups of people perceive certain activities and technologies in one risk and benefit fashion, and have quite a different set of perceptions about other activities and technologies.

Both of these questions require measurement of the cognitive map used by the subject for risk and benefit perceptions, and we will use the first two principal factors of a covariance matrix for this purpose.

Although it holds the promise of being both abstract and general, factor analysis has been widely criticized. Otway, for example, questions "the existence of underlying dimensions whose validity is dependent upon complex techniques, none of which are anything other than a function of the scales initially devised by the researcher." (8, p.

77) In his studies of attitudes toward energy technologies, he found that "the structure and interpretation of the dimensions varied with the number and combination of energy sources compared at any one time. (p. 78)

However, we believe that factor analysis results are stable under certain conditions, and that these conditions can be fulfilled within the framework of conventional survey methods at reasonable cost. If the survey instrument is designed so that the first few principal components of the covariance matrix focus on the desired cognitive dimensions, the results will, contra Otway[8], usually "generalize across populations or risk sources." But, if the first factor relates to only one of the cognitive dimensions or to an average response pattern of the subjects, so that the interesting features of the analysis occur in factors explaining less variance, the results are less likely to be reproducible. We simply note that although researchers around the world may give different names to the factors they elicit when studying similar data, and that the resulting factors may differ in detail, the principal factors corresponding to the largest eigenvalues contain many elements in common and support many of the same conclusions. (The same is not true, however, of factors with smaller eigenvalues.) Two major precautions are needed to assure such an outcome: First, one must either use a relatively large number of overlapping questions so that one is certain that all relevant cognitive dimensions have been tapped, and balance them so that no one dimension explains significantly more variance than any other; or one must use pretested questions that are known to tap these dimensions efficiently. Second, one must check the results for robustness by collecting data in different ways from different groups of respondents.

On the subject of the dimensionality of survey questions, it seems to us that Otway and Thomas are correct when they assert that "probabilistic information cannot be studied without reference to context. Meaning, relevance, and other motivational factors are crucial variables in 'judgment seeking' questionnaires." (8, p. 72-73) For this reason, we feel that many additional issues should be tapped when designing a survey instrument for risk/benefit studies. Not only should questions be designed to distinguish the different characteristics of risk, but equal numbers of questions should be included to tap the benefit, belief, and value dimensions of the respondents' perceptions.

In the study reported on below, the strategy outlined above has been followed to the extent that both risk and benefit dimensions have been included. But, having turned in this direction, it appears that further steps to broaden scope of risk/benefit/value studies would be worthwhile.

THE PRESENT RESEARCH

Samples and Method

Six groups of respondents living in the northeast are reported upon here. The first group is composed of 42 employees of a high technology manufacturing firm, including managers, technicians, scientists, and support staff. Samples 2 through 6 comprised respectively: 66 college students, 76 blue collar workers, 80 business people, 68 technologists, and 71 environmentalists. These latter samples were obtained in two different locations in roughly equal numbers, and were drawn in a set, non-random manner. This aspect of the procedure will be discussed further below.

Respondents in Sample 1 were asked to evaluate 30 activities and technologies nearly identical with those of Slovic et al.[3] Respondents in Samples 2 through 6 were given almost the same tasks, but only 17 technologies/activities were evaluated (see Table I). For all six samples, three task sets covering a total of nine cognitive dimensions were involved. The first task was to evaluate how the restrictions and standards applying to the activity or industry should be changed, where industry was loosely defined to include not only the direct products of the industry, but also the use of its products in pursuing other activities. A scale ranging from one (much more lenient) to five (much more strict) was used for evaluation. This question measures the perception of how well the risks in question are being managed at present.

The second set of tasks comprised evaluating four presumed benefits derived from operation of the technology or engaging in the activity. The four benefits--economic, contribution to basic human needs, ability to provide security or safety, and contribution to the higher well-being of the community and to the individuals within it--were assessed on seven-point scales. On these scales, a "1" represented "no contribution" and "7" represented a "very great contribution". Respondents in Samples 2 through 6 were asked to rate "overall benefit" on a 90 mm scale rather than "the contribution to well-being" on a seven-point scale.

The third set of tasks directed respondents to consider the hazards and risks accompanying the operation of a technology/use of its products, and engaging in the activities. Four risks were then assessed on seven-point scales, with "1" representing no risk, and "7" representing "very great" risk. The risks were: the amount of dread inspired by the technology or activity; the potential that operation of the industries and their products, or engaging in the activity had for causing catastrophic death and destruction; the extent to which the hazards associated with these technologies/products and activities were not known or understood by scientists; and finally, loss of life. Respondents were asked to estimate the annual number of fatalities in the United States on a log scale: 10, 100, 1000, etc.[9]

TABLE I. TECHNOLOGIES AND ACTIVITIES EVALUATED

SAMPLE 1	SAMPLE 2,3,4,5,6
ALCOHOLIC BEVERAGES *	--
BICYCLE RIDING	BICYCLE RIDING**
COMMERCIAL AVIATION	COMMERCIAL AVIATION
ELECTRIC POWER	ELECTRIC POWER
FIRE FIGHTING	--
FOOD PRESERVATIVES	--
GENETIC ENGINEERING	GENETIC ENGINEERING
HANDGUNS	HANDGUNS
H.S. & COLLEGE FOOTBALL	--
HOME APPLIANCES	--
HUNTING	--
LARGE CONSTRUCTION	--
MOTORCYCLES	--
MOTOR VEHICLES	MOTOR VEHICLES
MOUNTAIN CLIMBING	--
NUCLEAR POWER	NUCLEAR POWER
PESTICIDES	PESTICIDES
POLICE WORK (ON DUTY)	--
POWER MOWERS	POWER MOWERS
PRESCRIPTION ANTIBIOTICS	--
PRIVATE AVIATION	--
RAILROADS	--
SKIING	--
SMOKING	TOBACCO
SURGERY	--
SWIMMING	--
TOXIC WASTE DISPOSAL	--
VACCINATIONS	VACCINATIONS
WEAPONS (NUCLEAR)	WEAPONS (NUCLEAR)
X-RAYS (DIAGNOSTIC)	X-RAYS (DIAGNOSTIC)
--	SACCHARIN
--	COSMETICS
--	SPACE EXPLORATION
--	TRACTORS

*Toxic waste disposal, nuclear weapons, and genetic engineering were not on the original Slovic et al. list; alphabetically in their places were contraceptives, food coloring, and spray cans.
**Two different random presentation orders were used with Samples 2-6 to check order effects.

Statistical Summary of Results

The data were subjected to simple statistical procedures, yielding means and standard deviations for various groupings. Two questions motivated these analyses: first, how do perceptions differ between identifiable groups of people, and second, how do perceptions of a single case (or of a group of cases) differ with respect to different technologies and activities.

A summary of the means of the six sample groups and the standard deviation is given in Table II. The column labeled "standard deviation by individual" characterizes the distribution of all 409 responses, while "standard deviation by group" is the average of the standard deviations for each group.

TABLE II. SUMMARY OF RESULTS: MEANS AND STANDARD DEVIATIONS OF COGNITIVE DIMENSIONS (ACROSS ALL TECHNOLOGIES/ACTIVITIES) FOR SAMPLES 1 THROUGH 6

COGNITIVE DIMENSION Sample	MEANS 1*	2	3	4	5	6	STANDARD DEV indiv	group
Desired Restric.	3.42	3.49	3.49	3.25	3.18	3.58	0.92	0.45
Economic Benef.	4.04	2.99	3.45	3.25	3.09	2.99	2.00	1.19
Human Needs	3.74	2.83	3.32	3.28	3.10	3.05	2.12	1.52
Security Needs	3.25	2.74	3.37	3.31	3.01	2.79	2.11	1.41
Higher Well Being	3.86	--	--	--	--	--	1.91	1.13
Overall Benefit	--	4.57	5.19	5.13	5.08	4.59	2.99	1.97
Means	3.66	3.28	3.83	3.74	3.57	3.35		
Dread Risk	3.56	2.89	3.07	2.93	2.79	3.10	1.96	1.33
Catastr. Death	2.45	2.95	3.24	2.94	2.65	2.97	2.00	1.27
Known to Science	2.58	2.78	2.74	2.48	2.49	2.86	1.79	0.92
Fatality Est.	2.30	3.10	2.99	2.69	2.53	3.10	1.78	1.18
Means	2.72	2.93	3.01	2.76	2.61	3.01		

* Group 1 evaluated 30 technologies, while Groups 2-6 rated 17. Group 1 answered one benefit question dealing with a technology's contribution to "the higher well-being or joie de vivre of life," while in its place, Groups 2-6 rated "overall benefit."

Except for Group 1, each of the sample groups was comprised of two roughly equal subgroups which differed in substantial ways. Not only were the subgroups located in different areas, but one was about 50 percent men and 50 percent women while the other was 100 percent men. Despite their differences, the two subgroups were statistically similar in four cases out of five. Significant differences appeared between the two subgroups of Group 6, the Environmentalists, but this finding was not pursued further.

As can be seen in the above table, there is little difference between the means of the sample groups in comparison to the spread within each group. But since these are not random samples, significance tests are inappropriate and are not reported. Instead, let us note that the entire spectrum of perception appears to be represented in all of the groups with only modest distributional differences. The most obvious difference among the various groups is that Group 3 (Blue Collar Workers) is consistently higher than the others on all benefit dimensions and is also high on the average for risks.

Much is made of the supposed differences between "experts" with intimate knowledge of technology and lay persons lacking this knowledge. Technologists (Groups 1 and 5) consistently show the lowest concern about catastrophic potential, but their remaining perceptions do not differ greatly from those of others.

Results of Covariance Analysis

Before proceeding to our major results, let us note that the survey included questions on the overall benefit and overall risk of an ensemble of activities and is, therefore, amenable to comparison with the results of Slovic et al. Those authors found a slight negative correlation between these characteristics when individual perception measures were aggregated and correlated over activities, but expressed some concern as to the robustness of their finding. We find a similar result for all six groups in our sample, with correlation coefficients ranging from -.04 to -.22 (Slovic et al. found $r = -.20$).

We next turn to the results of the factor analyses. A major aim of the methodology to be described was to avoid committing the "ecological fallacy" by mixing levels of analysis (group level) with levels of interpretation (individual level). Consequently, correlations between the cognitive dimensions were first computed at the individual level using the ensemble of activities and technologies. Each of the resulting covariance matrices was then factor-analyzed, and varimax rotation was performed in two dimensions. This resulted in two sets of rotated factor loadings which, in almost all cases, could easily be identified as a risk theme and a benefit theme. That is, one set of loadings was consistently high on all risk dimensions and low on all benefits (risk theme), while the other set was consistently high on all benefit dimensions and low on all risks (benefit theme). Desired restrictions also loaded heavily on the risk theme.

The observed loadings were classified on the basis of the dot product between the observed rotated loadings and predetermined "prototype" themes, and class averages were computed. These averaged risk and benefit theme loadings are presented in Table III for each of the six sample groups in the study. Except for a few isolated exceptions, all groups yielded similar results. Note, for example, the light loading for Group 4, the Business people, on Desired Restrictions (.285) and the light loading for Group 5, Research Technologists, on the risk from lack of knowledge (.178). The loadings for all benefit dimensions and the Dread and Catastrophic Potential risk dimensions are consistently high and about equal, confirming that these characteristics are good indicators of the dimensions they are intended to tap. The fact that desired level of restrictions, risk due to lack of knowledge, and fatality estimates do not always load heavily on the risk theme, however, indicates that these dimensions are not highly correlated with the more specific risk dimensions for some groups, and that they are, therefore, not always reliable measures of perceived risk.

TABLE III. RISK AND BENEFIT THEME LOADINGS

A. AVERAGE LOADINGS ON RISK THEME:

GROUP	DES. RES.	$ BEN.	BASIC BEN.	SAFETY BEN.	WELL-BEING	DREAD RISK	CATAS. RISK	NOT KNOWN	FATAL EST.
1	.326	-.086	-.020	.051	-.041	.486	.613	.472	.217
2	.450	-.003	-.059	.065	-.018	.544	.563	.303	.289
3	.310	-.042	-.064	.100	.029	.507	.559	.273	.493
4	.285	-.053	-.045	.096	-.039	.528	.563	.322	.451
5	.339	-.069	-.056	.163	-.006	.562	.594	.178	.387
6	.449	-.042	-.062	.071	-.065	.515	.534	.274	.397

B. AVERAGE LOADINGS ON BENEFIT THEME:

GROUP	DES. RES.	$ BEN.	BASIC BEN.	SAFETY BEN.	WELL-BEING	DREAD RISK	CATAS. RISK	NOT KNOWN	FATAL EST.
1	.151	.532	.523	.438	.460	.061	-.076	.002	.091
2	.007	.456	.546	.384	.555	-.030	-.066	.065	.171
3	.113	.501	.563	.409	.487	-.028	-.075	.093	-.005
4	.139	.500	.555	.370	.529	-.004	-.049	.038	.034
5	.046	.525	.541	.360	.511	-.060	-.072	.109	.136
6	.027	.514	.540	.403	.511	.016	-.044	.090	.096

The procedure described above is robust because it uses only the first two principal factors of the covariance matrix which are generally more stable than the remaining ones. Furthermore, it automatically "rejects" any characteristic that does not contribute to the explanation of variance. There is, therefore, very little penalty for including irrelevant questions in the survey instrument as these will be assigned very low weights in both themes. Redundant questions are also handled gracefully; these will "share" the weight. But one must be careful not to unbalance the mix of questions, as a large number of questions tapping a single dimension can "create" a factor along that dimension.

Having found that the space spanned by the first two principal factors contains a substantially pure risk and a substantially pure benefit dimension, let us return to the subject of covariance. This space is schematically depicted in Figure 2a, which shows two orthogonal directions (vertical and horizontal) representing the first two principal factors as well as two rotated directions which point in the directions of the conceptual dimensions (risk theme and benefit theme). Each of the rotated dimensions can be viewed as a linear combination of the principal factors, and conversely, each of the principal factors can be thought of as a linear combination of the conceptual dimensions. This latter view is depicted in Figure 2b, where the risk and benefit themes now form the coordinate system. Since the first principal factor "explains" the largest amount of the variance of the individual's responses, we may ask what combination of the conceptual dimensions "explains" the most variance. That is, we can characterize the first principal factor of the covariance of each respondent in terms of its components along the risk theme and benefit theme directions. Alternatively, it can be characterized by its magnitude and the angles it forms with the risk and benefit theme directions. Conventionally, the principal component is normalized such that the square of its length on a vector plot is proportional to the variance it explains. Thus, the square of its projection onto any other axis can be thought of as the explanatory power of that axis. Close alignment of the first principal factor with the benefit theme direction, for example, means that that individual sees activities as having either many benefits at the same time or very few, while risks are perceived in a different manner. Conversely, alignment between the first principal factor with the risk theme direction indicates that an individual perceives activities as either having many risks at once or very few, while the benefits are not uniformly perceived.

Frequently, it is found that the direction of the principal factor lies approximately halfway between the directions of the risk and benefit themes. That is, its risk theme and benefit theme components are about equal. For cases where the loadings for risks and benefits have the same sign, risk and benefits predominantly vary together. That is, activities that are perceived as having above-average risks are often also perceived as having above-average benefits, while activities having less than the average risk are often

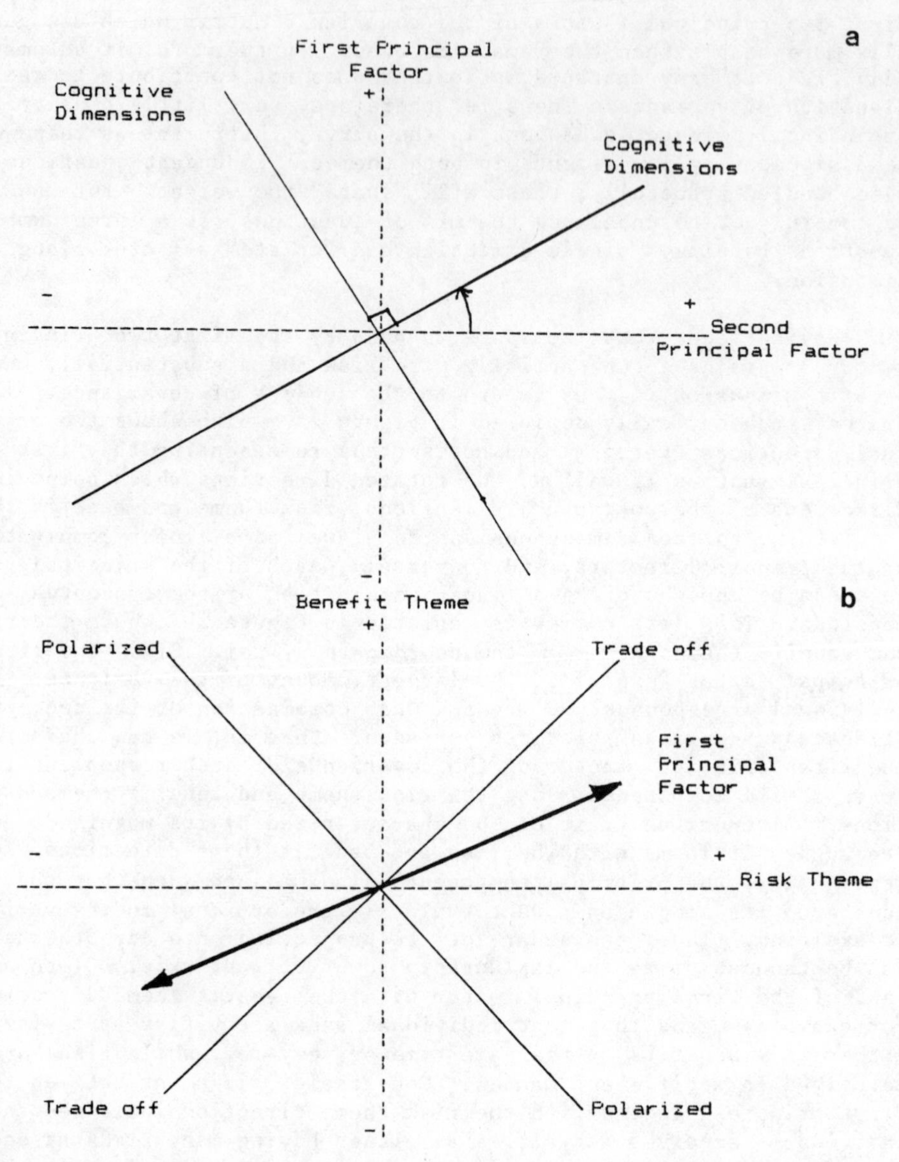

Figure 2a. Space of First Two Principal Factors and Cognitive Dimensions

Figure 2b: The First Principal Factor (which explains the most covariance) is viewed as a linear combination of the cognitive dimensions. The alignment of this Factor indicates the cognitive orientation of the subject.

perceived as having less than the average benefit. We have chosen to call this orientation a "Trade-off" approach. The opposite case, wherein a subject perceives activities as being both highly risky and low in benefits or highly beneficial and low in risks is termed a "Polarized" approach.

It is of interest to note that risk theme-weighted responses and benefit theme-weighted responses do not show the slight negative correlation exhibited by overall benefit and overall risk. The observed correlations for the six sample groups ranged from $r = -.05$ to $r = +.09$, and the mean was slightly positive. Since this result may be thought to be qualitatively different from the previously cited results of Otway et al. and Gardner et al., its origin was investigated. It was found that the increase in correlation was due to the inclusion of safety benefits in the benefit theme. In all of the sample groups, weapons-related technologies were rated high on both risks and on contribution to safety and security benefits, and this correlation was enough to cancel the negative correlation cited earlier. To some extent, this may be an artifact of the questionnaire, which directed the respondents to consider such security benefits as "preventing foreign invasion" in making their judgments.

Scatter plots of the risk theme and benefit theme components of the first principal factors of respondents are shown in Figure 3. Each of the six groups is shown as a separate plot, and in each plot the risk theme lies along the horizontal axis, while the benefit theme is vertical. Each case is plotted as a pair of small squares, and the distance of these squares from the origin is a measure of the amount of variance the principal factor explains. Two squares are used to represent each case because variance is a symmetrical function, and the eye would be misled by a single-sided display of the data. (This presentation is not suitable for many purposes, since it essentially maps the mean responses of all cases to the origin. It is not a substitute for a factor score plot.)

It is interesting to note that the radii of the (more or less circular) clusters of points are about equal for all groups, and that roughly the same amount of variance is explained no matter what angle the points fall on with respect to the theme axes. Upon closer examination, however, more variance is explained when the points are in the "Polarized" or "Trade-off" directions than when they occur along the "Risk" or "Benefit" directions. That is, the points can be seen to fall along a slightly squared off curve rather than a true circle. This means that people who form cognitive maps oriented in the "Trade-off" or "Polarized" direction utilize these maps more often than those who adopt "Risk" or "Benefit" oriented maps, a finding which is similar to that of Renn.[11]

The wide spectrum of perceptions displayed by each of the groups in the descriptive statistics also characterizes the factor analysis

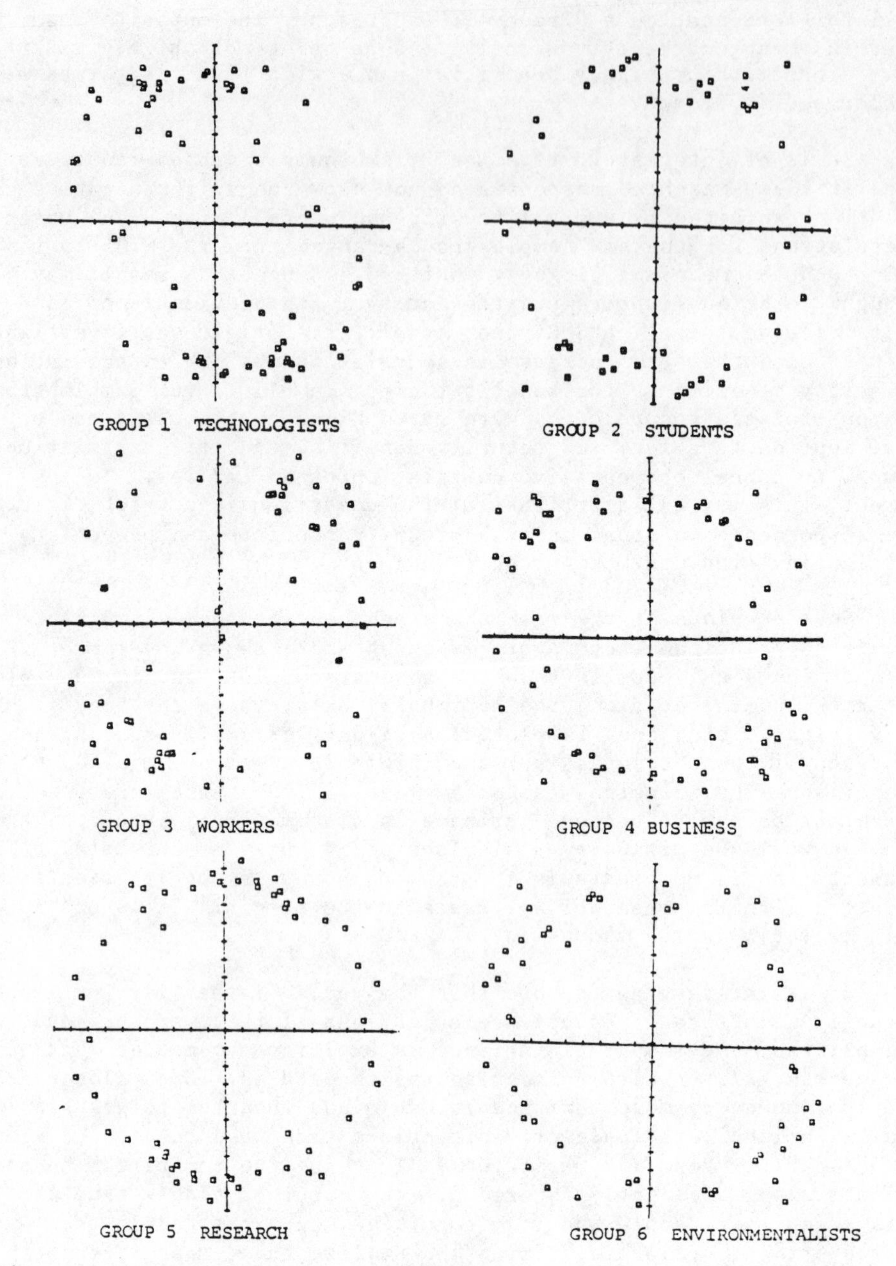

Figure 3. Scatter Plots of the Risk and Benefit Theme Components of the First Principal Component of Each Individual's Covariance Matrix by Occupational Group.

results. However, different distributional characteristics can be seen in the scatter plots as a clustering of the points along particular angles for some of the groups. Both Groups 1 and 5 (Technologists) show strong clustering toward the direction of the "Benefit" axis, while Groups 2, 3, and 6 (Students, Blue Collar Workers, and Environmentalists) are clustered between the "Polarized" and the "Risk" directions. Since these observations conform to predicted orientations, it is felt that the methodology outlined above is effective as a measurement tool for cognitive maps. Further study involving the predictive power of the orientation of an individual's principal factor appears justified, but is beyond the scope of the present paper.

One of the major applications for the understanding of people's perceptions of risks and benefits is the evaluation of the acceptability of a particular activity or technology. We have, therefore, applied the methodology outlined above to analyze the covariance over each group between cognitive dimensions for each of the survey activities and technologies. Matrices for the covariance of the nine risk and benefit dimensions discussed above were computed for each technology using each of the groups as a basis for the correlation, and each of these matrices was factor analyzed as described above. The largest principal factor of each matrix was projected onto the risk and benefit theme axes for the relevant group. Because each of the groups produced a full spectrum of technology locations, and, because of space limitations, the results are pooled into a single plot in Figure 4.

The scatter plot shows clear evidence of clustering, with one dense group of technologies lying between the "Benefit" and the "Polarized" directions, and a broader group lying between the "Benefit" and the "Trade-off" directions. We take these clusters as evidence that people do, in fact, use cognitive maps in forming their perceptions of risks and benefits. The dense cluster of technologies near the "Polarized" direction, for example, may represent those technologies of which people either approve or disapprove.

SUMMARY AND CONCLUSIONS

A methodology has been presented which will hopefully provide new insight into the covariance between responses to survey questionnaires. In contrast to existing methods which focus on a single numerical measure (a correlation measure, such as r, or Dyx), this method provides a multi-dimensional distribution showing the location of each case with respect to the major conceptual dimensions of the study. If the linkages between conceptual dimensions are analyzed two at a time, the distributions can be plotted graphically to permit visual analysis of the data. The method has the advantage of being relatively more independent of specific assessment questions than previous approaches, and consequently questions which do not require complex judgments on the part of respondents can be used.

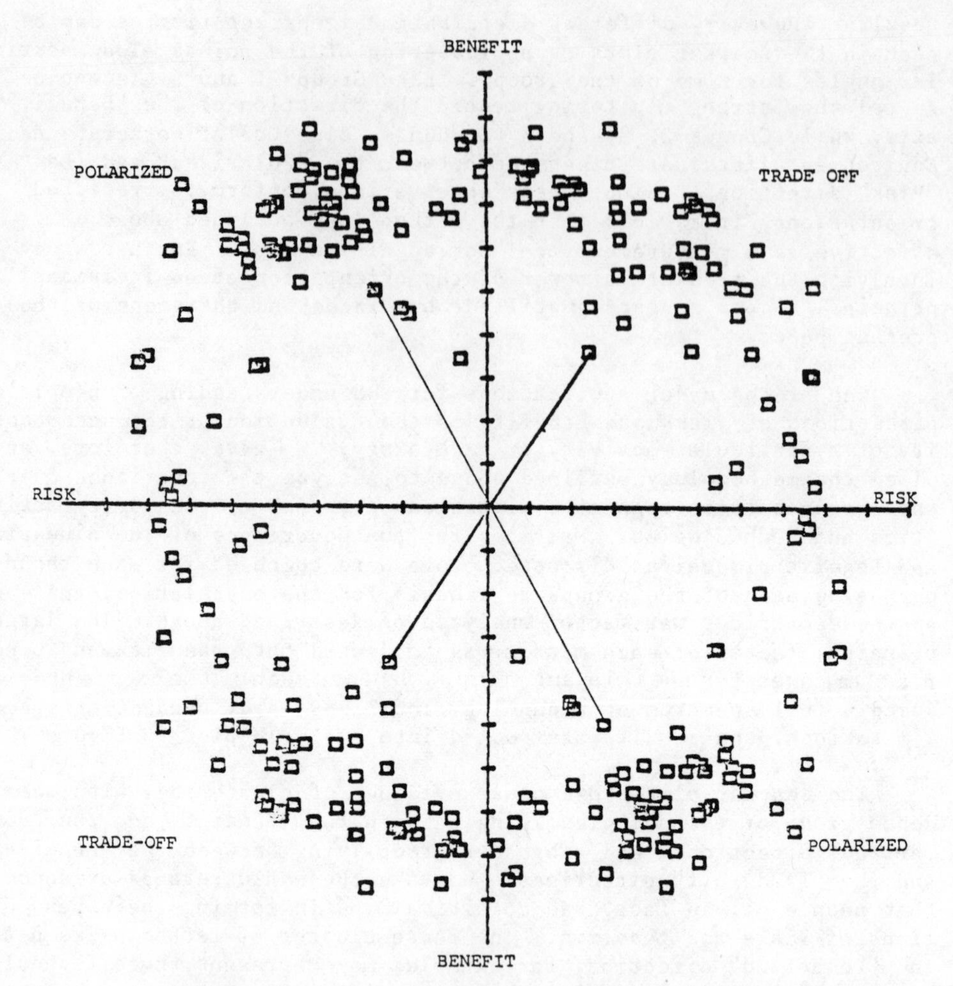

Figure 4. Scatter Plot of Risk and Benefit Theme Components of the First Principal Component of the Covariance Matrix for Each Technology. The covariance was computed for each occupational group, and results for all 6 groups are plotted. Two lines are drawn for illustrative purposes connecting the squares representing two of the cases. Note the clustering of the points on either side of the benefit direction.

Generally speaking, questionnaire items must be replicated over an ensemble of instances which can serve as the basis for correlation at the individual level, and a sufficient number of respondents must be obtained to provide a basis for the correlation for each instance. Thus, in this study of the linkages between perceptions of risk and benefit, the survey instrument replicated the same questions for 17 (or 30) different technologies, and the survey covered at least 40 or 50 individuals from any one group. Ideally, the questionnaire should tap the relevant dimensions through a large number of (possibly redundant) characteristics, but if this results in a prohibitively complex task for the respondents, smaller numbers of pretested characteristics can be used. Replication of the results for different groups of respondents is important, and the research design should include at least four or five independently chosen groups.

The covariance matrix for the chosen characteristics is computed either for each individual (using the ensemble of questionnaire objects as the basis for correlation), or for each object in the ensemble (using the respondents in a particular group for correlation). The correlation matrices are next factor analyzed, and both unrotated and rotated factor loadings are obtained. The rotated loadings are examined to verify that the conceptual dimensions are well represented, and if they are, the analysis can proceed. Assuming the conceptual dimensions are well captured by the rotated loadings, the first principal factor (i.e., the one that "explains" the most variance) is projected onto the conceptual axes. In the two dimensional case, the principal factor can be represented by a vector whose projections on the x- and y- axes correspond to the projections onto the two conceptual dimensions under consideration. The ensemble of survey objects or the ensemble of respondents can then be presented graphically as a scatter plot of the corresponding vectors, and various cluster analysis procedures can be applied using either cartesian or polar coordinates as discriminant variables.

In this paper, the survey objects are various activities and technologies, the conceptual dimensions are perceptions of risks and benefits, and the groups of respondents are purposively sampled so as to cover occupations and levels of social status that are thought to be relevant to those dimensions. Even though the responses of these groups differ only modestly in terms of simple statistical measures, the distribution of the linkages between their responses at the individual level are quite different. Thus, measures of cognitive maps are more dependent on occupational group than other measures such as mean responses to particular questions or means of factor scores. More importantly, when the survey objects are studied, clustering can be seen that could be explained by the assumption that approval or disapproval of a particular activity influences perception of its risks and benefits. Or, as Cummings' perceptive comment suggests, major decisions can have hidden agendas, and "risk may become the surrogate for all that the stakeholder feels is right or wrong about a particular choice and the lifestyle...it represents."[10] Studies could now be undertaken, based on measurements of cognitive maps, to explore the extent to which risk and benefit perceptions represent a vote for or against a technology or activity.

The results presented above are encouraging, but are obviously preliminary in nature. Further work will quantify the predictive power of "cognitive orientation" and its relationship to the measurement of "acceptability."

ACKNOWLEDGEMENTS

This work was supported in part by grants from The General Electric Company, Northeast Utilities, and the National Science Foundation, Grant PRA 8014194. The authors are deeply grateful to Gerald T. Gardner who, together with the senior author, was responsible for the major design of the survey instrument, and for obtaining half of the respondents in Groups 2-6. Further appreciation is expressed to J. Stolwijk, D. DeLuca, L. C. Gould and L. Doob of Yale University who also participated in the design of the survey instrument and analysis of related results.

REFERENCES

1. Chauncey Starr, Social Benefit vs. Technological Risk. Science 165, (1969), 1232-1238.
2. Harry Otway and J. J. Cohen, Revealed Preferences. Comments on the Starr Benefit-Risk Relation. RM 76-80, International Institute for Applied Systems Analysis, Laxenburg, Austria, (1975).
3. Paul Slovic, Baruch Fischoff, and Sarah Lichtenstein, Rating the Risks. Environment 21, (1979), 14-20, 36-39.
4. Gerald T. Gardner et al., Risk and Benefit Perceptions, Acceptability Judgments, and Self-Reported Actions Toward Nuclear Power. The Journal of Social Psychology 116, (1982), 179-197.
5. Mary Douglas and Aaron Wildavsky, How Can We Know the Risks We Face? Why Risk Selection is a Social Process. Risk Analysis 2(2), (1982), 49-58.
6. Douglas MacLean, Risk and Consent, Philosophical Issues for Centralized Decisions. Risk Analysis (2)2, (1982), 59-67.
7. Harry Otway and Kerry Thomas, Reflections on Risk Perception and Policy. Risk Analysis (2)2, (1982), 69-82.
8. Paul Slovic, Baruch Fischoff, and Sarah Lichtenstein, Why Study Risk Perception? Risk Analysis (2)2, (1982), 83-93.
9. Adrian R. Tiemann, Technologists and Risk Assessment. ONRL-Sponsored Conference on Risk Assessment, Eugene, OR, December (1980).
10. Robert B. Cumming, Risk and the Social Sciences. Risk Analysis 2(2) (1982), 47-48.
11. Otto Renn, Man, Technology and Risk. Kernforschlungsanlage Juelich GmbH, June (1981).

WARNING SYSTEMS AND RISK REDUCTION

M. Elisabeth Pate-Cornell

Department of Industrial Engineering
and Engineering Management
Stanford University
Stanford, CA 94305

ABSTRACT

A probabilistic method is proposed to assess the efficiency of warning systems in terms of costs and risk reduction. This evaluation is based both on the quality of the signal and the human response to warnings. The quality of the signal is described by probabilities of correct warnings, Type I errors, and Type II errors. Human response depends on the memory that people have kept of past warnings and is described here by a Markov model. On our hypothetical example concerning a fume detector in a chemical plant, the results show that the optimum sensitivity is at an intermediate position between extreme levels.

KEY WORDS: Warning Systems, Risk Analysis, Probability, Signals, Memory, Cost-Benefit Analysis

INTRODUCTION

A warning system that is too sensitive issues so many false alarms that after some experience people may cease to respond. On the contrary, if it is not sensitive enough, it may fail to issue a signal when an event is about to occur, or might do it with such a short lead time that no protective measure can be taken to reduce losses.

In this paper, we propose a probabilistic analysis of costs and benefits of a warning system as a function of (1) the quality of the signals issued, and (2) the response of the people to whom warnings are directed, as a function of their memory of the system's past performance. These results allow evaluation of a warning system, including the effects on the system's overall efficiency of true alerts, missed predictions (Type I errors), and false alerts (Type II errors)

VALUE OF INFORMATION OF A WARNING SYSTEM

In previous work, the author analyzed the costs and benefits of monitoring dams, (1) and predicting earthquakes for the San Francisco Bay Area.(2,3) The value of the information provided by the system is determined by the losses that can be avoided by appropriate actions following early warnings. These benefits are computed in probabilistic terms as the difference of losses (human casualties and/or financial loss) incurred with and without the warning system.

The model proposed here is an extension of this work to include the fact that the proportion of people that will respond to a new warning depends on the pattern of true and false alerts that have been observed in the past. To represent this phenomenon, we shall use a simple Markov model in which the system's states are the patterns of the previous alerts that determine future response, and the quality of the present alert that determines the warning's benefits.

As a concrete support for a more abstract model, we shall assume that our problem is the following: we want to set up the sensitivity of a fire alarm system in a student dormitory in such a way that the rate of false alerts does not discourage response, but that the lead time is sufficient for evacuation. We shall assume, for this purpose, that we can set a mobile hand on the fire alarm device in an optimal position between maximum and minimum sensitivity (see Figure 1). We shall also assume, for simplicity, that our goal is to maximize the expected value of the number of people saved, that is, of students who might otherwise have been hurt or killed in fires and are protected, thanks to early warnings. In general, the goal depends on the risk attitude of the decisionmaker, and we might want to maximize the expected value of other utility functions including both costs and benefits of the system.

Elements

The proposed model includes the following elements:

- The initiating events (e.g., fires) are noted E, and are assumed to occur following a Poisson process at a mean rate λ per time unit (4), and to have only one level of magnitude. $\overline{E}$ means that E does not occur.

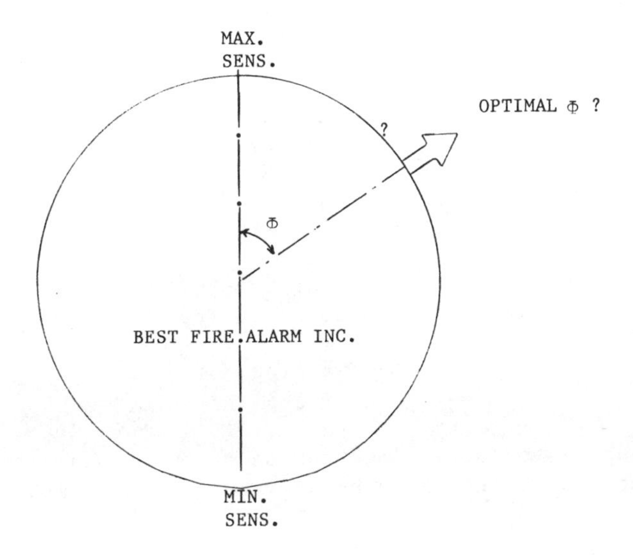

- IF Φ IS TOO SMALL, TOO MANY FALSE ALERTS → LOW RESPONSE.
- IF Φ IS TOO LARGE, LEAD TIME TOO SHORT FOR EFFECTIVE RESPONSE.

Figure 1

The Fire Alarm Decision

o The signals are assumed to be of two types:

--<u>Signals 1</u> are independent false alerts, assumed to occur following a Poisson process at a mean rate λ_1 per unit of time. For example, they can be alerts triggered by a spider, or another foreign object unrelated to smoke, entering the alarm device.

--<u>Signals 2</u> may lead to false alerts or to real ones, but they are linked to the occurrence of events by an underlying phenomenon (e.g., the smoke level). They are also assumed to occur following a Poisson process at a mean rate λ_2 per time unit. They can be fire alerts triggered by smoke from cooking or cigarette smoking, as well as the beginning of a real fire in the building.

o The warnings are caused by signal 1 and signal 2. When caused by signal 1, they are always false alerts; when caused by signal 2 followed by "no event", they are false alerts; when caused by signal 2 followed by an event, they are true alerts.

- The final outcomes (warnings and events) are, therefore, of three types:

  --Event A is a warning followed by an event (true alert)

  --Event B is no warning followed by an event (missed prediction or Type I error)

  --Event C is a warning followed by no event (false alert or Type II error)

Figure 2 shows an illustration of the process and the different scenarios (signals, events, warnings, and outcomes) along a time axis.

Variables

The variables of the model are the following:

λ: mean rate of occurrence of event E per time unit.
λ_1: mean rate of occurrence of signal 1 per time unit.
λ_A: mean rate of occurrence of true alerts per time unit.
λ_W: mean rate of occurrence of warnings per time unit.
I: occurrence of signal 2.

Occurrence of Signals and Warnings

$$\lambda_2 = \lambda \frac{p(I|E)}{p(E|I)} \quad \text{(eq. 1)}$$

The occurrence of warnings (all warnings, true alerts only, and false alerts only) follows Poisson processes. The mean rates of occurrence are respectively:

$$\begin{cases} \lambda_W = \lambda_A + \lambda_C = \lambda_1 + \lambda_2 \\ \lambda_A = \lambda_{2\&E} = \lambda_2 \times p(E|I) = \lambda \times p(I|E) \\ \lambda_C = \lambda_1 + \lambda_{2\&\overline{E}} = \lambda_1 + \lambda_2 \times p(\overline{E}|I) \end{cases} \quad \text{(eq. 2)}$$

$p(E|I)$: characterizes the setting of sensitivity of the device measured by the probability of event E conditional of signal I; ($p(E|I)$ gets higher as sensitivity is lower)

$p(I|E)$: characterizes the reliability of the device measured by the probability of signal conditional on event E;

α: rate of response, is the proportion of people willing to respond to warning given pattern of past alerts. We shall assume here that all people willing to evacuate can actually evacuate;

C_W: cost of response to a warning; depends on the rate of response;

N_E: is the number of victims per event that one should expect without the warning system.

The results of the model are:

EV(L): Is the expected value of losses per time unit if the warning system did not exist;

EV(ΔL): is the expected value per time unit of the benefits (i.e., loss reduction) of the warning system;

EV(CO): is the expected value per time unit of the cost of warnings (total costs of the system also include costs of monitoring device and system's operation.

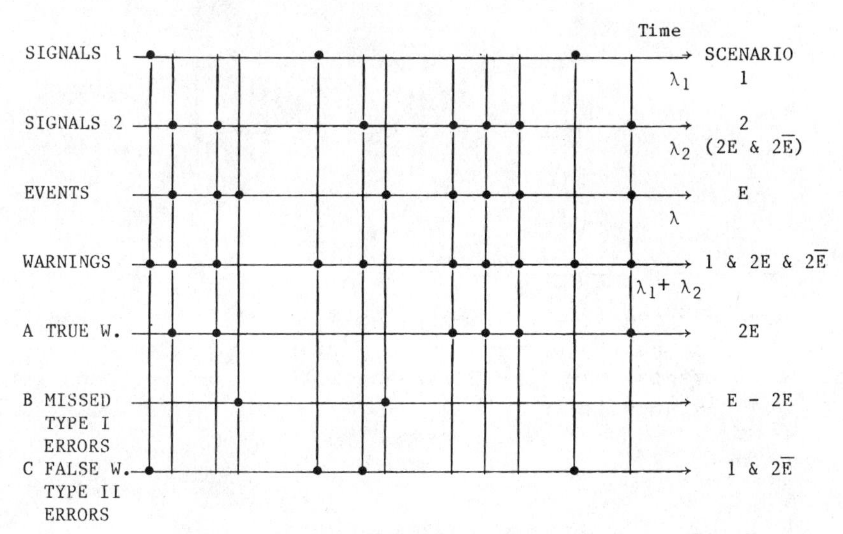

Figure 2

Events and Warnings: The Process

Response Without Memory

Assume first that no memory is involved in the response. The rate of response is independent of previous performance of the system. People are saved in case of true alerts only and the benefits are:

$$\circ \quad EV(\Delta L) = \alpha N_E \lambda_A$$

$$= \alpha N_E \lambda p(I|E) \qquad \text{(eq. 3)}$$

$$\circ \quad EV(\Delta L)/EV(L) = \alpha N_E \lambda_A / N_E \lambda$$

$$= \alpha p(I|E) \qquad \text{(eq. 4)}$$

Costs are incurred whether warnings are true or false alerts, and their expected value per time unit is:

$$\circ \quad EV(CO) = C_W \times \lambda_W$$

$$= C_W \left[\lambda_1 + \lambda \frac{p(I|E)}{p(E|I)} \right] \qquad \text{(eq. 5)}$$

Response With One Step of Memory

Assume now one step of memory, that is, the rate of response depends on the quality of the previous alert (A or C). The following sequence of events on a time axis shows the successive occurrences of alerts and response:

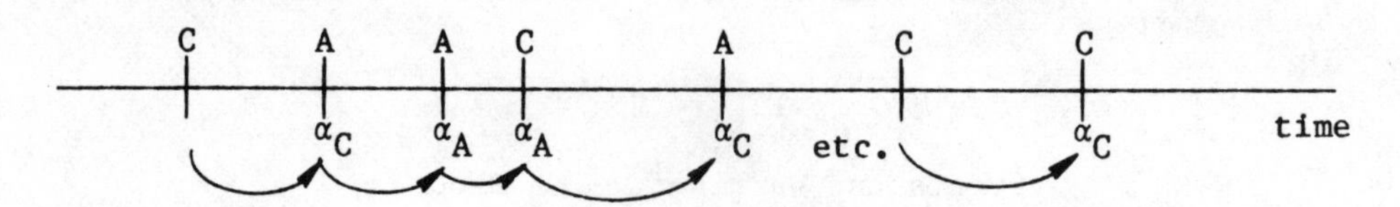

This very simple model does not account for the severity of the previous event or the time elapsed since it occurred, but only for the quality of the previous warning.

One can model the process using a Markov model in which each state represents the quality of the previous warning (that determines the response to the new one), and the quality of the new warning (that determines the actual benefit). In this case, we therefore have four states: AA, AC, CA, and CC. Let p_A and p_C be the probabilities of true and false alerts conditional on warning. The transition matrix is the following:

$$
\begin{array}{c}
\\
\Pi =
\end{array}
\begin{array}{cccc}
AA & AC & CA & CC \\
\end{array}
\leftarrow \text{States}
$$

$$
\Pi = \begin{bmatrix} p_A & p_C & 0 & 0 \\ 0 & 0 & p_A & p_C \\ p_A & p_C & 0 & 0 \\ 0 & 0 & p_A & p_C \end{bmatrix}
\begin{matrix} AA \\ AC \\ CA \\ CC \end{matrix}
$$

(Columns: AA, AC, CA, CC ← States; rows ↓: AA, AC, CA, CC)

For example, the probability of transition from State 2 (AC) to State 3 (CA) is the probability p_A that the new warning is a true alert. The probability of transition from State 3 (CA) to State 4 (CC) is 0 because State 3 ends in an A and State 4 begins with a C.

The steady state probabilities p_i^* represent the limit of the proportion, in the very long run, of warning sequences corresponding to each state i. Costs and benefits depend both on the response rate and the outcome of the warning.

The transition probabilities are the following:

$$
\begin{cases}
p_A = \dfrac{\lambda_A}{\lambda_W} = \dfrac{\lambda\, p(I|E) \times p(E|I)}{\lambda_1\, p(E|I) + \lambda\, p(I|E)} \\[2ex]
p_C = \dfrac{\lambda_C}{\lambda_W} = \dfrac{\lambda_1\, p(E|I) + \lambda\, p(I|E) \times [1 - p(E|I)]}{\lambda_1\, p(E|I) + \lambda\, p(I|E)} \\[2ex]
p_A + p_C = 1
\end{cases}
\qquad \text{(eq. 6)}
$$

The steady state probabilities are:

$$\begin{cases} p^*(AA) = p_A^2 \\ p^*(AC) = p^*(CA) = p_A\, p_C \\ p^*(CC) = p_C^2 \end{cases} \qquad \text{(eq. 7)}$$

The benefits are measured by the expected value of the number of people saved per time unit. In state AA the benefit is $\alpha_A N_E$, in state CA, $\alpha_C N_E$, and 0 in states AC and CC that end up in false alerts. N_i, the number of potential victims, is equal to N_E for i = 1 or 3 and 0 for i = 2 or 4. The results are therefore the following:

° $$EV(\Delta L) = \lambda_W \sum_i \alpha_i\, p_i^*\, N_i$$

$$= (\lambda_1 + \lambda_2)(\alpha_A\, p_1^*\, N_E + \alpha_C\, p_3^*\, N_E)$$

$$= \left[\lambda_1 + \lambda \frac{p(I|E)}{p(E|I)}\right](\alpha_A\, p_A^2 + \alpha_C\, p_A p_C)\, N_E \qquad \text{(eq. 8)}$$

° $$EV(\Delta L)/EV(L) = \frac{(\lambda_1 + \lambda_2)(\alpha_A\, p_A^2 + \alpha_C\, p_A\, p_C)N_E}{\lambda\, N_E}$$

$$= \left[\frac{\lambda_1}{\lambda} + \frac{p(I|E)}{p(E|I}\right](\alpha_A\, p_A^2 + \alpha_C\, p_A p_C) \qquad \text{(eq. 9)}$$

Let C_A be the cost of a high level response warning, and C_C be the costs of a low level response warning. The expected value of the costs per time unit is thus:

$$\text{EV(CO)} = \lambda_W \sum_i C_i p_i^*$$

$$= \lambda_W \left[C_A(p_A^2 + p_A p_C) + C_C(p_A p_C + p_C^2) \right]$$

$$= \left[\lambda_1 + \lambda \frac{p(I|E)}{p(E|I)} \right] (C_A P_A + C_C P_C) \qquad \text{(eq. 10)}$$

Generalization to n Step Memory

Assume now that people remember the quality of the n previous alerts which will determine their response to a new one. The previous model can be generalized by defining 2^{n+1} states of the following type:

AAAC...A|A ← outcome of the present alert (unknown at the time of response)

↓

n previous alerts

The transition matrix Π has a generic term p_{ij} which is the probability of transition from state i to state j conditional on the occurrence of a new alert. p_{ij} is equal to zero unless the last elements of state i are identical to the first n elements of state j. If states i and j have these elements in common, the probability of transition is that of the last element of state j. Let X and Y be either A or C.

$p_{ij} = p_A$ if state i is of type: X|YXY...XY|

and state j is of type: |YXY...XY|A

$p_{ij} = p_C$ if state i is of type: Y|XYY...XX|

and state j is of type: |XYY...XX|C (eq. 11)

$p_{ij} = 0$ otherwise

Steady state probabilities are solutions of the system:

$$\begin{cases} p_j^* = \sum_i p_i^* \, p_{ij} & \text{(for } n-1 \text{ values of } j \text{ between 0 and } n) \\ \sum_j p_j^* = 1 \end{cases} \quad \text{(eq. 12)}$$

Assuming that state j involves k_1 A's and k_2 C's, the solution to this system is:

$$p_j^* = p_A^{k_1} \, p_C^{k_2} \quad \text{(eq. 13)}$$

Both the response ratio in state i (α_i) and the costs of an alert corresponding to state i (C_i) depend on the n first digits of state i. There are at least three possible approaches to obtaining this data: to observe past (or past related) behaviors, to use questionnaires to predict behaviors, or to assume a "rational behavior" model and compute a rate of response, given a distribution of risk attitudes in the considered population.

One can then generalize the results shown in equations 8, 9, and 10. Let {k} be the set of indexes of states i ending in a true alert. The costs and benefits of the system are:

$$\circ \; EV(\Delta L) = \lambda_W \sum_{i \in \{k\}} \alpha_i \, p_i^* \, N_E$$

$$= \left[\lambda_1 + \lambda \frac{p(I|E)}{p(E|I)} \right] \sum_{i \in \{k\}} \alpha_i \, p_i^* \, N_E \quad \text{(eq. 14)}$$

$$\circ \; EV(\Delta L)/EV(L) = EV(\Delta L)/N_E \lambda$$

$$= \left[\frac{\lambda_1}{\lambda} + \frac{p(I|E)}{p(E|I)} \right] \sum_{i \in \{k\}} \alpha_i \, p_i^* \quad \text{(eq. 15)}$$

$$° \; EV(CO) = \lambda_W \sum_i C_i \, p_i^*$$

$$= \left[\lambda_1 + \lambda \frac{p(I|E)}{p(E|I)} \right] \sum_i C_i \, p_i^* \qquad \text{(eq. 16)}$$

ILLUSTRATION: A HYPOTHETICAL EXAMPLE

Consider the application of this model to the assessment of the efficiency of a hypothetical detector of fumes in a chemical plant. Assume the following data:

The unit of time is the year.
The mean rate of occurrence of incidents (chemical spills) is 3 per year.
The mean rate of irrelevant alerts (signal 1) is 2 per year.
The reliability (availabililty on demand) of the detector device $p(I|E)$ is equal to 0.9.
The sensitivity of the device is characterized by $p(E|I)$ is equal to 0.7 (a high sensitivity).
Without the early warnings given by the sensor, the expected number of persons who get injured or sick per incident is $N_E = 5$.
The number of steps of memory is 3.
The response to a new alert is given by the following (non-parametric) function (X is the outcome of the present alert, A or C, unknown at the time of response):

States:	AAAX	CAAX	ACAX	AACX	CCAX	CACX	ACCX	CCCX
Response α:	0.9	0.8	0.7	0.6	0.5	0.4	0.3	0.2

Costs: 50\$ x α_i x 100 (employees involved).

This does not look like a particularly efficient system, it has a very high rate of irrelevant alerts and a high probability of false alerts from signal 2. Indeed, the results of the model developed above show that the expected value of the number of people effectively protected by the system is only seven per year, out of the 15 potential victims. The mean of annual costs attributable to alerts is \$15,000. The base case is therefore:

$$\begin{cases} EV(\Delta L) = 7 \\ EV(\Delta L)/EV(L) = 47\% \\ EV(CO) = \$15{,}000 \end{cases}$$

Several improvements can be considered. Assume first that someone has a method to decrease the number of irrelevant alerts (signal 1) to a mean rate of one per year without changing the other characteristics of the problem. The system's performance is now:

$$\begin{cases} EV(\Delta L) = 9.5 \\ EV(\Delta L)/EV(L) = 63\% \\ EV(CO) = \$13,468 \end{cases}$$

Simply by eliminating irrelevant alerts one not only decreases the costs of alerts as expected, but also increases the number of people protected that would otherwise have been injured. Because of a higher probability of correct warning given an alert, the rate of response to each alert is higher. Of course, we have not considered here the effect of regulations and police actions forcing people to evacuate the plant, or various incentives (punishment or reward) that can be applied to increase the evacuation rate. It is simply assumed that after a certain number of false alerts there will be a tendency from all parties involved to question the necessity to respond to the signals.

Assume now that starting from the improved base case, one decides to act on the sensitivity of the device (the angle ϕ of Figure 1) and that one wants to evaluate the system's performance at different levels of sensitivity. We shall not get into the details here of how $p(E|I)$, $P(I|E)$, and λ_1, depend on this angle ϕ because it would involve studying the physical characteristics of the alarm device. We shall simply assume that if ϕ increases from maximum to minimum sensitivity, $p(E|I)$ increases with ϕ, whereas the rate of alerts, therefore λ_1 and $p(I|E)$, decrease when ϕ increases (the more sensitive the system, the higher the probability of getting signal 2 conditional or not on E, and of getting signal 1 at any time).

A case of high sensitivity (ϕ very small) can correspond to the following characteristics:

$$\text{Data} \quad \begin{cases} p(E|I) = 0.1 \\ p(I|E) = 0.99 \\ \lambda_1 = 2 \end{cases}$$

$$\text{Results} \quad \begin{cases} EV(\Delta L) = 4 \\ EV(\Delta L)/EV(L) = 27\% \\ EV(CO) = \$20,028 \end{cases}$$

A case of medium-high sensitivity corresponds to the improved base case described above, whose characteristics are:

$$\text{Data} \quad \begin{cases} p(E|I) = 0.7 \\ p(I|E) = 0.9 \\ \lambda_1 = 1 \end{cases}$$

Results $\begin{cases} EV(\Delta L) = 9.5 \\ EV(\Delta L)/EV(L) = 63\% \\ EV(CO) = \$13{,}468 \end{cases}$

A case of medium-low sensitivity can be described by the following data:

Data $\begin{cases} p(E|I) = 0.9 \\ p(I|E) = 0.85 \\ \lambda_1 = 1 \end{cases}$

Results $\begin{cases} EV(\Delta L) = 8.6 \\ EV(\Delta L)/EV(L) = 57\% \\ EV(CO) = \$13{,}190 \end{cases}$

A case of low sensitivity can be characterized by:

Data $\begin{cases} p(E|I) = 0.99 \\ p(I|E) = 0.5 \\ \lambda_1 = 0. \end{cases}$

Results $\begin{cases} EV(\Delta L) = 6.7 \\ EV(\Delta L)/EV(L) = 45\% \\ EV(CO) = \$6{,}760 \end{cases}$

This example shows that if we restrict the range of possible sensitivities to these four positions, the optimum is not reached at either extreme, but in the medium-high sensitivity level.

CONCLUSION

The function EV(ΔL) that represents the benefits of the fire alarm system is a continuous and bounded function over the continuous range of possible degrees of sensitivity (angle ϕ). One can, therefore, use this analysis to find the optimum position of ϕ, which maximizes the benefits (or other objective function). This optimization, however, requires a more complete physical analysis of the relationship between the position of the handle, the density of smoke that will trigger the alarm, the rate of irrelevant signals, and the conditional probabilities p(I|E) and p(E|I). The same method can also be applied to compare the costs and benefits of two warning systems (e.g., a fire alarm triggered by human intervention and an automatic smoke detector, or to compare the cost-effectiveness of a warning system with that of other means of risk reduction.)

REFERENCES

1. Pate, M. Elisabeth, Risk-Benefit Analysis for Construction of Ne Dam Sensitivity Study and Real Case Applications. Research Report #R81-26. Department of Civil Engineering, Massachusetts Institute of Technology, Cambridge, MA, July 1981.
2. Pate, M. Elisabeth, and Haresh C. Shah, "Public Policy Issues: Earthquake Prediction." Bulletin of the Seismological Society of America, Vol. 69, No. 5, October 1979, pp. 1533-1547.
3. Pate, M. Elisabeth, "Analysis of Warning Systems: Application t Earthquake Prediction." Earthquake Prediction Research 1. Tokyo, Japan: Terra Scientific Publishing Company, 1982.
4. Benjamin, J. R., and C. A. Cornell, Probability, Statistics, and Decision for Civil Engineers, McGraw-Hill, New York, 1970.

ACKNOWLEDGEMENT

This study was funded by the Stanford Center for Economic Policy Research whose support is gratefully acknowledged.

ON DETERMINING PUBLIC ACCEPTABILITY OF RISK

George Oliver Rogers

University Center for Social and Urban Research
University of Pittsburgh

ABSTRACT

In recent years public opinion polls and social surveys have come to be more widely used by the media, government, and private industry to track public sentiment regarding a variety of issues. Risk analysts are simultaneously beginning to rely on such data in the tracking of attitudes and opinions bearing on the acceptability of risks. This research seeks to demonstrate how social structural analysis may be used to enhance our understanding of trends in attitudes concerning the acceptability of risk as reflected in social surveys.

When social surveys are used to track trends in public sentiment, attitudes on the acceptability of risk are often presented as univariate distributions of key indicators, such as the favorability of nuclear energy over time. However, variations in such distributions may also reflect changes in methodological approach, social structural changes, and variation in social processes. By developing a social structural model of the trend, a modicum of separation is achieved, allowing the researcher to distinguish between some methodologically induced shifts, socially-based shifts, and genuine shifts in attitudes bearing on acceptability over time. The log-linear approach used here to model these data also allows for prediction of future trends, or the prediction of acceptability of risk for different population structures. By this method we are able to interpret survey research data on the public acceptability of risk in more comprehensive and effective ways.

KEYWORDS: Perceived Risk, Acceptable Risk, Lay-Estimation, Lay-Evaluation, Social Structural Location, Social Surveys

INTRODUCTION

Survey data are widely used to track public sentiment and reported behavior regarding a variety of issues by all sectors--the media, the government at all levels, as well as private industry. In many cases risk analysts rely on such data with respect to risk in assessing the potential for public acceptability. This paper seeks to demonstrate how social structural analysis may be used to enhance our understanding of expressed attitudes reflecting the acceptability of risk. The approach allows the risk analyst to most fruitfully take advantage of large archives of existing social surveys, conduct their own dedicated surveys, or "piggy-back" some relevant questions concerning risk on existing surveys to better understand the social processes involved in lay risk assessment.

Based upon the physical requirements of nutrition, safety, and reproduction, the fundamental need for society to survive is a primary function of society.(1-4) Hence, the examination of both formal and informal mechanisms within society through which these requirements are met is nontrivial. To the extent that risk analysts are becoming sensitive to revealed social preferences for risk (5,6) or attitudes bearing on the potential acceptability of risk, an understanding of the social processes involved in the differential assessment of risk is essential. By placing these preferences and attitudes concerning risk into their social structural context, a more comprehensive understanding of their meaning is obtained.

Social scientists in general, and sociologists in particular, have typically been interested in the associations among properties, attitudes, and behavior.(7) It is precisely this perspective that addresses the various kinds of people with particular attitudes and preferences for risk. For example: Do men and women assess risk differently? If so, what seems to account for the differences? How do people with different educational backgrounds assess risk? And, if so, why? Does differential power, say in the form of resources, (7) affect risk preferences and attitudes? To the extent that people are characterized by various sociocultural and demographic properties, what is the multiple effect on their risk assessment? In short, how are differential social positions associated with risk preferences and attitudes, and indeed, with risk-related actions?

The social meaning of risk attitudes and preferences is more fully understood when the "how they are different" and the "why they are different" are addressed. Hence, the aim of this research is to address how and why risk attitudes and preferences are different among various social strata, in so far as they are, in fact, different. A social structural model of perceived and acceptable risk is developed in terms of lay-estimated risk and attitudes bearing on acceptability. It is in the context of this model that recent trends in existing social survey data are analyzed.

This research conceptualizes risk assessment as hazard identification, involving the reduction of uncertainty through recognition; risk estimation, consisting of the measurement of risk potential; and risk evaluation, involving the determination of the social acceptability of risk.(5,6,9) Furthermore, we find no a priori reasons to believe that lay-people assess risk through significantly different conceptual processes when compared with those of experts. In fact, lay-people probably use a system of evaluation similar to that of experts, albeit less sophisticated, less rigorous, and less quantified.(10-13) The lay-system treats hazards recognition as a lay identification, of hazard, perceived risk being a lay estimation of likelihood of actualization and magnitude of effects, (14) and attitudes bearing on acceptability representing a social evaluation of risk, in which a trade-off between risk and benefits is often at least implied. The public assesses risk through a procedure that roughly parallels the system used by risk analysts. This does not imply that the results are, or even need be, similar as differences in results have been frequently reported (15,16) and thus well documented.

THE EMPIRICAL TREND

Secondary survey data concerning attitudes associated with nuclear energy are available. Public opinion polling agents have frequently assessed such attitudes and much of the data has been made available to researchers.(17) Among these data are regional, national, and local surveys. Spanning just slightly more than four years, the Louis Harris and Associates (18, 19-22) surveys analyzed here were conducted in an important period in the development of nuclear power as a source of electric energy. Several events in the 1975 to 1979 period were significant, including the Browns Ferry fire; several state referenda to ban nuclear development; and perhaps most significant, the Three Mile Island incident. The surveys conducted in March 1975 and July 1976 involved face-to-face interviews concerning the many salient issues associated with nuclear power, including: the perceptions of the seriousness and meaning of the energy crisis, alternative methods of meeting future demand for energy, and attitudes about nuclear power, its development, and safety. The survey, conducted in April 1977, consisted of personal interviews addressing more typical issues of public opinion, economic and political interest, energy conservation, and safety. While the March and April 1979 surveys once again focused more succinctly upon issues associated with energy and nuclear power, the former was a face-to-face interview and the latter was conducted by telephone.[1] The 1975 and 1976 Harris surveys are household samples of the entire continental United States. They are limited to adults 18 years of age and over and to the non-institutionalized population. "Scientifically random sampling techniques guaranteed to each household in the country an equal chance of being drawn into the sample."(19:p.vii) Based on the Harris approach to sampling, (23) we presume that a stratified multistage cluster sample of about 100 primary sampling units (PSUs) was used with random selection within these PSUs. The PSUs are drawn with probabilities proportional to population size, and random sampling within PSUs assures the overall equal probability of household selection. While no specific sampling information is available for the April 1977 and March 1979 surveys, these procedures appear to hav remained unchanged. (That is, the procedures used in the 1975 and 1976 surveys were also utilized in April 1977 and March 1979.) The April 1979 survey, being a telephone interview, requires special mention. There is no specific sampling information, but Harris telephon procedures are of the random-digit-dialing variety, (24) and thus assure the inclusion of both unlisted and non-listed telephones.(25) The sample size of 1,200 is consistent with a ± three percent samplin error at the 95 percent confidence level.

Lay risk assessment rests on the foundation of evaluation and estimation of risk for identified hazards, because unrecognized risks cannot be estimated or evaluated. Furthermore, since evaluation and estimation cannot be measured directly, they must be cast in terms of their components or elements. In the case of nuclear power, the risk have been recognized for some time. Thus, indicators of social evaluation and risk estimation are sought for the general public. In each of the Harris surveys, respondents were asked to assess their general attitudes concerning building more nuclear power plants in terms of being favorably disposed or not. In addition, respondents were asked in each survey, to evaluate nuclear power plant safety. The favorability associated with building nuclear power plants may be viewed as bearing on potential acceptability, while perceived safety reflects a lay estimate of risk.

The distribution of responses concerning favorability associated with nuclear power across these five surveys (Table I) indicates continued support for nuclear power in general terms (i.e., there are more favorable responses than any other in each survey). However, this support erodes from 63.4 percent favoring nuclear power in March 1975 to a low of 44.5 percent immediately following the Three Mile Island accident in March 1979. Secondly, the opposition response mor than doubles over the period from 18.7 percent in March 1975 to 43.6 percent in April 1979. Finally, the percentage of people remaining unsure is reduced from 17.9 percent in March 1975 to 6.9 percent in April 1979. As the primary shifts in these responses occur between April 1977 and March 1979, long after other significant events, but more in the immediate aftermath of the Three Mile Island incident, th shift in attitudes concerning nuclear power appears to be attributabl to TMI and its aftereffects. For example, the Browns Ferry fire and the public debate surrounding the various State referenda concerning nuclear energy would have been expected to yield attitude shifts reflected in the April 1977 data; however, this does not occur. Base on this, it is tempting to speculate that the erosion of support for nuclear energy comprises people shifting to opposition responses from

TABLE I. PUBLIC ATTITUDES ABOUT BUILDING NUCLEAR POWER PLANTS IN UNITED STATES 1975-1979

Louis Harris and Associates:

In general, do you favor or oppose the building of more nuclear power plants in the United States?

Survey Date:	N	Responses		
		Favor	Oppose	Not Sure*
March 1975	1537	63.4%	18.7%	17.9%
July 1976	1497	60.8%	22.3%	16.9%
April 1977	1547	60.4%	23.9%	15.7%
March 1979	1510	44.5%	42.1%	13.4%
April 1979**	1200	49.5%	43.6%	6.9%

*A small proportion of the not sure response is comprised of don't know, no answer and non-response missing data codes.

**The April 1979 survey was conducted by telephone.

both favorable and unsure responses because of TMI. The explanation for such a shift might be that the accident and its aftermath served to increase the lay estimated risk and thereby induced such a shift in attitudes. However, the data regarding perceived safety (Table II) are to some extent inconsistent with this decision. First, the percent finding nuclear power very safe declines only modestly, from a high of 25.8 percent in March 1975 to a low of 20.7 percent immediately following the accident in March 1979. Second, the proportion reporting nuclear power as not safe (or dangerous in 1975 and 1976) increased from a low of 18.1 percent in March 1975 to a high of 28.8 percent in April 1979. This shift is significantly smaller than the magnitude of shifts found in opposition to nuclear power. The proportion of people finding nuclear power somewhat safe increased from 35.8 percent in March 1975 to 46.1 percent in April 1979 reflecting growing ambivalence. Finally, and perhaps most significantly, the proportion of people remaining unsure is reduced dramatically, from 20.3 percent in March 1975 to 2.3 percent by April 1979. This seems to indicate that the Three Mile Island accident served primarily to help people decide (perhaps only temporarily) about nuclear energy. To those people who found nuclear power plants relatively safe, the accident, or the lack thereof, proved just how effective the safety systems associated with nuclear power plants can be. For people who were apprehensive about the safety of nuclear energy, the accident served to confirm the risky nature of nuclear energy production systems.

TABLE II. PUBLIC PERCEPTION OF NUCLEAR POWER PLANT SAFETY 1975-1979

Louis Harris and Associates:

All in all, from what you have heard or read, how safe do you think nuclear power plants that produce electric power are--very safe, somewhat safe, or not safe?

Survey Date:	N	Responses			
		Very Safe	Somewhat Safe	Not Safe*	Not Sure**
March 1975	1537	25.8%	35.8%	18.1%	20.3%
July 1976	1497	25.4%	38.3%	23.3%	13.0%
April 1977	1547	28.7%	36.7%	23.0%	11.6%
March 1979	1510	20.7%	47.8%	27.0%	4.5%
April 1979	1200	22.8%	46.1%	28.8%	2.3%

*The 1975, 1976 and 1977 surveys allowed a voluntary "dangerous" response, however, in both 1979 surveys this response category was not allowed. The "dangerous" responses are collapsed with the "not safe" response in this analysis.

**A small proportion of the not sure response is comprised of don't know, no answer and other non-response codes.

The historical effect associated with the Three Mile Island accident and its aftermath is only one of several potential sources of attitude change. First, changes and modifications in survey instruments themselves can result in apparent attitude shifts. Furthermore if the sampling or survey methods employed changed, apparent attitude shifts might also result. This would include question wording, question placement, interviewer instruction, and mode of data acquisition (e.g., face-to-face, telephone, and the like). Certainly, the variety of question contexts in this analysis and the shift to a telephone interview in the April 1979 survey are among the critical consideration for the interpretation of the observed shifts in attitudes. Some of the trends presented in Tables I and II are sufficiently small so that sampling error alone might account for most or all of the apparent changes. Second, changes in the social processes that guide our behavior and attitudes may acount for the attitude shift. For example, the apparent shift in support for nuclear power plants might be concomitant with a more fundamental shift away from technology and technological solutions.(26) In essence, the norms that determine what is acceptable may change, which would induce change in specific risks as well. Third, changes in demographic character can significantly alter attitudes, particularly those concerning risk. This arises because of the distribution of various risks among geographic locations, age groups, and the sexes. Recent population trends such as the growing number of Hispanics in the United States and the in-migration to Southern and Western states (Snowbelt to Sunbelt) may affect attitudes about specific risks through differential values and altered saliency. Nuclear power plants are currently concentrated more in the Northeastern United States than elsewhere. Hence, large outflows of population to Southern and Western States may make nuclear energy both less needed and desirable, as greater proportions of people come to rely on alternative sources of electrical energy such as coal, hydro-electric power and solar power. Finally, because people are associated with various groups and organizations in society, and each of these associations is characterized by a role, which, in turn, is guided by some norms quite specific to the organization's needs, attitudes are likely to vary among social strata. The value systems created by these various role sets form the foundation for assessing attitudes concerning acceptable risk. The social evaluation of risk rests comfortably within the domain of such values.

The social structural analysis discussed in this paper examines attitudes concerning perceived and acceptable risk in terms of some aspects of social structural location. That is, there are groups or classes of people that estimate and evaluate the risks associated with nuclear power plants similarly. The social structural approach used herein draws both on the social theories of stratification and the risk assessment process, by explicitly (wherever possible) including both social structural elements and attitudes reflecting risk assessment.

SOCIAL STRUCTURAL LOCATION

Human societies are marked by inequality in the sense of access to, and distribution of, scarce resources. Since risk may be thought of as "negative resources" or, conversely, protection (safety) as goods and services, it follows that risk is also likely to be unequally distributed. Whether stratification is functionally necessary (27-29) or simply universally observed (30-33), people within various strata maintain different values, beliefs, attitudes, and opinions. Attitudes concerning risk are also expected to mirror some of these important differences.

The concept of social structural location includes elements of vertical and horizontal status and the associated roles, norms, values, and experiences. The social structure comprises a web of relationships among roles or positions in the social system.(8,34-36) Patterns of association among positions reflect status clusters, which are characterized by frequent in-group relations and less frequent between-group relations.(36,37) Describing these status clusters by their multivariant status indicators, the primary status clusters are examined. While hierarchical ranking among status clusters is termed stratification, social structural location reflects explicitly both the horizontal and vertical dimensions of stratification. Further, it is static, frozen in time, and as such does not reflect a change of role or position in the social structure in the form of social mobility. Social structural location has elements of both horizontal and vertical stratification as reflected in a static position in the social structure. Further, it is conceptually broader in that it is more closely associated with the patterned web of relationships in the social structure and reflects differential values, norms, roles, and experience.

A Louis Harris and Associates survey (24) conducted from December 1979 to March 1980, finds that women, the elderly, and ethnic minorities are more sensitive to risks "associated with living today" than are their counterparts. Hence, differential risk sensitivity is associated with various strata or positions in the social structure, and people in the social structure view risk and potentially risky situations differently. Simultaneously, risks are not distributed equally. For example, older people may be more susceptible to some types of risk than other adults, while some risks affect only the very young or predominantly women. Because risks are not equally distributed, death rates are nearly always adjusted by age and sex. Furthermore, research concerning the generation of electricity in nuclear power plants (17,38,39) indicates that individuals differ with respect to their attitudes at least in terms of acceptability of the risk associated with nuclear power.

This variation is associated with a variety of indicators of structural location (status), including sex, education, income, occupation, and age. When Brody (40) re-examined the data from two Harris surveys (18,19), he posited that women find nuclear power plants less acceptable than men because they maintain different values with respect to technology, the value of life and future generations, and the risks associated with nuclear power facilities. It was not argued so much that women know less about technology (although this was considered a modest factor), but rather that a different value system is employed in the social evaluation of risk due to the more marginal location of women in the social structure (at least historically so). Because men are more "centrally" located in "economic, political, and technical spheres," they will recognize more of the associated benefits of nuclear power and, hence, be more likely to support continued development of nuclear energy, despite such risks as they may themselves perceive. Women, on the other hand, "are either excluded or occupy less central ('marginal') positions in these spheres" (40:p.14) and are thus less likely to recognize the benefits of nuclear energy, so that "risks", in a way, "loom larger."

Perceived risk, in terms of estimated likelihood of nuclear power plant accidents or of nuclear war, has been shown to be associated with structural location in terms of education, age, and sex.(11,12) The association of age with estimated risk is most consistent, and particularly so among young adults (18 to 29 years of age). Young people tend to consistently estimate risks at higher levels than middle-aged (30 to 49 years of age) or older Americans. Previous work (12) suggests that this effect is due to differing roles and values at various age groups rather than effects associated with growing older (age cohort rather than simply maturation effect).

Finally, social structural location plays an integral role in determining response to realized hazards.(41-50) The fundamental position is that because of the conflicting loyalties associated with a complex stratification system in a complex industrial and urban society, response to realized hazard, say in the form of disaster, is a function of social identification and social structural location. In essence, who the social actor is in terms of social identification, and what responsibilities are associated with that particular role, determine to a large extent the nature of disaster response. For example, an adult in a household with children (a parent) will be more likely to take adaptive action in the form of relocation than a non-parent.(51)

MEASUREMENT

This research develops a simple model of social structural location, in terms of descriptive characteristics, using both perceived risk in terms of lay-estimated likelihood and perceived safety, and acceptability of risks associated with nuclear power plants, in terms of favorable and non-favorable attitudes. Descriptive characteristics of social structural location include indicators of both achieved and ascribed status. The status indicators of age, sex, and education are considered in this research due to:

- Their prominence in the status literature, (35-37, 52-57)

- The unequal distribution of risks among people as demonstrated by age and sex adjusted life tables,

- Their relatively consistent relationship to views regarding nuclear power, (17, 38, 39, 40) and the reported associations with risk sensitivity, (24) o Their role in determining response to an actualized risk in the form of disaster.(41, 44, 59)

Hence, the social structural characteristics of age, sex, and education are selected not only because they reflect differing social experience (7) but also because they have been shown to be associated with differential beliefs, attitudes, and behaviors regarding risks o various kinds--and, in particular, nuclear risks.

The effect of age in determining the acceptability of nuclear risks (17) and risk in general, (24) has been reasonably well documented. In addition, the effect of age seems to be predominantly a function of differential value maintenance, (12) perhaps associated with the roles that are typically being played by persons in various age groups, or the flows of generations in terms of "critical experience." Hence, these differential value systems produce patterned variation in the estimation of likelihood associated with various risks. The consistent propensity for young adults to estimate the likelihood of nuclear war as "likely" or "very likely" is primarily a function of the values associated with a position in life which may b thought of as a product of the roles associated with that position, a opposed to the experience of growing older through the aging process. Three categories of age are used throughout this research: young adults--those 18 to 29 years of age; middle-aged adults--those 30 to 49 years of age; and older Americans--those 50 years of age or older. These three categories of age are thought to represent the differential roles associated with position in life better than alternative age groupings.

Education is related to nuclear power attitudes (17) with more than compulsory education (High School) being associated with a tendency to favor the technology. One explanation suggests that enhanced knowledge associated with higher than compulsory levels of education tends to reduce perceived risk which, in turn, makes nuclea power plants more acceptable. Another explanation suggests that higher education is associated with higher social status, which, on the one hand, reflects greater personal resources which can be relied upon in the event the risk is actualized, and, on the other, may reflect an underlying propensity to take risk. The latter effect reflects the observed association between risk taking, say in the for of innovative behavior, and social status.(61-66) Figure 1 summarize this relationship.

PERCEIVED SAFETY AND FAVORABILITY

This analysis examines the association between perceived nuclear power plant "safety" and "favorability" associated with nuclear technology in terms of the age, education, and sex indicators of social

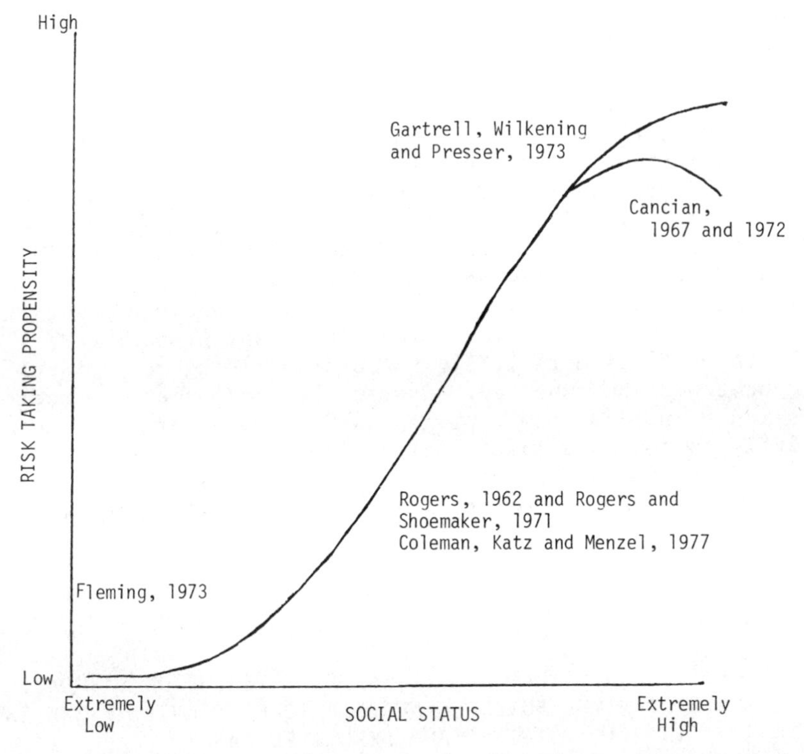

Figure 1. A Suggested Curvilinear Relationship Between Social Status and Propensities to Take Risks

structural location. It is undertaken in terms of contingency table analysis using log-linear techniques.(67-70) This analytic technique is appropriate for cross-classified or contingency data. It essentially searches for a set of subtables that account for systematically patterned variation in contingency tables once "fitted" in accordance with an underlying conceptual model. This technique allows the researcher to systematically account for variation in the contingency[2] tables and isolate specific associations in the context of the model. A parsimonious model fits the data in the Harris surveys with probabilities of Type I error of .107 and explains nearly 93 percent of the variation in the table. The subtables in the model are represented as {FS} {FX} {FY} {SE} {SA} {SX} {SY} {EA} {EY} {FA} {AX} and {AY}, where F represents favorability, S safety, E education, A age, X sex, and Y year of the survey. Notice that the AX and AY terms reflect a slightly disproportionate distribution of females in the 30 to 49 years of age grouping. The latter effect could be viewed as a model adjustment which accounts for the use of telephone survey methodology in the April 1979 survey, which would be more likely to produce female respondents in the middle age group.

The resulting (best fitting) model includes only two factor terms. Hence, the majority of the variation in the five survey contingency table is accounted for (i.e., R^2 analogs greater than .92) with a two factor model. The log-linear effects for this model are presented in Table III. The most important association in the data, and, hence, in the model, is the association between the favorability associated with "building more nuclear power plants in the United States" and perceived safety associated with nuclear power plant operation. Respondents finding nuclear power plants "very safe" are nearly three times more likely to favor building more plants than would otherwise be expected. Conversely, respondents perceiving the operation of nuclear power plants as "not safe" are less than one-half as likely to favor building nuclear power plants. In the sense that the standardized values are quite large in this analysis, the association between perceived safety and favorableness is a very robust relationship.

TABLE III

DIRECT AND MARGINAL LOG-LINEAR EFFECTS (λ) ON FAVORABILITY ASSOCIATED WITH "BUILDING NUCLEAR POWER PLANTS IN THE UNITED STATES" AND PERCEIVED SAFETY

	Marginal	Very Safe	Somewhat Safe	Not Safe	Favor
Marginal	2.22	-.158	.913	.119	.063
Favor	--	1.08	.207	-.845	--
More than High School Education	-.258	.261	.037*	.003*	NA
Age 18 to 29 years	-.141	-.138	.142	.050*	-.091
Age 50 or more years	.065*	.027*	-.193	-.037*	.077
Males	-.036*	.323	-.012*	-.084	.118
March 1975 Survey	.144	-.197	-.235	-.245	.202
July 1976 Survey	.142	-.197	-.212	-.014*	.158
April 1977 Sruvey	.145	-.018*	-.206	-.050*	.098
April** 1979 Survey	.382	.258	.349	.272	-.171

*Not significant at .05 level comparing standardized linear value and students. NA indicates that the effect is not estimated as it is not part of the model.

**April 1979 Survey was conducted by telephone.

The most important social structural component is the simultaneous effect being male or being female has on both perceived safety and associated favorability. Females[3] are nearly 12 percent less likely to "favor" nuclear technology and simultaneously more than 30 percent less likely to find nuclear power plant operation "very safe",and around 8 percent more likely to judge it "not safe" when compared to males. In addition, women are about 23 percent more likely to remain "unsure" with regard to nuclear power plant safety than their male counterparts. Hence, this analysis is consistent with both the findings of Melber et al. (17) and Brody.(40) Previous analysis of these data (12) indicated that females were over 40 percent less likely to judge nuclear power plants "very safe," approximately 15 percent more likely to find them "not safe," and about 25 percent more likely to remain "not sure" when only perceived safety is considered. Hence, the substantial proportion of the association between perceived safety and sex is accounted for when nuclear power plant favorability is introduced into the analysis. This reflects the added interpretive value we get by modeling the social structural characteristics and the risk assessment attitudes simultaneously.

As previously mentioned, age and favorability are significantly associated in the model, with young adults being about nine percent less likely to favor building more nuclear power plants than other age groups. Young adults are less likely to find nuclear power plants "very safe" than would otherwise be expected. They are about 14 percent more likely to find nuclear power "somewhat safe." In addition, four Harris surveys (18, 19, 21, 22) were employed in an analysis of only perceived safety, which suggests a different model and data structure, and in that analysis also, young adults are 14 percent more likely to find nuclear power "somewhat safe" than would be expected otherwise.(12) Comparisons between the four survey model (without favorability) and the five survey model (with favorability) reflect remarkably consistent associations between age groups and perceived safety response. These consistent results point to a tendency among young adults to be somewhat apprehensive about nuclear power plant safety. Adults aged 30-49 years tend to be more or less neutral regarding nuclear power plant safety. This tendency is reflected in the fact that in these models, the association for middle age group is not significant at the .05 level for any category of perceived safety.

Also reflected in the reciprocal association is the finding that adults 30-49 years of age are less "unsure" than other age groups, when nuclear power plant safety is considered. Adults in the middle age category are simultaneously more likely to have made a decision concerning nuclear power plant safety, and are more or less equally distributed among the other categories. Hence, their neutrality is not just a lack of decision or ignoring of the issues in an indifferent sense, but rather reflects a decisive neutrality among the respondents of the middle age group.

Older Americans (age 50 years or more) are simultaneously less likely to find nuclear power "somewhat safe," which is also found in the four survey model, (12) and more likely to remain "unsure" regarding nuclear power plant safety. These findings are consistent with regard to the model employed and may reflect a certain indifference, perhaps in the sense of an inability to decide about complex technological systems. However, having made a decision, older adults are less likely to choose the least committed "somewhat safe" response, indicating a "polar" response tendency among older Americans.

While education is not significantly associated with attitudes concerning acceptability of nuclear power in terms of favorableness (even enough to be included in the model), education is significantly associated with perceived safety. Respondents having completed more than a high school education are consistently less likely to make an "unsure" response, and at the same time, more likely to respond that nuclear power plant operations are "very safe." The effects are remarkably consistent across the three survey model of community and United States favorability reported by Rogers (13) where three surveys were used and the current five survey model, as well as the four survey model where favorability is not considered and income is introduced.

Viewing Rogers' (12) four survey model as an elaboration of the present model with regard to the association between education and perceived safety, income and education are confirmed as having independent effects. Given that education and income are strongly associated, this suggests an important differential role being highlighted by the two status indicators. Higher income presumably reflects a propensity to take risks and enhanced resources with which to deal with any consequences of actualized risks, and the reverse descripton may apply to the lower income groups.(71) Education, on the other hand, presumably reflects a "knowledge base," which is also reflected in the particularly robust association between education and the "not sure" response in the model.

The historical trend, reflected in the models, with regard to the perceived safety and favorability associated with "building nuclear facilities in the United States" is best examined in this context. This is true for the most part because variations associated with changes in the empirical distributions for any given model characteristic (e.g., age, sex, or education) are accounted for in the model. Hence, the log-linear effects for "favorability and year of survey" in the model decrease from .202 in the March 1975 survey to -.171 in the April 1979, indicating that holding age, sex, education, and perceived safety constant, respondents were much less likely to favor nuclear power in March of 1979 than in March of 1975.

Furthermore, by examining the intervening survey effects and the historical events of the period, it is recognized that this shift does not appear to be associated with the fire at Browns Ferry or its aftermath (72) or the state referenda of 1976 (19, 40) in California, Colorado, Arizona, Montana, Oregon, Washington, and Ohio.[4] The shift is most pronounced as a by-product of the most visible event of the period, the nuclear power plant accident of March 1979 at Three Mile Island (TMI) and its aftermath.(21, 22, 48-50, 73) While the duration of these rather dramatic shifts, in terms of how lasting the effect might be, is not clear from these data, it may be suggested that the impact of TMI has been primarily to solidify opposition and support and reduce the number of people holding no opinion or remaining undecided.

Accounting for any variation associated with the structural components of the model--age, education, or sex--and the risk estimation component, perceived safety in the favorability of nuclear power plants allows the risk analyst to isolate the effect associated with the intervening social/historical event, such as TMI. Table IV presents the odds ratios for the proportion favoring nuclear energy to all other responses, and the odds ratios for finding nuclear power plants somewhat to very safe. While the adjusted ratios are a direct consequence of the model, the adjusted ratios reflect the expected historical trend after accounting for the effects of age, sex, and education. Hence, any shifts in survey or sampling techniques which might be accounted for by any or all of these social structural components are accounted for by the model. Support for nuclear power, as reflected in the adjusted ratios, is highest in 1975 at 1.699 favorable response per those not favoring. In the March 1979 survey, immediately following TMI, the adjusted ratio is lowest at .640. This adjusted ratio is considerably lower than the observed ratio, which

TABLE IV

OBSERVED AND ADJUSTED RATIOS FOR FAVORING BUILDING NUCLEAR POWER PLANTS IN THE UNITED STATES TO ALL OTHER RESPONSES AND FINDING NUCLEAR POWER PLANTS SOMEWHAT SAFE TO VERY SAFE--MARCH 1975 TO APRIL 1979

	Favor/Non Favor		Somewhat/Very Safe	
	Observed	Adjusted	Observed	Adjusted
March 1975	1.732	1.699	1.388	2.810
July 1976	1.551	1.556	1.508	2.875
April 1977	1.525	1.380	1.279	2.418
March 1979	.802	.640	2.309	3.397
April 1979	.980	.806	2.022	3.197

tends to reflect the importance of the conceptual model of the trend that includes theoretically important factors, both concerning risk and the social structure. Furthermore, the adjusted ratio fails to recover in the April 1979 survey to a nearly one to one ratio as the observed ratio does, which also reflects the added statistical control associated with the approach. The smoother trend represented by the adjusted ratio points out the potential pitfalls of using unadjusted, uncontrolled surveys. While this research adjusts for the social characteristics of age, sex, and education, and the risk factor, perceived safety, other social structural and risk factors may also be needed to adequately adjust such data for use in the evaluation of risk in terms of acceptability.

Perceived safety may be compared across surveys in a similar manner. This examination shows that for the current model of favorability, and particularly for the "Community" favorability, (13) respondents were 25.8 percent more likely to find nuclear power plants "very safe" in March of 1979 than otherwise expected. Simultaneously, respondents were approximately 34 percent more likely to find nuclear power "somewhat safe" in 1979 than otherwise expected, while the variation in the period for the "not safe" category was only marginally significant, with only the March 1975 and April 1979 surveys being significantly associated with the "not safe" response. This is a slightly different pattern than is observed for the four survey model where favorability is not considered. While the primary shifts associated with the "somewhat safe" and "not sure" categories remain very similar, the shifts in the "not sure" category are substantially reduced and a shift is substantially introduced into the "very safe" responses. Hence, we conclude that by elaborating the data in terms of acceptability, a confirmatory role of the TMI incident is illustrated. That is, among those who favor nuclear power, TMI was interpreted as confirming its safety; while among those who do not favor nuclear power, TMI served to underscore the lack of safety in nuclear power plant operation.

Isolating perceived safety by accounting for the social structural characteristics and the acceptability associated with nuclear power demonstrates the utility of the modeling approach. Not only do the observed ratios change with the adjustment, but even the overall pattern is altered. In every survey, except the April 1977 survey, the ratio is near the three to one mark. The observed ratios never exceed 2.309 to one and go as low as 1.388 to one in March 1975. Hence, by modeling the five survey trends, we find that the ambivalence associated with responding that nuclear power plants are somewhat safe is accentuated. While it is highest immediately following TMI and lowest in the April 1977 survey (in both observed and adjusted ratios), the adjusted ratios are considerably higher. Furthermore, because the difference between observed and adjusted ratios is more pronounced in the somewhat to very safe ratio than in the favor to not favor ratio, most prominent causal direction between lay-estimated risk and lay-evaluated risk is illuminated. That is, while perceived risk affects the manner in which risk is evaluated, it appears, from this research, that the degree of acceptability affects the estimation of risk even more. Hence, the prominent causal direction is from lay-evaluated to lay-estimated risk, rather than the other way around.

CONCLUSION

As the trends in survey data are used by risk analysts and policymakers to guide public policy concerning risk, our approach must reflect accurately our best understanding of the problem. It cannot be limited to merely technical aspects of the risk any more than such decisions can be based on purely social considerations. In order for the risk analyst to appropriately utilize public opinion data in determining the potential for acceptability of any given risk, they must first begin to understand the social processes involved in lay risk assessment. If policies concerning risk are to appropriately reflect the value system in which they rest, we must incorporate these values and the people that hold them into the information base that guides these policies. This research takes a first step in this direction by analyzing risk attitudes in terms of social structural location in order to gain insight into the processes involved in lay risk asessment.

Substantively this analysis seems to support the marginal location explanation for differential risk attitudes in terms of perceived and acceptable risk. Women, the less educated, and particularly young adults (but to some extent older Americans) tend to estimate risk at higher levels and be less likely to find, at least nuclear, risks acceptable. This analysis not only confirms Brody's (40) explanation for women's attitudes concerning nuclear power, it expands the explanation to other marginal positions of the social structure. This analysis is consistent with the posited explanation that marginal locations in social, economic, political, and scientific spheres will be likely to have lower perceived self-efficacy and be less likely to occupy key or powerful positions than more central locations. Hence, individuals occupying marginal locations are less likely to find the risks associated with nuclear energy acceptable and more likely to estimate those risks at higher levels than more central positions.

This research seems to confirm that lay people use a similar risk assessment process to that of experts, although perhaps with less sophistication and rigor, and certainly with differing results.(13) Lay people tend to treat risk estimation, in the form of perceived safety, and risk evaluation, in the form of favorability, independently, and simultaneously as related. Furthermore, this research suggests that while perceived safety affects favorability, the prominent causal relation is in terms of how potential acceptability shapes the perception of risk. It is almost as if the predominant question is "whether the risks are worth taking," as opposed to "how risky they may be." Hence, for lay people the conceptual process consists of hazard recognition, risk estimation, and social evaluation, just as for risk analysts. For lay people, however, the emphasis rests on acceptability and how that bears on social evaluation, while risk analysts place greater emphasis on estimation of risk and occasionally upon identification. This social structural approach to determining the acceptability of risk places greater emphasis on the value system and how its differential nature across the social system affects risk assessment. By doing so, we not only gain insight into the social processes affecting the way risk is assessed, but we also enhance our ability to determine the potential for acceptability across various risks and among people with different value systems.

FOOTNOTES

1. Harris survey data provided through the courtesy of Louis Harris Data Center, University of North Carolina, Chapel Hill.
2. Actual analysis was conducted using BMDP Biomedical Computer Program: P-series software package, specifically program P3F.
3. The female association with favorability is the reciprocal of the male association with favorability presented in Table III.
4. All of which were defeated, as reported in the November 11, 1976 issue of Nucleonics Week.

REFERENCES

1. Parsons, Talcott, "Age and Sex in the Social Structure of the United States", American Sociological Review, 7, 604-616 (1942).
2. Spencer, Herbert, Essays, Scientific, Political, and Speculative, Appleton, New York (1892).
3. Davis, Kingsley, Human Society, The McMillan Co., New York (1949).
4. Durkheim, Emile, Division of Labor in Society, Free Press, New York (1933).
5. Otway, H. J., "Risk Estimations and Evaluation," Proc. of IIASA Planning Conference on Energy Systems, International Inst. for Applied Systems Analysis, Schloss Laxenburg, Austria (1973).
6. Rowe, William D., An Anatomy of Risk, John Wiley & Sons, New York (1977).
7. Rosenberg, Morris, The Logic of Survey Research, Basic Books, New York (1968).
8. Parsons, Talcott, The Social System, The Free Press, New York, (1951).
9. Kates, Robert W., Risk Assessment of Environmental Hazard: Scope 8, John Wiley & Sons, New York (1978).

10. Nehnevajsa, Jiri, "Low-Probability/High-Consequence Risks: Issues in Credibility and Acceptance", in Low-Probability/-High-Consequence Risk Analysis, V. Covello and R. Waller (eds.), New York: Plenum Press (In Press).
11. Rogers, George Oliver, "Life-Experience as a Determinant of Perceived and Acceptable Risk," in Low-Probability/High--Consequence Risk Analysis, V. Covello and R. Waller (eds.), New York: Plenum Press (In Press).
12. Rogers, George Oliver, "Social Status and Perceived Risk: Some Social Processes of Risk Perception," 10th World Congress of Sociology, Mexico City, Mexico (August 1982b).
13. Rogers, George Oliver, "Toward A Sociology of Risk: Values, Experience, Perceived and Acceptable Risk," Dissertation submitted to the Graduate Faculty of Arts and Sciences, University of Pittsburgh, Pittsburgh, PA (1983).
14. Rogers, George O. and Jiri Nehnevajsa, "Behavior and Attitudes Under Crisis Conditions: Selected Issues and Findings," University Center for Social and Urban Research, University of Pittsburgh, Pittsburgh, PA (July 1983).
15. Slovic, P., B. Fischoff, and S. Lichtenstein, "Cognitive Processes and Societal Risk Assessment: How Safe is Safe Enough," R. C. Schwing and W. A. Albers, Jr. (eds.), 1-38, Plenum Press, New York (1980).
16. Burton, I., R. W. Kates, and G. F. White, The Environment as Hazard, Oxford University Press, New York (1978).
17. Melber, Barbara D., et al., "Nuclear Power and the Public: Analysis of Collected Survey Research," Battelle Human Affairs Research Centers, Seattle (November 1977).
18. Harris, Louis and Associates, Inc., A Survey of Public and Leadership Attitudes Toward Nuclear Power Development in the United States, EBASCO Services, Inc., New York (1975).
19. Harris, Louis and Associates, Inc., A Survey of Public and Leadership Attitudes Toward Nuclear Power Development in the United States, EBASCO Services, Inc., New York (1976).
20. Harris, Louis and Associates, Inc., "The President's Energy Program: A Public View," Harris Report #35, New York (1977).
21. Harris, Louis and Associates, Inc., "Americans Unwilling to Declare Moratorium on Nuclear Power Despite Increasing Worries About Its Safety," ABC News - Harris Survey, I, 54 (May 3, 1979a).
22. Harris, Louis and Associates, Inc., "Americans Divided on Future of Nuclear Power," ABC News - Harris Survey, I, 66 (May 31, 1979b).
23. Martin, Elizabeth, Diana McDuffee, and Stanley Presser, Source Book of Harris National Surveys' Repeated Questions 1963-1976, University of North Carolina, Institute for Research in Social Science, Chapel Hill, NC (1981).
24. Harris, Louis and Associates, Inc., Risk in a Complex Society: A Marsh & McLennan Public Opinion Survey, Marsh & McLennan, New York (1980).

25. Dillman, Donald, Mail and Telephone Surveys: The Total Design Method, New York: John Wiley (1978).
26. Tanfer, Koray, et al., "National Survey of the Attitude of the U.S. Public Toward Science and Technology," Institute for Survey Research, Temple University (May 1980).
27. Davis, Kingsley and Wilbert E. Moore, "Some Principles of Stratification," American Sociological Review, 10, 242-249 (1945).
28. Moore, Wilbert E., "Comment," in Class, Status, and Power, R. Bendix and S. M. Lipset (eds.), The Free Press, New York (1953).
29. Davis, Kingsley, "Reply to Tumin," American Sociological Review, 18, 394-397 (1953).
30. Tumin, Melvin M., "Some Principles of Stratification: A Critical Analysis," American Sociological Review, 18, 387-393 (1953a).
31. Tumin, Melvin M., "Reply to Kingsley Davis," American Sociological Review, 18, 672-673 (1953b).
32. Wesolowski, W., "Some Notes on the Functional Theory of Stratification," The Polish Sociological Bulletin, 3-4, 28-38 (1962).
33. Stinchcomb, Arthur L., "Some Empirical Consequences of the Davis-Moore Theory of Stratification," American Sociological Review, 28, 805-808 (1963).
34. Nadel, S. F., A Theory of Social Structure, Cohen & West Ltd., London (1957).
35. Merton, Robert K., Social Theory and Social Structure, The Free Press, New York (1968).
36. Berger, Joseph, et al., Status Characteristics and Social Interaction: An Expectation-States Approach, Elsevier, New York (1977).
37. Blau, Peter M., Inequality and Heterogeneity, The Free Press, New York (1977).
38. Rankin, William L. and Stanley M. Nealey, "The Relationship of Human Values and Energy Beliefs to Nuclear Power Attitudes," Battelle Human Affairs Research Centers, Seattle (November 1978).
39. Nealy, Stanley M. and William L. Rankin, "Nuclear Knowledge and Nuclear Attitudes: Is Ignorance Bliss?," Battelle Human Affairs Research Centers, Seattle (October 1978).
40. Brody, Charles J., "Nuclear Power: Sex Differences in Public Opinion," Dissertation submitted to the Department of Sociology, University of Arizona, University Microfilms International (1981).
41. Mileti, Dennis, Thomas E. Drabek and J. Eugene Haas, "Human Systems in Extreme Environments: A Sociological Perspective," Inst. of Behavioral Science, University of Colorado (1975).
42. White, Gilbert F. and J. Eugene Haas, Assessment of Research on Natural Hazards, M.I.T. Press, Cambridge (1975).

43. Form, W. H. and C. P. Loomis, "The Persistence and Emergence of Social and Cultural Systems in Disasters," American Sociological Review, 21, 180-185 (1956).
44. Mack, Raymond W. and George W. Baker, "The Occasion Instant," National Academy of Sciences/National Research Council Disaster Study #15, Washington, DC (1961).
45. Drabek, Thomas E., "Social Processes in Disaster: Family Evacuation," Social Problems, 16, 336-349 (1969).
46. Danzig, E. R., P. W. Thayer, and L. R. Galanter, "The Effects of a Threatening Rumor on a Disaster-Stricken Community," Disaster Study #10, Disaster Research Group, National Academy of Sciences, Washington, DC (1958).
47. Brunn, S. D., J. H. Johnson, Jr., and D. J. Zeigler, Final Report on a Survey of Three Mile Island Area Residents, Department of Geography, Michigan State University, East Lansing (1979).
48. Flynn, C. B., Three Mile Island Telephone Survey, U.S. Nuclear Regulatory Commission, Washington, DC, NUREG/CR-1093 (1979).
49. Flynn, C. B. and J. A. Chalmers, The Social and Economic Effects of the Accident at Three Mile Island: Findings to Date, U.S. Nuclear Regulatory Commission, Washington, DC, NUREG/CR-1215 (1980).
50. Barnes, Kent, et al., Responses of Impacted Populations to the Three Mile Island Nuclear Reactor Accident: An Initial Assessment, Department of Geography, Rutgers University, New Brunswick, NJ (1979).
51. Rogers, George Oliver, "Presidentially Directed Relocation: Compliance Attitudes," University Center for Social and Urban Research, University of Pittsburgh, Pittsburgh, PA (May 1980).
52. Lenski, Gerhard E., "Status Crystallization: A Non-Vertical Dimension of Social Status," American Sociological Review, 19, 398-404 (1954).
53. Lenski, Gerhard E., Power and Privilege: A Theory of Social Stratification, McGraw-Hill Book Co., New York (1966).
54. Linton, Ralph, "Age and Sex Categories," American Sociological Review, 7, 5, 589-603 (1942).
55. Parsons, Talcott, "On the Concept of Political Power," Proc. of the American Philosophical Society, 107, 232-262 (1963).
56. Blalock, Hubert M., "The Identification Problem and Theory Building: The Case of Status Inconsistency," American Sociological Review, 31, 52-61 (1966).
57. Blalock, Hubert M., "Status Inconsistency, Social Mobility, Status Integration and Structural Effects," American Sociological Review, 32, 790-801 (1967).
58. Mills, C. Wright, "The Middle Classes in Middle-Sized Cities," American Sociological Review, 11, 520-529 (1946).
59. Lachman, Roy, Maurice Tatsuoka and William Bonk, "Human Behavior During the Tsunami of May 1960," Science, 133, 1405-1409 (1961).

60. Friedsam, H. J., "Older Persons in Disaster," in Man and Society in Disaster, George Baker and Dwight W. Chapman (eds.), Basic Books, New York (1962).
61. Fleming, John J., Jr., "Social Position and Decisionmaking Involving Risk," Human Relations, 26, 1, 67-76 (1973).
62. Rogers, Everett, The Diffusion of Innovation, The Free Press, New York (1962).
63. Rogers, Everett and Floyd Shoemaker, Communication of Innovations: A Cross-Cultural Approach, The Free Press, New York (1971).
64. Coleman, J., E. Katz and H. Menzel, "The Diffusion of an Innovation Among Physicians," in Social Networks: A Developing Paradigm, Samuel Leinhardt (ed.), Academic Press, New York (1977).
65. Cancian, Frank, "Stratification and Risk-Taking: A Theory Tested on Agricultural Innovation," American Sociological Review, 32, 912-927 (1967).
66. Cancian, Frank, Change and Uncertainty in a Peasant Economy, Stanford University Press, Stanford (1972).
67. Goodman, L. A., "A General Model for the Analysis of Surveys," American J. of Sociology, 77, 1035-1086 (1972).
68. Goodman, L. A., Analyzing Qualitative/Categorical Data, Abt Books, Cambridge (1978).
69. Knoke, David and Peter J. Burke, Log-Linear Models, Sage Publications, Beverly Hills, CA (1980).
70. Fienberg, S. E., The Analysis of Cross-Classified Data, MIT Press, Cambridge (1977).
71. Wildavski, Aaron, "Richer is Safer," The Public Interest, 60, 23-39, Summer (1980).
72. Nuclear Regulatory Commission, Annual Report (1975).
73. Nuclear Regulatory Commission, Annual Report (1979).

INDEX